Megaregionalism 2.0

Trade and Innovation within Global Networks

World Scientific Studies in International Economics
(ISSN: 1793-3641)

World Scientific Studies in International Economics includes works dealing with the theory, empirical analysis, and evaluation of international economic policies and institutions, with topics covering international macroeconomics and finance, international trade theory and policy, as well as international legal and political economy. Monographs and edited volumes will comprise the core of the publications.

Vol. 68 *International Trade, Capital Flows and Economic Development*
edited by Francisco L. Rivera-Batiz (Columbia University, USA) &
Luis A. Rivera-Batiz (University of Puerto Rico, USA)

Vol. 67 *Megaregionalism 2.0: Trade and Innovation within Global Networks*
edited by Dieter Ernst (East-West Center, USA &
The Centre for International Governance Innovation/CIGI, Canada) &
Michael G. Plummer (The Johns Hopkins University, SAIS, Italy)

Vol. 66 *International and Interregional Migration: Theory and Evidence*
edited by Francisco L. Rivera-Batiz (Columbia University, USA)

Vol. 65 *International Trade Theory and Competitive Models: Features, Values, and Criticisms*
by Ronald W. Jones (University of Rochester, USA)

Vol. 64 *Information Spillovers and Market Integration in International Finance: Empirical Analyses*
by Suk-Joong Kim (University of Sydney, Australia)

Vol. 63 *Dimensions of Trade Policy*
by Martin Richardson (Australian National University, Australia)

Vol. 62 *International Trade and Labor Markets: Welfare, Inequality and Unemployment*
by Udo Kreickemeier (TU Dresden, Germany)

Vol. 61 *Achieving Financial Stability: Challenges to Prudential Regulation*
edited by Douglas D. Evanoff (Federal Reserve Bank of Chicago, USA),
George G. Kaufman (formerly Loyola University Chicago, USA),
Agnese Leonello (European Central Bank, Germany) &
Simone Manganelli (European Central Bank, Germany)

The complete list of the published volumes in the series can be found at
http://www.worldscientific.com/series/wssie

67 World Scientific Studies in International Economics

Megaregionalism 2.0

Trade and Innovation within Global Networks

Editors

Dieter Ernst

East-West Center, USA
The Centre for International Governance Innovation/CIGI, Canada

Michael G Plummer

The Johns Hopkins University, SAIS, Italy

World Scientific

NEW JERSEY • LONDON • SINGAPORE • BEIJING • SHANGHAI • HONG KONG • TAIPEI • CHENNAI • TOKYO

Published by

World Scientific Publishing Co. Pte. Ltd.
5 Toh Tuck Link, Singapore 596224
USA office: 27 Warren Street, Suite 401-402, Hackensack, NJ 07601
UK office: 57 Shelton Street, Covent Garden, London WC2H 9HE

Library of Congress Cataloging-in-Publication Data
Names: Ernst, Dieter, 1942– editor. | Plummer, Michael G., 1959– editor.
Title: Megaregionalism 2.0 : trade and innovation within global networks /
edited by Dieter Ernst (East-West Center, USA & The Centre for
International Governance Innovation/CIGI, Canada), Michael G. Plummer (The
Johns Hopkins University, SAIS, Italy).
Description: New Jersey : World Scientific Publishing Co., [2018] | Series:
World scientific studies in international economics | Includes
bibliographical references and index.
Identifiers: LCCN 2017051204 | ISBN 9789813229822 (hardcover)
Subjects: LCSH: Regional economics. | International trade. | Commercial policy. |
Technological innovations--Economic aspects.
Classification: LCC HT388 .M44 2018 | DDC 330.9--dc23
LC record available at https://lccn.loc.gov/2017051204

British Library Cataloguing-in-Publication Data
A catalogue record for this book is available from the British Library.

For any available supplementary material, please visit
http://www.worldscientific.com/worldscibooks/10.1142/10705#t=suppl

Desk Editor: Alisha Nguyen

Typeset by Stallion Press
Email: enquiries@stallionpress.com

Printed in Singapore

Contents

Acknowledgements

An endeavor of this magnitude requires generous support from a variety of sources, for which we are truly grateful and without which the project would be much diminished.

We are indebted to the National Science Foundation, which sustained us financially (via a grant) and with its encouragement and input. We gratefully acknowledge the insight and contributions by Dr. Maryann Feldman (Heninger Distinguished Professor in the Department of Public Policy at the University of North Carolina, and former Science of Science and Innovation Policy (SciSIP) Program Director at the National Science Foundation).

This project owes a debt of gratitude to the following participants of the January 2016 workshop who contributed substantially to this project: Shigeyuki Abe (Center for Contemporary Asian Studies, Doshisha University, Tokyo); Staffan Laestadius (Department of Industrial Economics and Management, Royal Institute of Technology, Stockholm); Lan Xue (Tsinghua University, Beijing); William Lazonick (UMass Lowell/ The Academic-Industry Research Network, Cambridge/Mass); Mark Wu (Harvard Law School, Cambridge/Mass); and Klaus Ziegler (China Representative, DIN German Institute for Standardization, Beijing). G.M.P. (Peter) Swann (Nottingham University Business School, UK) provided substantial guidance on the role of standards for international trade.

We are very grateful to Ms. Alissa Tafti for the invaluable support she provided via extensive comments on second drafts of the papers. Her help improved significantly the quality of the chapters (including our own). We are also grateful to Mr. Evan Frondorf (IP Research Fellow at University of California Hastings College of Law), and to Ms. XIE Lincan (Research Fellow, International Cooperation Center at China's National Development and Reform Commission) for taking extensive notes during our Workshop in Honolulu in January 2016.

We would like to thank the leadership and staff of the East-West Center (EWC, Honolulu/USA) and the Centre for International Governance Innovation (CIGI, Waterloo/Canada) for their continued support and encouragement throughout the life of the project. While so many were involved in helping us, we would particularly like to underscore our appreciation to Richard Vuylsteke and Charles Morrison, the current and former EWC Presidents; to Rohinton P. Medhora, CIGI President; to Paul Jenkins, Interim Director of CIGI's Global Economy Program; and to Jefferson Fox and Nancy Lewis, the current and former EWC Directors of Research. At the EWC, we are very grateful to Ms. Ketty Loeb, Ms. June Kuramoto, Ms. Tina Tom, Ms. Elisa Johnston, Ms. Ann Hartman, Ms. Carol Wong, Ms. Sharon Shimabukuro, and Ms. Laurel Pikcunas for helping us to implement this complex project. Last but not least, Ms. Barbara Wiza at the Johns Hopkins University, SAIS (Bologna/Italy) provided excellent logistical and editorial support.

Dieter Ernst and Michael G. Plummer

Preface

> "Trade helps strengthen the middle class, but for it to work we must ensure that the benefits are shared with all our citizens, not just the few. The Comprehensive and Progressive Trans Pacific Partnership is a new step on that path." Canada's Prime Minister Pierre Trudeau, at Davos, 23 January 2018

The Trans-Pacific Partnership (TPP) negotiations, in progress since 2008, had been finalized less than three months before the January 2016 conference at which the papers for this volume were first presented. An agreement was signed by the 12 negotiating partners a month later. This seemed to be the right time to assess how trade and innovation links would be affected by this new push towards "mega-regionalism" in the Asia-Pacific.

While some conference participants were skeptical about certain aspects of the TPP (especially with regard to intellectual property and standards), most expected that it would bring significant gains to the region, advance non-economic aspects of cooperation including security relationships, protect labor rights, and improve institutions. Importantly, it was widely assumed that the TPP would strengthen US engagement in the Asia-Pacific region.

What a difference a year makes: in its third day in office, the Trump Administration withdrew the United States from the TPP, which technically terminated the agreement (since the agreement required that at least six of the 12 members constituting 85 percent or more of the combined GDP would have to ratify the agreement). This abrupt dumping of a major trade agreement, without any consultation with trading partners, certainly constituted a major change in US policy: the TPP was nurtured by a Democratic Administration and was supported by most of the Republican Party and, hence, was bipartisan in nature. Negotiators for the United States fought hard to ensure that US interests would be protected, from tighter restrictions on the protection of intellectual property to disciplines on state-owned enterprises. Indeed, opposition to the TPP outside the United States derived to no small degree from the belief that the agreement was overly biased in favor of US interests, a sentiment that emerged during our January conference. It was ironic, therefore, that the Trump Administration pulled out of the agreement because it believed that TPP would badly hurt US interests. No serious research supporting this assertion was ever produced.

As many studies of the TPP conclude and those included in this book would suggest, the US withdrawal from the TPP constitutes a colossal strategic mistake, not just because of foregone economic benefits but also because of the perception, exacerbated by the "America First" doctrine, that the United States is no longer a reliable ally in the process of Asia-Pacific economic integration. Meanwhile, China's leadership is being given a big boost by the US withdrawal and the country is "doubling down" on integration, from the Regional Comprehensive Economic Partnership (RCEP) agreement, which is likely to be concluded in 2018, to the Belt and Road Initiative and the new Asian Infrastructure Investment Bank.

The US withdrawal, however, did not doom the TPP: shortly after the country left, the remaining 11 TPP partners began to move forward with revisions to the agreement that would allow it to proceed without the United States. After several rounds of negotiations, the remaining members reached an agreement in principle on the "Comprehensive and Progressive Trans-Pacific Partnership" (CPTPP), announced on the sidelines of the November 2017 APEC members in Viet Nam. The CPTPP agreement is remarkably close to the original TPP agreement, with the exception of 20 "suspensions". These suspensions are very minor except in the area of intellectual-property protection (11 out of the 20) and express delivery services (2 out of the 20), which are areas of key importance to US interests. The CPTPP has suspended these clauses in lieu of leaving them out completely in hopes that the United States will eventually return to the agreement.

While the chapters in this book were written before the US withdrawal from the TPP, the analysis continues to be pertinent even without the participation from the agreement's largest member. At the very least, these chapters provide a fascinating case study of how a resurgent economic nationalism under the banner of "America First" can disrupt the critically important trade and innovation nexus. More importantly, the chapters demonstrate why it is now time to move forward to an Enlightened Globalization Agenda.

Dieter Ernst
East-West Center and
Centre for International Governance Innovation/CIGI

Michael G. Plummer
Johns Hopkins University, SAIS, and East-West Center

About the Contributors

Knut Blind has been holding the Chair of Innovation Economics at the Technical University of Berlin since 2006 and is a Senior Researcher at the Fraunhofer Institute of Open Communication System FOKUS. Between 2008 and 2016, he also held the Endowed Chair of Standardization at the Rotterdam School of Management at the Erasmus University. His research interests include innovation and regulation, including standardization.

Carlos A. Primo Braga is an Associate Professor, Fundação Dom Cabral, Brazil. He is also a Visiting Professor at IMD, Switzerland. He was Professor of International Political Economy at IMD and Director of the Evian Group (2012–15). Before that he was the Special Representative and Director for Europe, The World Bank. He was also Director, Economic Policy and Debt (2008–10) and, in 2010, he was the WBG's Acting VP and Corporate Secretary and Acting Executive Secretary of the Development Committee. He was a Fulbright Scholar (1988/89) at the Paul Nitze School of Advanced International Studies (SAIS) where he also taught as a visiting professor (1988–98). He was also an Assistant Professor of Economics, University of São Paulo (USP), and Senior Researcher at FIPE in the 1980s. He received a degree in Mechanical Engineering (1976), Instituto Tecnológico de Aeronáutica and an MSc (1980) in Economics from USP. He holds a PhD (1984) in Economics from the University of Illinois, Urbana-Champaign, USA.

Ronald Brown is a law professor at the University of Hawai'i Law School and has served as the University's Director of the Center for

Chinese Studies from 2000–2006. His experience includes working as an attorney with the National Labor Relations Board, acting as private impartial arbitrator in labor-management disputes, and serving as state-appointed public fact-finder in Hawai'i public sector disputes.

Professor Brown's teaching specialties include labor and employment law, employment discrimination law, arbitration, Chinese law, Asian International and Comparative Labor Law, and U.S.-China Labor & Employment Law. He has authored numerous articles on free trade agreements in Asia and the labor provisions of their social dimension provisions and has recently published *East Asian Labor and Employment Law: International and Comparative Context* (Cambridge 2012), *Understanding Labor & Employment Law in China* (Cambridge 2010), and a book entitled *Understanding Chinese Courts and Legal Process: Law with Chinese Characteristics*.

Professor Brown has worked in China under the USIA's professional-in-residence program, has served as a Consultant with the World Bank, published with the ILO, and has lectured throughout Asia on comparative labor law topics. He has taught at Beijing University Law School and currently serves as a foreign advisor to BEIDA on graduate law programs. He also serves as an Editorial Board Member on the Hague Institute for Global Justice, International Labor Rights Case Law Journal. He conducts legal exchange and international training programs for Chinese lawyers, judges, law drafters, and prosecutors under arrangements with the key government legal agencies. In 2004–2005, Professor Brown was in China as a Fulbright Distinguished Scholar, teaching U.S. and Chinese Labor & Employment law at both Peking University Law School and Tsinghua University Law School.

Tain-Jy Chen is Professor of Economics at National Taiwan University. He previously worked at Chung-Hua Institution for Economic Research, a think tank specializing in economic policy studies. His research interests are industrial and trade policies. His recent research work focuses on industrial development of Taiwan and China.

Colleen Chien is a Professor at Santa Clara University School of Law where she teaches, writes, and mentors students and a visiting Scholar at the Stanford Computational Policy Lab, where she works on leveraging innovation to advance access to justice. From 2013–2015 she served in the Obama White House as a Senior Advisor, Intellectual Property and Innovation. Professor Chien is nationally known for her research and publications on domestic and international patent law and policy issues. She has testified on multiple occasions before Congress, the DOJ, the FTC, and the US Patent and Trademark Office. In 2017, Professor Chien was awarded the American Law Institute's Early Career Medal, awarded every two years to up to two law outstanding professors; she also has received the Vanguard Award for Public Service, and has been named Eric Yamamoto Emerging Scholar, a Tech Law Leader, one of Silicon Valley's "Women of Influence," one of the 50 Most Influential People in Intellectual Property in the world. Chien is a graduate of Stanford (Engineering) and Berkeley Law Schools and lives in Oakland with her husband and their two sons. She is the founder of civic engagement projects Wall of Us and ActLocal and an advisor to ClearAccessIP, an AI-driven patent analytics and enterprise software firm and CitizenBe, which works to create and sustain impactful political participation.

Jorge L. Contreras is a Professor of Law at the University of Utah and a Senior Fellow at the Centre for International Governance Innovation in Ontario, Canada. He has previously served on the law faculties of American University and Washington University in St. Louis. Before entering academia, Professor Contreras was a partner at the international law firm Wilmer Cutler Pickering Hale and Dorr LLP, where he practiced transactional and IP law in Boston, London and Washington DC. His research focuses, among other things, on the development of technical standards and the use, dissemination and ownership of data generated by scientific research. He has published more than 100 scholarly articles and chapters and is the editor of four books relating to technology law and technical standards, including the Cambridge Handbook of Technical Standardization Law (2017). He has been quoted in the NY Times, Wall Street Journal, Economist, Washington Post, Korea Times, has been a guest on NPR, BBC and various televised broadcasts, and has been cited by the U.S. Federal Trade Commission, European Commission and courts in the U.S. and Europe. He currently serves as Co-Chair of the

Interdisciplinary Division of the ABA's Section of Science & Technology Law, and as a member of the National Institutes of Health (NIH) Council of Councils and the IPR Policy Committee of the American National Standards Institute (ANSI). He has previously served as Co-Chair of the National Conference of Lawyers and Scientists, and as a member of the National Academy of Sciences (NAS) Committee on IP Management in Standard-Setting Processes. He is a graduate of Harvard Law School (JD) and Rice University (BSEE, BA).

Dieter Ernst is a Senior Fellow at the East-West Center/EWC (Honolulu, Hawaii, USA) and at the Centre for International Governance Innovation/CIGI (Waterloo, Canada). He is an authority on trade, global production networks and the internationalization of research and development in high-tech industries, with a focus on standards and intellectual property rights. His research examines corporate innovation strategies and innovation policies in the United States and in China, India, Taiwan, Korea, and Malaysia. Dr. Ernst has provided testimony to US Congress, and has served as a member of the United States National Academies "Committee on Global Approaches to Advanced Computing"; Senior Advisor to the Organisation for Economic Co-operation and Development, Paris; Research Director of the Berkeley Roundtable on the International Economy at the University of California at Berkeley; Professor of International Business at the Copenhagen Business School; and Scientific Advisor to governments, private companies, and international institutions.

Robin Feldman received a Bachelor's Degree from Stanford University and a J.D. from Stanford Law School, graduating the Order of the Coif and receiving the Urban A. Sontheimer Award for graduating second in the class. Professor Feldman also served in the Articles Department of the Stanford Law Review. After graduation, she clerked for The Honorable Joseph Sneed of the U.S. Court of Appeals for the Ninth Circuit. Professor Feldman holds the Harry & Lillian Hastings Chair. She has received the Rutter Award for Teaching Excellence, the Visionary Award from the UC Hastings Board of Directors, and the 1066 Foundation Award for Scholarship. In 2015, she was honored by the American Lawyer Publications as one of the Women Leaders in Law & Tech, and was the sole academic to receive the honor. Professor Feldman has published three books, *Drug Wars: How Big Pharma Raises Prices and Keeps Generics Off the Market* (Cambridge 2017), *Rethinking Patent Law* (Harvard 2012) and IZZ (Oxford 2009), as well as an extensive number of articles in journals including Yale, Stanford, Columbia, Virginia, and Georgetown law schools, the New England Journal of Medicine, and the American Economic Review. Her work has been cited by the White House and by numerous federal and state agencies and members of Congress. She has provided testimony and commentary for committees of the US Congress and the California legislature, as well as the Federal Trade Commission, the Department of Justice, the Patent & Trademark Office, and the National Academy of Sciences. Professor Feldman has chaired the Executive Committee of the Antitrust Section of the American Association of Law Schools and has served as the Herman Phleger Visiting Professor of Law at Stanford Law School. She was elected to the American Law Institute in 2012 where she serves as an advisor to the ALI's Restatement of Copyright Project. Professor Feldman is also the Director of the UC Hastings Institute for Innovation Law, which is dedicated to data-driven law making and the active classroom. As part of the Institute, Professor Feldman runs the Startup Legal Garage, which has been named one of the most innovative law school programs in the country and recognized by the American Association of Law Schools as one of the nation's "Innovative & Other Outstanding Programs." More information on the Institute, including disclosures on funding, can be found at http://innovation.uchastings.edu/.

Susan Helper is the Carlton Professor of Economics at the Weatherhead School of Management, Case Western Reserve University. She served as the Chief Economist of the U.S. Department of Commerce from 2013–2015, and as Senior Economist at the White House Council of Economic Advisers in 2012–3. Dr. Helper has been a visiting scholar at University of Oxford, the University of California (Berkeley), Harvard University and the Massachusetts Institute of Technology (MIT). Her research focuses on the globalization of supply chains, and on impacts of automation on productivity and job quality. She has published in journals such as Sloan Management Review, American Economic Review, and the Journal of Economic Perspectives. Dr. Helper received her PhD in Economics from Harvard and her B.A. from Oberlin College.

Derek Hill currently works as an economist at the National Science Foundation on indicators of science and technology focusing on high-technology industry and trade of major economies, commercialization of science, and innovation. Between January 2016 and August 2016, he worked on detail at the National Renewable Energy Laboratory conducting research in the co-location of innovation and production of clean energy technologies. Before joining NSF in 1999, Derek worked at the US Department of Treasury as an economist on international financial issues. Derek obtained a masters' degree in Applied Economics at the University of Michigan, Ann Arbor in 1991 and a Bachelor's in Political Economy at the University of California, Berkeley in 1989.

Brian Kahin is Digital Economy Fellow at OECD's Directorate of Science, Technology and Innovation; Research Scholar at the Berkeley Roundtable on the International Economy (BRIE); and Senior Fellow at the Computer & Communications Industry Association. He was Founding Director of the Information Infrastructure Project at Harvard's Kennedy School of Government (1989–1997), the first academic program to address the economic and policy implications of the internet. From 1997 to 2000, Kahin was Senior Policy Analyst at the White House Office of Science and Technology Policy, where he initiated a study on patent quality and business effect; he also led an interagency working group on the digital economy on behalf of the National Economic Council. He has since been deeply engaged in patent reform efforts in both Europe and the U.S. Kahin has edited ten books on the internet and the digital economy, including *Understanding the Digital Economy* (with Erik Brynjolfsson), and *Advancing Knowledge and the Knowledge Economy* (with Dominique Foray). He has taught at Harvard, Maryland (College Park), Michigan (Ann Arbor), Colorado (Boulder), and Washington (Seattle). He holds a BA from Harvard College and a JD from Harvard Law School.

A globally recognized leader in standards development and intellectual property, **Dr. Ing. Konstantinos Karachalios** is managing director of the IEEE Standards Association and a member of the IEEE Management Council. As managing director, he has been enhancing IEEE efforts in global standards development in strategic emerging technology fields, through technical excellence of staff, expansion of global presence and activities and emphasis on inclusiveness and good governance, including reform of the IEEE standards-related patent policy. As member of the IEEE Management Council, he championed expansion of IEEE influence in key techno-political areas, including consideration of social and ethical implications of technology, according to the IEEE mission to advance technology for humanity. Results have been rapid in coming and profound; IEEE is becoming the place to go for debating and building consensus on issues such as a trustworthy and inclusive Internet and ethics in design of autonomous systems. Before IEEE, Konstantinos played a crucial role in successful French-German cooperation in coordinated research and scenario simulation for large-scale nuclear reactor accidents. And with the European Patent Office, his experience included establishing EPO's patent academy, the department for delivering technical assistance for developing countries and the public policy department, serving as an envoy to multiple U.N. organizations. Konstantinos earned a Ph.D. in energy engineering (nuclear reactor safety) and masters in mechanical engineering from the University of Stuttgart.

Tim Krueger is a consultant at Newry Corporation, which advises manufacturing firms on strategy and innovation. He has also worked as a Research Assistant for Policy Matters Ohio and as Program Manager for Case's Great Lakes Energy Institute. Tim has an MBA from Case Western and a B.A. from Cornell.

Douglas Lippoldt is Chief Trade Economist¸ HSBC Global Research. He joined HSBC in 2014 coming from the OECD in Paris where he had worked for 22 years in various roles as a senior economist. Prior to that, he served seven years as an international economist with the US Department of Labor in Washington, DC. His central focus is on international trade issues and he has published extensively on trade topics as well as related aspects of economic development, labour market adjustment, innovation and intellectual property. Doug holds a PhD in Economics from the Institut d'Etudes Politiques de Paris (Sciences Po), a MA in International Studies from the University of Denver, and a BA in International Studies from Washington College in Maryland, USA. He was a Fulbright Scholar at the University of Cologne, Germany. He represents HSBC as the deputy delegate on the B20 Trade and Investment Taskforce.

Keith E Maskus is Professor of Economics and former Associate Dean for Social Sciences at the University of Colorado, Boulder, USA. He has been a Lead Economist in the Development Research Group at the World Bank. He is also a Research Fellow at the Peterson Institute for International Economics, a Fellow at the Kiel Institute for World Economics, and an Adjunct Professor at the University of Adelaide. He has been a visiting professor at the University of Bocconi and a visiting scholar at the CES-Ifo Institute at the University of Munich and the China Center for Economic Research at Peking University. He serves also as a consultant for the World Bank and the World Intellectual Property Organization and is currently chairing a panel of the National Academy of Sciences on intellectual property management in standards-setting organizations.

Maskus received his PhD in economics from the University of Michigan in 1981 and has written extensively about various aspects of international trade. His current research focuses on the international economic aspects of protecting intellectual property rights. He is the author of *Intellectual Property Rights in the Global Economy*, published by the Institute for International Economics, and co-editor of *International Public Goods and the Transfer of Technology under a Globalized Intellectual Property Regime*, published by Cambridge University Press. A new volume, *Private Rights and Public Problems: The Global Economics of Intellectual Property in the 21st Century*, was published in 2012 by the Peterson Institute for International Economics.

Karen McCabe is a Senior Director of Technology Policy and International Affairs at IEEE, where she leads efforts to build and connect communities working in the technology policy ecosystem and where her focus is on the development and promotion of collaborative strategies that produce impactful outcomes among multi-stakeholders to help advance technology for society and humanity benefit. Karen has over 20 years experience working at the nexus of mission-driven organizations, industry, NGO and government bodies to raise awareness, to educate and build capacity among stakeholders in the technology sphere. Karen focuses on projects and initiatives to expand global footprint, build communities, develop and execute integrated communications and global outreach programs and build and nurture relationships. During the last 20 years, Karen has held various leadership and senior management positions in the technology sector, with a specific focus in the global standards and technology development domain. Karen in engaged in efforts in Internet governance, ethics and technology and global standards in trade and policy. She leads the organization's engagement with the UN (SDGs, WSIS etc.) and other international bodies with a focus on engagement of technical community and a focus on the intersection of policy and technology in ICTs. Karen is a member of the OECD Internet Technical Advisory, the Internet Society and works with an array of organizations and bodies to connect technologists, industry leaders and policy makers where she develops partnerships and builds alliances across stakeholder communities.

Wang Ping, Senior Researcher of CNIS, former Deputy Chief Engineer of China National Institute Standardization (CNIS), has been long engaged in CAD/CAM Standardization and Manufacturing Informatization Standardization. He was involved in ISO/TC184/SC4 Standardization of Industrial Data as a formal representative of China from 1991 to 2002, and involved as the first Project Leader in several National Key Technologies R&D Programs and Soft Science Research Projects of the Ministry of Science and Technology (MOST) in the eighth, ninth and tenth Five-Year-Plan. From 2000, he started to do research on standards theory and history, standards strategy, institutional reform and innovation. He has been employed as Advisory Research Fellow, Policy Research Center for China's Science and Technology, Tsinghua University, from 2012.

Michael G Plummer is Director of SAIS Europe and the Eni Professor of Economics at the Johns Hopkins University, as well as (non-resident) Senior Fellow at the East-West Center. Previously he was Head of the Development Division of the Organization for Economic Co-operation and Development (OECD) in Paris and a tenured associate professor at Brandeis University. He was Editor-in-Chief of the *Journal of Asian*

Economics (Elsevier) from 2007–2015 and President of the American Committee on Asian Economic Studies (ACAES) from 2008–2015. A former Fulbright Chair in Economics and Pew Fellow in International Affairs at Harvard University, he has been an Asian Development Bank (ADB) distinguished lecturer on several occasions and team leader of projects for various organizations including the Association of Southeast Asian Nations, the United Nations, the OECD, the ADB and the World Trade Organization. He has taught at more than a dozen universities in Asia, Europe, and North America. He has advised several governments on the Transpacific Partnership (TPP) negotiations and Asian economic integration, particularly in the ASEAN context. He is author/co-author of over 100 journal articles and book chapters.

Florian Ramel focused on the effects of standards on trade during his doctoral studies at the Chair of Innovation Economics at the Technical University of Berlin. He currently holds a position at a boutique investment firm in Berlin.

Gregory Tassey currently works at the Economic Policy Research Center, University of Washington. He was formerly Senior Economist for the National Institute of Standards and Technology (NIST). His major fields of research are economics of innovation: investment behavior, industry expansion path analysis, technology life cycle analysis, public-private growth models; roles and impacts of technology infrastructure; technology-based economic growth policies, including R&D investment models and appropriate policy tools such as direct funding and tax incentives, public-private technology research consortia; public-sector R&D program impact analysis. He has published papers in policy and economics journals and four books on the technology-based economy and relevant growth policies. He also worked with White House and Congressional staff on innovation policy development. Tassey participated in national advisory bodies, including the National Academy's National Materials and Manufacturing Board, the National Research Council's Committee on Assessing the Value of Federal R&D; co-chaired a White House Committee that advised Congress on advanced manufacturing technology strategies (developed National Strategic Plan). He consults with several agencies of the federal government and nonprofit research groups. Gregory Tassey obtained his B.A. in Physics from McDaniel College and a Ph.D. in Economics from The George Washington University.

Maximilian von Laer is a PhD student at the Chair of Innovation Economics at the Technical University of Berlin. Prior to his PhD studies he has founded a startup and holds a Bachelor's degree from Humboldt University Berlin as well as a Master's degree from the Barcelona Graduate School of Economics (GSE). His research focuses on standardization, emerging markets and trade.

Liang Zheng now serves as Associate Professor of the School of Public Policy and Management, Tsinghua University, as well as Research Fellow and Deputy Director of China Institute for Science &Technology Policy at Tsinghua University (CISTP). The main areas of his research now focus on Science, Technology and Innovation Policy, Globalization of R&D, IPRs and standardization, National Innovation System etc. Besides academic publishing, Dr. Liang also participated in some important research projects such as the Strategic Research for National Medium and Long Term Science and Technology Development Program (MLP) of China. Now he is a member of National Innovation Survey Expert Group, as well as a member of Sino-US Innovation Dialogue Expert Group.

Introduction

Megaregionalism 2.0: Trade and Innovation in Global Networks — Towards an Enlightened Globalization Agenda

Dieter Ernst

East-West Center and Centre for International Governance (CIGI)

Michael G. Plummer

Johns Hopkins University, SAIS, and East-West Center

1. Capturing gains from trade for innovation: Pathways to Megaregionalism 2.0?

In November 2016, the United States elected a president who, for the first time in the post-war era, promised to openly embrace protectionism to "make America great again".[1] This unabashed inward-looking approach of US trade diplomacy raises new challenges for the study of trade and innovation.

Prospects for gains from trade and investment liberalization are arguably greater than ever before, but realizing them has never been more complicated. The globalized economy requires not only lower border restrictions — such as tariffs and non-tariff barriers — but also behind-the-border policy harmonization and rules for the digital economy. "Megaregionalism" in the Asia-Pacific region has emerged as a promising potential vehicle for promoting innovation and productivity — both essential to raising living standards — via trade and investment links. Moreover, governments need to adopt a new set of well-funded social and other support policies (including innovation policies) to facilitate structural adjustment in efficient and equitable ways. We might call this two-dimensional set of policy challenges — i.e., maximizing gains from globalization and effectively managing the process — "Megaregionalism 2.0".

This book, *Megaregionalism 2.0: Trade and Innovation within Global Networks,* presents new research insights for policy debates on how to strengthen the gains from trade for innovation through an inclusive trading environment that promotes productivity, while at the same time facilitating access to knowledge for all. The goal is to open up space for new agenda-setting research on this critically important policy challenge. This opening chapter introduces these issues by: (1) emphasizing the role that economic policy needs to play in ensuring that gains from trade for innovation are shared by all and that losses are minimized (Section 2); (2) evaluating the risks involved in the emerging "economic nationalism" that has been growing for some time but, with Brexit and the election of Donald Trump as US President, is now shaping policy (Section 3); (3) casting these issues in light of the new megaregionalism trend (Section 4); and (4) reviewing how the innovation and trade nexus (Section 5) might be affected by the alternative scenarios of megare-

---

[1]Tumulty, Karen, "How Donald Trump came up with 'Make America Great Again'", *Washington Post*, January 18, 2017. To be sure, other administrations used various forms of protectionism to pursue political objectives; still, they never advocated it as a development strategy.

gionalism in the future (Section 6). The final section reviews some of the salient messages from the book's 19 chapters that lay out key building blocks of Megaregionalism 2.0.

2. Gains from trade for innovation — An evolving research agenda

With the backlash against globalization, "protectionist bilateralism" describes the current direction of US trade diplomacy. For example, the US government has withdrawn from the Transpacific Partnership agreement (TPP), which was considered to be the gold standard of the new "megaregionalism" trend, vowing to only undertake bilateral, rather than regional, agreements. It is reopening NAFTA negotiations and has threatened to do the same with the Korea–US Free-trade Agreement (KORUS). It has promised an "America First" trade agenda that threatens countries experiencing bilateral trade surpluses with the United States with punitive tariffs. It has been especially critical of China, which it has accused of myriad unfair practices including dumping, currency manipulation, intellectual property theft, and discrimination against foreign investors.

It is thus time to highlight once again the critical role that international trade and investment partnerships can play in fostering sustainable growth and prosperity, decreasing poverty, reducing inequality, and lowering risks of international conflict. Economic policy is key in promoting international economic integration. Equally important however is the critical role that economic policy plays in ensuring that gains and losses from trade for innovation are shared by all. As digital technology and geopolitical shifts rapidly transform the global economy, all the more important are domestic support policies that seek to foster economic growth, competitiveness and prosperity through productivity-enhancing innovation.[2] Greater attention to such

[2]For a theoretical foundation, see Chapter 12. For key contributions to the economics of innovation policy, see Perez, Carlota, 2002. *Technological Revolutions and Financial Capital: The Dynamics of Bubbles and Golden Ages*, Edward Elgar, Cheltenham, UK, 198 pages, and the contributions by Richard R. Nelson, David C. Mowery and Jan Fagerberg in *Oxford Handbook of Innovation,* 2002, edited by Jan Fagerberg, David C. Mowery, and Richard R. Nelson.

domestic support policies underscores the essence of Megaregionalism 2.0. The relevance of the corresponding research agenda becomes clear when one considers the following long-term developments that are eroding the innovation capacity of the United States:

- The secular decline of public funding for basic and applied research needed to improve competitiveness and the creation of quality jobs;
- The long-term decline of corporate research in the United States, while other countries (especially China) are rapidly increasing their investment in R&D and in the development of frontier technologies that are critical for, *inter alia*, economic growth, employment, and defense in the 21st century[3];
- The first budget proposal from the White House (May 2017) which recommends a drastic decline in funding for government agencies with responsibilities related to supporting investment, education, and knowledge diffusion for critical frontier technologies and for the provision of public goods related to the environment, health, economic development, and conflict resolution.

The distribution of gains from trade for innovation is of even greater concern for emerging and developing countries. While this is a highly divergent group, access to foreign technology (as manifested in licensing fees and conditions) is of critical importance for all of these countries.[4] They lag well behind the United States and other advanced countries given their stage of development, human capital, and other resources and capabilities, and hence pursue quite different growth models and development strategies. Most importantly, barriers to innovation remain substantial in these countries, ranging from severe quality shortcomings in education to

[3]The drastic decline in corporate research is documented in a series of conferences, e.g., Conference on Innovation — The Decline in Corporate Research: Should We Worry? — March 31, 2017, at Duke University's "Duke in DC" offices in Washington, DC.

[4]See Kim, Linsu and Richard R. Nelson, eds, 2000, *Technology, Learning, and Innovation: Experiences of Newly Industrializing Economies*; and Ernst, D., L. Mytelka and T. Ganiatsos (1998a), 'Learning, Technological Capability Building and Sustainable Export Growth', in D. Ernst, L. Mytelka and T. Ganiatsos, eds, *Technological Capabilities and Export Success in Asia*. London: Routledge.

plagiarism in science, a weak and fragmented innovation system, and constraints to entrepreneurship and private R&D investment. These differences in economic structure are acknowledged in the World Trade Organization (WTO)'s "special and differential treatment" provisions for trade agreements which, in principle, give developing countries special rights.[5]

Research on trade and innovation in the context of megaregional trade agreements cuts across many disciplinary boundaries.[6] To address this challenge, this book brings together chapters from trade economists, policymakers, and trade negotiators, as well as from experts on innovation, intellectual property rights, competition law, technical standards, labor standards, and industrial development.

Gains from trade have been the subject of intense debate for centuries. Trade theory has focused on "the proposition that freedom of [international] trade is on the whole economically more beneficial than protection".[7] In the last half of the 20th century, globalization culminated in the emergence of international corporate networks that integrate dispersed production, engineering, product development, and research across geographic borders. Knowledge-sharing is the glue that keeps these networks growing, strengthening the potentially powerful link between trade and innovation.[8]

New research has documented that international trade today is closely intertwined with foreign direct investment (FDI) as part of increasingly complex global value chains or global production networks (GPNs).[9]

---

[5] WTO. n.d.(a). Special and Differential Treatment Provisions. Trade and Development Committee. Available at www.wto.org/english/tratop_e/devel_e/dev_special_differential_provisions_e.htm.

[6] See the important contribution by Feldman, M. and N. Lowe. (2017). "Evidence-Based Economic Development Policy." *Innovations: Technology, Governance, Globalization, 11*(3), 34–49.

[7] Johnson, H. (1977, p. 187), as quoted in Irwin, D. A. (1996). *Against the Tide. An Intellectual History of Free Trade*. Princeton University Press, Princeton, New Jersey: p. 3.

[8] See Chapter 4 (Beyond value capture — Exploring innovation gains from global networks)

[9] On the proliferation of these networks, see Ernst, D., 1997, From Partial to Systemic Globalization: International Production Networks in the Electronics Industry, Berkeley Roundtable on the International Economy (BRIE) Working Paper 98, April: 113 pages; Ernst, D. and P. Guerrieri (1998). "International Production Networks and Changing Trade

Trade theory demonstrates how trade and global networks can increase global incomes substantially via efficiency-enhancing structural adjustment, economies of scale and scope via an international — rather than national — marketplace, improved incentives for investment in technology, and greater competition. There is broad consensus among trade and development economists that, in principle, a country's integration into global networks may facilitate access to best-practice technology and management approaches.

Integration into global networks could thus enable developing countries to overcome "barriers to exporting by accommodating specialization in narrow business functions and niche activities and [to] limit dependence on the degree of industrial development and broader skills set in the country" (OECD and World Bank, 2015). A new "gains from trade" doctrine for economic development has emerged that emphasizes the role of global network integration as "the 21st century's fast lane to industrial development" (Baldwin, 2013, 2014). By joining GPNs, it is argued that "developing countries can benefit from foreign-originated intellectual property, trademarks, managerial and business practices, marketing expertise, and organizational models" (Taglioni and Winkler, 2014).

This is in line with important insights from Paul Romer's "endogenous growth theory", which states that trade provides access to "new types of goods and new types of productive activities being introduced from abroad" and thus stimulates productivity and technology diffusion.[10] To facilitate such global knowledge sourcing, countries have engaged in various rounds of trade liberalization and are searching for ways to

---

patterns in East Asia: The Case of the Electronics Industry". *Oxford Development Studies, 26*(2); Ernst, D. and L. Kim (2002). "Global Production Networks, Knowledge Diffusion and Local Capability Formation". *Research Policy*, special issue in honor of Richard Nelson and Sydney Winter; Ernst, D., "Complexity and Internationalization of Innovation: Why is Chip Design Moving to Asia?". *International Journal of Innovation Management*, March 2005, special issue in Honour of Keith Pavitt. See also most recently, Baldwin, R. (2016). *The Great Convergence. Information Technology and the New* Globalization, The Belknap Press of Harvard University, Cambridge, Mass.

[10] Romer, P. (1994). "New Goods and the Welfare Costs of Trade Restrictions". *The Journal of Development Economics, 43*(1):6.

improve compliance with international patent rules and standards. With but a few exceptions, emerging and developing countries have been especially active in pursuing this new pathway to industrialization.[11]

However, the distribution of gains from trade and global networks remains highly unequal, both within countries and between them. According to the OECD, large sections of the middle classes in advanced countries have not experienced gains in their real incomes over the past 25 years or so.[12] Branko Milanovic's pioneering study on "global inequality" shows that, between 1988 and 2008, two groups experienced the greatest real per capita income gains: (1) the global top richest 1% concentrated in the advanced countries; and (2) median income groups in emerging economies, predominantly China, but also India, Thailand, Vietnam, and Indonesia. The result is a major increase in global inequality, with erstwhile median income households in advanced economies being the main relative losers.[13] Recent OECD research shows that wage inequality has increased in recent years and that globalization and digitalization are widening the gap between the growth of wages and productivity. Equally important, the polarization of labor demand threatens those with average skills. The perceived threat is that we are heading toward a "winners take all" economy.

Of great concern for the distribution of gains from trade for innovation is evidence that points to growing corporate concentration. Global oligopolies are spreading, especially in "industries characterized by network externalities (such as firms in the expanding digital sector) with 'winner-take-most' features, and the integration of different economies can help to allow a small number of dominant firms (like *Apple*, *Facebook*, *Google* in the US, and *Baidu, Alibaba, Tencent* in China) in each industry to emerge and earn growing rents. There is certainly a widespread perception that some multinational enterprises (MNEs) have become excessively large and powerful and, in some cases, use their

---

[11] Baldwin, R. (2016). "The World Trade Organization and the Future of Multilateralism". *Journal of Economic Perspectives,* 30(1): 95–116.

[12] OECD, 2017, *Fixing Globalisation: Time to Make it Work for All*, OECD, Paris, April: p. 9.

[13] Milanovic, B. (2016). *Global Inequality. A New Approach for the Age of Globalization.* Cambridge, MA: Harvard University Press.

power to entrench their position (e.g., pharmaceutical companies pushing for stronger protection of intellectual property rights in developing economies). Tax evasion and aggressive tax avoidance by some MNEs, together with bribery and corruption, further add to the problem" (OECD, 2017: *ibid.*).[14]

All of this suggests that liberalization needs to be accompanied by well-resourced and targeted active government policies to facilitate structural adjustment, provide training and compensation to displaced workers, and support the most vulnerable. Such policies are necessary for both economic efficiency and equity-related reasons. The above OECD study, for instance, makes a strong case for "complementing advances in globalization by policies to ensure that the benefits of trade, investment and digitalization are more widely and more equally distributed and that all feel they contribute to the dynamic environment created by these forces. Key policy tools include social protection and labor market activation policies, measures to ensure healthy financial markets, and strategic investments in education, skills, health, innovation and physical infrastructure." (OECD, 2017: Key Messages).

Even before the rise of the new "Economic Nationalism", existing trade-adjustment-assistance programs were woefully under-funded and subject to other problems (e.g., bureaucracy) that prevented them from achieving the goals of compensation and facilitating re-integration of displaced workers. Rather than expanding public policies to mitigate trade-adjustment costs, a resurgent economic nationalism appears committed to scaling back such policies even further. Of course, if economic nationalism leads to trade wars, there will be fewer workers displaced by import competition. Still, workers in exports industries will be hit and economic growth could suffer severely. Such was the result of the economic nationalism manifested in the Smoot–Hawley tariff of 1930, which led to retaliation by foreign partners and an exacerbation of the Great

[14] For similar findings about unequal distribution of gains from trade, see *Making Trade an Engine of Growtrh for All. The Case for Trade and for Policies to Facilitate Adjustment,* a report prepared for discussion at the meeting of G20 Sherpas, March 23–24, 2017, Frankfurt, Germany, prepared by the staff of the International Monetary Fund, The World Bank, and the World Trade Organization.

Depression. The US economy was far less exposed to foreign trade back then; the impact of a trade war could be greater this time around.

In short, economic policy is essential in ensuring that gains and losses from trade are shared by all. Equally important however is the role that economic policy plays in fostering a wide distribution of gains from trade for innovation. After all, a country can use, absorb, and further develop imported ideas and knowledge only if its firms (both producers and users) can tap into strong domestic technological and management capabilities.

This may be less of a concern for the United States and other advanced countries whose innovation systems have developed over more than a century.[15] However, for countries like China or Malaysia, developing such capabilities as a latecomer requires supporting industry and innovation policies.[16] This is why Romer's proposition needs to be balanced with the "infant economy" argument, as formalized by Greenwald and Stiglitz (2006).

Philippe Aghion (2014) summarizes the argument as follows: For developing countries with a nascent industrial sector, "full trade liberalization will make it very costly for domestic industrial sectors to invest in learning by doing … since domestic costs are initially higher than foreign

[15] But even for the US, Ralph Gomory, a former IBM SVP for R&D, has documented that "the self-interests of our corporations, especially our large global corporations, can diverge from the interests of our country, We need to find ways to fundamentally change the motivation of corporations to align with those of our country." (Gomory, Ralph, 2009, "Country and Company: Part I — Divergent Goals", *The Huffington Post*, March 16. Updated March 25, 2011, http://www.huffingtonpost.com/ralph-gomory/country-and-company-part_b_174875.html). For a detailed analysis, see Berger, S. (2013). *Making in America: From Innovation to Market*, Cambridge, MA: MIT Press.

[16] See for instance, Ernst, D. (2017). *Advanced Manufacturing and China's Future for Jobs,* in Paus, Eva (editor), Emerging Dystopias? *The New Technological Revolution and the Future of Work*, Cornell University Press, Ithaca, NY 14850; Ernst, D., 2011, Indigenous *Innovation and Globalization: The Challenge for China's Standardization* Strategy, UC Institute on Global Conflict and Cooperation; La Jolla, CA and East-West Center, Honolulu, HI., 123 pages http://www.EastWestCenter.org/pubs/3904 ; and Ernst, D., 2004, "Global Production Networks in East Asia's Electronics Industry and Upgrading Perspectives in Malaysia," chapter 3 in Shahid Yusuf, M. Anjum Altaf and Kaoru Nabeshima (eds.), Global Production Networking and Technological Change in East Asia, The World Bank and Oxford University Press.

costs [and] the social benefits from learning by doing are not fully internalized."[17] At the same time, innovation theory tells us that strengthening national innovation capabilities improves a country's ability to engage in and benefit from the international trading system" (Kim and Nelson, 2000; Ernst *et al.*, 1998). In fact, the gains from trade through global network integration are contingent — or a country's capacity to capture those gains depends — "on the structure of specialization and the level of development" (Kowalski *et al.*, 2015).

3. A resurgent economic nationalism creates new uncertainty

Rising economic nationalism, as embodied in the "America First" doctrine of the Trump Administration, has added new challenges to this enlightened globalization agenda. In a fundamental break with established US trade diplomacy, the new US trade dogma is ostensibly predicated on the fundamental arguments of *mercantilism* that trade is essentially a zero-sum game in terms of its welfare effects, and hence an optimal policy is to ensure that exports exceed imports.[18]

At his inauguration, President Trump laid out a vision of unabashed protectionism: "We must protect our borders from the ravages of other countries making our products, stealing our companies and destroying our jobs. Protection will lead to great prosperity and strength." Moreover, "We will follow two simple rules: buy American and hire American." Success is measured by the following criteria: any trade deal "must increase the

---

[17] Aghion (2014: 498) argues that "adequately targeted sectoral intervention, e.g., to more skill-intensive or to more competitive sectors, can enhance growth," especially for developing countries who seek to benefit from trade and global network integration.

[18] For Thomas Mun, a director of the *East India Company* and one the most well-known of the mercantilist writers, the best way to increase the wealth of a nation was to "sell more to strangers yearly than we consume of theirs in value." (Mun, Thomas, 1664, *England's Treasure by Foreign Trade,* republished in 2013 in *Thomas Mun, The Complete Works: Economics and Trade*, Newton Page, Jul 29, 2013 — Business & Economics — 296 pages).

growth rate [of the economy], decrease the trade deficit and strengthen the US manufacturing base".[19]

While some observers had hoped that over time harsh rhetoric would yield to pragmatism, no such change was in evidence during the first months of the Trump Administration. By May 2017, "... [t]he administration has not distanced itself from the doctrine that bilateral trade balances are to be considered the litmus test to determine whether particular trade agreements are benefitting the United States. In addition, bilateral agreements are to be the path forward, not multi-party agreements. There is currently no positive trade agenda for the United States participating in, much less leading, multi-party agreements — nothing for the WTO, nothing for the Trade in Services Agreement (TiSA), nothing on replacing the Trans Pacific Partnership Agreement and nothing with respect to continuing negotiations with the European Union through the Trans-Atlantic Trade and Investment Partnership (TTIP)."[20]

Such "protectionist bilateralism" creates pervasive uncertainty with unknown consequences for the rules-based system of global governance established after World War II. To many foreign observers, the "America First" doctrine is tantamount to a declaration of economic warfare. Specifically regarding China, a declared intention of the Trump administration, repeated frequently on the 2016 campaign trail, has been to impose an across-the-board 45% tariff on Chinese imports. In addition, the Trump administration is threatening to implement broad sanctions under Section 301 of the 1974 US Trade Act, restrictions on steel and aluminum imports due to "national security" considerations, and aggressive screening of inward Chinese FDI.

The United States remains the predominant economic and military power, but it is increasingly unable to dictate the rules of international trade.

---

[19]Wolf., Martin, 2017, "Trump and Xi battle over globalization", *Financial Times, February 1.*

[20]Wolff. Alan Wm., 2017, *The U.S.–Japan Relationship and American Trade Policy, Observations of a Former Trade Negotiator*, presented at the conference on Japan's Global and Domestic Futures and the U.S. Relationship, The Center on Japanese Economy and Business, Columbia Business School, Tokyo, Japan, May 25. The author is Deputy Director General of the World Trade Organization and a previous Chairman of the Board of the U.S. National Foreign Trade Council (NFTC).

Both the United States and China would pay a heavy price in the case of a trade war, with severe knock-on effects in other areas as well. The Chinese government has already begun to issue warnings that a US-led trade war will hurt both sides.[21]

The Chinese government may have a strong desire to negotiate a Grand Bargain, "partly because it needs time to identify the bottom line of the Trump administration and partly because of the critical importance of the United States as its largest export market."[22] However, if no compromise can be worked out, China can very effectively retaliate and can do considerable damage to US exports (agriculture, Boeing, semiconductors). In particular, China could target companies in critical Midwestern swing states that enabled Donald Trump's victory. China is prepared for a long battle of attrition, ranging from a campaign of silence and non-cooperation to retaliatory open-trade warfare if it came to that. In the open US system, it is much harder to manage discontent and political blow back than in the authoritarian Chinese system. A US-engineered trade war could easily backfire.

Equally important, a return to mercantilism plays havoc on the domestic policy front as well. Proposed changes in healthcare and for tax reform have the potential to be highly regressive, exacerbating the inequality problem in the United States. In 2014, the US income distribution (as measured by a Gini coefficient of more than 40 on a scale of 100) is now more like that of a developing country than an advanced country."[23] In addition, the White House has proposed drastic cuts in the budgets for programs on trade-adjustment-assistance, social welfare, climate change, as well as major science and technology programs. Taken together, these policies are likely to weaken even further the scope for structural adjustment policies needed to compensate the losers from trade, to provide

[21]For example, Wall Street Journal, "China Warns the US on Trade: We Are Both Hurt in a Fight, August 3, 2017, Available at: https://www.wsj.com/articles/china-pushes-back-on-u-s-accusations-on-trade-urges-cooperation-1501751592.

[22]Zhai, Fan, 2017, "Trade Cooperation and Conflicts between the United States and China: Risks and Realities", chapter 16, in Posen A.S. and Ha Jiming, editors, *US-China Cooperation in a Changing Global Economy*, PIIE Briefing 17-1: 151.

[23]Wolf, Martin, 2017, "A clash of civilisations or community?", *Financial Times*, 12 July: p. 9.

training for displaced workers, and to invest in new advanced manufacturing and service industries.

4. Rethinking the framework for Megaregionalism 2.0

Writing new rules and updating old ones is of the essence in order to accommodate unprecedented change. Equally important are transparent processes of negotiating and implementing new framework conditions.

4.1. *Limitations to multilateralism*

Developing such rules at the multilateral level would be best in theory, but the success in expanding the membership of the WTO to 164 economies with very different policy priorities has rendered progress difficult: the Doha Development Agenda (DDA), launched in November 2001, has been at an impasse since 2013 and is unlikely to move forward in the short-term (if at all).[24] The last time that comprehensive new trade rules were put into place in the global system was in 1995, with the implementation of the Uruguay Round. The new international economy, in which globalization plays an essential role, has changed radically since then but without global rules to accommodate and facilitate it.

Hence, bilateral and regional trade agreements have emerged as key commercial policy initiatives designed to bring down barriers to trade and update trade-related rules, regulations, and standards. These agreements are mostly in the form of bilateral free-trade areas (FTAs), as they tend to be the easiest to negotiate as "coalitions of the willing" between like-minded economies. While subject to the (mild) exigencies of Article XXIV of the GATT/WTO, these FTAs can vary significantly in terms of depth, coverage, and ambition. Bilateral and regional FTAs negotiated by OECD economies tend to be the "deepest", but FTAs

[24]For the US, the decline of the Doha round was definitive, even before the Trump administration came to power. For former USTR Michael Froman, "moving beyond Doha… [and] freeing ourselves from the strictures of the Doha framework" was the main theme of his remarks to the 10th Ministerial Conference of the WTO in Nairobi, Kenya (USTR 2015b).

among developing economies are also increasing in ambition, with the ASEAN Economic Community (AEC), established in December 2016, being a salient example.

To date, empirical estimates of the economic effects of existing FTAs tend to be positive (see WTO 2011 and Gilbert 2016 for surveys). Still, given the importance of global and regional production networks as sources of 21st century trade and investment, FTAs with a limited number of partners will inevitably fall short of their potential for many reasons, from trade and investment diversion to costs associated with rules of origin. Moreover, few are large enough to boast the possibility of creating global industrial standards.

4.2. *A focus on new international benchmark standards for trade*

"Megaregional" trade agreements, such as the TPP and the Regional Economic Comprehensive Partnership (RCEP), are ambitious trade agreements devised to address these problems by including many partners at varying levels of economic development and embracing wide coverage and depth. The TPP in particular was intended to be a "21st century" trade agreement with the aim of setting a "gold standard" for regional cooperation. Signed by its 12 negotiating member states in February 2016, the TPP included cutting-edge chapters on the digital economy, state-owned enterprises engaged in trade, trade in services, intellectual property rights (IPR) protection, non-tariff barriers, and protection of labor and environmental standards, while at the same time cutting virtually all tariffs to zero (except in a few notable sectors, such as agriculture).

Megaregional trade agreements seek to bypass the stalemate in multilateral negotiations under the WTO. While its original intentions were ambitious when it was launched in November 2001, the DDA has yielded little in terms of concrete progress: the "Aid for Trade" initiative, the expansion of the Information Technology Agreement and the Government Procurement Agreement, and the Trade Facilitation Agreement are the main fruits of more than a decade and a half of negotiations. The main purpose of megaregionals, composed of a group of like-minded economies, is to find ways to create and harmonize new international benchmark standards for trade, for instance for IPR, technical standardization, government procurement, taxation, and competition policy, in a way that is not possible in the context of the wider WTO membership, as well as to bring down traditional tariff and non-tariff barriers. The move toward megaregionalism is driven by this need for deep integration in order to foster knowledge-sharing within international corporate networks of production and innovation.

The new architecture of these megaregional trade agreements is shaped by the emergence of international corporate networks that integrate dispersed production, engineering, product development, and research across geographic borders.[25] While the proliferation of global production networks dates back to the late 1970s, a more recent development is the rapid expansion of global innovation networks (GINs), driven by the relentless slicing and dicing of engineering, product development, and research.[26]

Megaregional trade agreements attempt to provide new international benchmark standards for international trade and investment within such complex global networks of production and innovation. And yet, we still know little about how the megaregionalism trend might change the link between trade and innovation.

---

[25] On the proliferation of these networks, see Ernst, D., 1997, From Partial to Systemic Globalization: International Production Networks in the Electronics Industry, Berkeley Roundtable on the International Economy (BRIE) Working Paper 98, April: 113 pages; Ernst, D. and P. Guerrieri, 1998, "International Production Networks and Changing Trade patterns in East Asia: The Case of the Electronics Industry", *Oxford Development Studies*, Vol. 26, No. 2; Ernst, D. and Linsu Kim, 2002, "Global Production Networks, Knowledge Diffusion and Local Capability Formation", *Research Policy*, special issue in honor of Richard Nelson and Sydney Winter; Ernst, D., "Complexity and Internationalization of Innovation: Why is Chip Design Moving to Asia?", *International Journal of Innovation Management*, March 2005, special issue in Honour of Keith Pavitt.

[26] Ernst, D. (2012). "Production and innovation networks, global", in Encyclopedia of Global Studies, Sage Publications (pp. 1393–1397). Thousand Oaks, CA; Ernst, D., A New Geography of Knowledge in the Electronics Industry? Asia's Role in Global Innovation Networks, Policy Studies #54, August 2009, East-West Center, Honolulu, USA.

4.3. *Potential economic effects*

Nevertheless, empirical economic analysis provides estimates of overall potential economic and structural effects. As noted in Chapter 1, the economic gains of these agreements are estimated to be large. For example, Petri and Plummer (2016) estimate that the TPP would increase global incomes by about a half-trillion dollars, with the United States being the biggest winner ($131 billion per year relative to the baseline beginning in 2030). The economic impact of RCEP could increase global incomes by even more; Petri *et al.* (2012) estimate the gains to be on the order of 1.9% of member-country GDP by 2025 ($664 billion), again relative to the baseline scenario.[27] Importantly, this income growth will always be led by trade; for example, exports of RCEP economies are projected to increase by 14% due to the agreement. Since many of the most important variables driving these gains can be traced to non-discriminatory measures (e.g., improvements in trade facilitation, investment- and service-related measures, digital economy), non-partner countries are only marginally affected and, in some cases, gain due to these megaregional accords. Indeed, regionalism of the 21st century is very different to that of the 20th century, which relied on the liberalization of traditional parameters — such as tariffs — that lead to trade diversion. The Trump Administration pulled the United States out of the agreement, but the remaining 11 countries continue to negotiate a TPP without the United States,[28] with the hope that it may join in the future. Also, several other economies (to date, Indonesia, the Philippines, South Korea, Taiwan, and Thailand) have expressed interest in possibly joining the TPP.

There have been legitimate worries that trade has a deleterious effect on income distribution, a salient contemporary policy concern, megaregional accords included. By improving efficiency via structural change, trade increases economic welfare, but the "winners" and "losers" of this process are different agents. The same is true of often disruptive transformations imposed on manufacturing and services by automation, smart robots,

[27] For details, see www.asiapacifictrade.org .

[28] Given the size of the United States, the TPP in its present form could not move forward without the United States, according to Chapter 30. To move forward, at a minimum, the remaining 11 economies must revise that chapter. On TPP-11, see Preface.

3D printing, artificial intelligence, and the Internet of Everything. Certainly, it is a major public policy challenge to compensate the losers by promulgating compensation (e.g., through a strong general safety net, targeted programs to help trade-displaced labor, wage-insurance policies) and spearheading measures to facilitate quick re-entry into the job market (e.g., via training programs, job-market information, and job-matching programs).

An even more fundamental challenge is raised by the renewed debate about what role a "guaranteed minimum income" could play in creating a level playing field for all participants in global trade, including the current losers. While this is not the place for a detailed discussion, it is worth noting that the idea has a long history and has been supported by a broad coalition of leading economists, including Paul Samuelson, James Tobin, and Kenneth Galbraith in the late 1960s.[29] Today, the search for viable mechanisms of guaranteed minimum income has moved to center stage in debates about compensating the losers of job-displacing technical change.[30] It is an approach that could play an important role for an enlightened globalization agenda and requires further study.

In short, a combination of these support policies is necessary to ensure equity and efficiency, as well as to reassure the public that trade is not the enemy of the working class, a theme embraced by populists of the left and the right. Globalization needs to be managed, not thwarted, which would be to the economy's severe detriment.

Still, the effects of trade on income distribution are less significant than is often conjectured in the media. In fact, trade liberalization can actually improve income distribution. For example, in the case of the TPP,

---

[29] See for instance "*Economists' Statement on Guaranteed Annual Income*, 1/15/1968-4/18/1969 folder, General Correspondence Series, Papers of John Kenneth Galbraith, John F. Kennedy Presidential Library. Cited in: Jyotsna Sreenivasan, "Poverty and the Government in America: A Historical Encyclopedia." (Santa Barbara: ABC-CLIO, 2009), page 269. It is noteworthy that the Nixon Administration responded with a policy initiative *The Politics of a Guaranteed Income: The Nixon Administration and the Family Assistance Plan*,1973, by Daniel Patrick Moynihan.

[30] See for instance, Ford, M., "Rise of the robots: technology and the future of work", in Paus, Eva (editor), Emerging Dystopias? *The New Technological Revolution and the Future of Work*, Cornell University Press, Ithaca, NY 14850.

Petri and Plummer (2016b) estimate that, because labor gains more than capital and the lower prices of goods from TPP countries would disproportionately help the less prosperous in the United States, income distribution in the United States should improve.[31] These estimates are consistent with those of Lawrence and Moran (2016).

How these megaregional trading agreements will affect innovation and invention will clearly determine, to no small degree, their ultimate success. Sustainable economic growth and prosperity for all increasingly depends on innovation and invention, and this is true for developed, emerging, and developing economies. For example, global innovation networks are evolving hand-in-hand with the global production networks in the Information and Communications Technology (ICT) industry; in turn, updated and improved rules and standards in the ICT industry, as well as lower barriers to trade, expand significantly the potential of these networks and the ability of emerging and developing economies to participate in them.

5. The Two-way Nexus between Innovation and Trade

Innovation reshapes international trade and production

Innovation has reshaped the structure of international trade and production, *inter alia,* through generic technology platforms such as information and communication technology, new materials, and biotechnology.[32]

---

[31] http://voxeu.org/article/economics-tpp-winners-and-losers .

[32] According to recent MIT research, these generic technology platforms encompass, for instance: Synthesized new materials (e.g., nano-engineering), as well as custom-designed and recycled materials; continuous manufacturing of pharmaceuticals and bio-manufacturing; green sustainable manufacturing; mass customization, for instance through Additive Manufacturing (3DP) and reconfigurable robotics which might enable Continuous Manufacturing in small batch sizes and break down the boundaries between fabrication and assembly; and integrated solutions through bundling of physical products with services and software. (Berger, S., 2013, Making in America. From Innovation to Market, Cambridge, MA: The MIT Press.)

Richard Baldwin and colleagues have captured the transformation of trade through innovation.[33] For Baldwin:

> Trade in today's world is radically more complex. The information and communications technology revolution has internationalized supply chains, which has created a tight supply-side linkage between trade and FDI: the "trade–investment–service–IP nexus". Today's international commerce comprises complex, two-way flows of goods, services, people, ideas and investments in physical, human and knowledge capital — in addition to trade in raw materials and final goods. These connections make it almost irrelevant to talk about trade without also talking about FDI — at least for many products and markets....As a result,...trade and investment are neither complements nor substitutes — they are simply two facets of a single economic activity: international production sharing. [34]

The electronics industry has been in the vanguard of this transformation; production is increasingly characterized by fragmented trade at the regional level — i.e., a regional division of tasks across multiple countries in the production of a final good — and is being driven by technological innovation.[35] But this modularization of all stages of the

[33]Baldwin, R., 2013, "Global supply chains: why they emerged, why they matter, and where they are going", chapter 1 in: D.K. Elms and P. Low, eds., 2013, Global value chains in a changing world, WTO, Geneva: pages 13–60; Baldwin, Richard and J. López González (2013) "Supply-Chain Trade: A Portrait of Global Patterns and several testable hypotheses" NBER Working Paper 18957 http://www.nber.org/papers/w18957.pdf.

[34]Baldwin, R. 2013, "The New Relevance of FDI: The GVC Perspective", in *World Economic Forum, Foreign Direct Investment as a Key Driver for Trade, Growth and Prosperity. The Case for a Multilateral Agreement on Investment,* Geneva: p. 13.

[35]Ernst, D., 2009, A New Geography of Knowledge in the Electronics Industry? Asia's Role in Global Innovation Networks, Policy Studies #54, August 2009, East-West Center, Honolulu, USA. For empirical research on China's global network integration, see Watanabe, M., ed., 2014, *The Disintegration of Production. Firm Strategy and Industrial Development in China*, Edward Elgar, Cheltenham, UK, for the Insitute of Developing Economis (IDE), JETRO; and Ernst, D., 2014, *From Catching Up to Forging Ahead?*

value chain has moved well beyond the electronics industry. Today, the dispersion of global networks of production, product development, and research is driving production and trade in myriad goods and services sectors.

5.1. *Trade as an enabler of innovation*

Equally important however is the other side of the trade nexus with innovation: the role of FDI as a channel of knowledge spillovers and co-invention, as articulated in Lee Branstetter's pioneering work.[36] Economic research typically focuses on three specific channels through which trade could strengthen a region's innovation capacity: (1) imports, FDI, and technology licensing; (2) learning-by-exporting, which both exposes the region to foreign technology and intangible knowledge as a source of product and process innovation; and (3) competition, which may reduce monopoly rents from innovation and create pressure to increase productivity.[37]

Thus, the link between trade and innovation is critical for trade laws and diplomacy as well as for innovation policy. Equally important is that trade and innovation are mutually interdependent: innovation has increased the benefits of trade due to its salutary effects on costs and the generation of a greater variety of inputs to firms and goods to consumers. At the same time, international trade facilitates the application of new innovations by providing new markets and economies of scale.

China's Prospects in Semiconductors, East-West Center Working Papers: Innovation and Economic Growth Series, no. 1 (November 2014).

[36]Branstetter, L., 2006, "Is Foreign Direct Investment a Channel of Knowledge Spillovers: Evidence from Japan's FDI in the United States," Journal of International Economics, vol. 68, February 2006, pp. 325–344. See also Branstetter, L., Guangwwei Li and Francisco Veloso, 2014, "The Rise of International Co-invention", Chapter in forthcoming NBER book *The Changing Frontier: Rethinking Science and Innovation Policy*, Adam Jaffe and Benjamin Jones, editors.

[37]Kiriyama, N., 2012, Trade and Innovation: Synthesis Report, OECD Trade Policy Papers, No.135, OECD, Paris, and Onodera, O., 2008, Trade and Innovation: a Synthesis Paper, OECD Trade Policy Working Paper No.72, August 7.

6. Alternative futures for Megaregionalism 2.0

6.1. *Impacts of the US withdrawal from the TPP*

To assess the stakes involved in the TPP, which was the first megaregional agreement to be concluded in the Asia-Pacific region, let us begin with some general economic context. Of the world's $19 trillion in trade in 2013, $12 trillion involve countries of the Asia-Pacific Economic Cooperation (APEC) organization[38] as either an exporter or importer or both (World Bank, WITS). Its share of FDI flows are no less impressive; in 2014, APEC accounted for 53% of global FDI inflows (UNCTAD 2015).

These figures reflect in part the explosive growth of the Chinese economy; China is now the world's largest trader, with a total trade of $3 trillion in 2013 up from $280 billion in 1995 (World Bank, WITS). It was the largest recipient of FDI flows in 2014 and the third largest source of FDI flows, behind the United States and Hong Kong (UNCTAD 2015). It is also the first or second most important export market of nearly all other Asia-Pacific economies, and its share in their exports has been rising in all cases since 1995 (World Bank, WITS).

Indeed, the structural transformation of the Asia-Pacific economy has been dramatic as well. The region is the testbed of the new global value chains made possible by the spread of ICT and improvements in connectivity. Asia has become a global manufacturing hub; it collects components from many countries and bundles them into products destined for European, US, and other markets. China is now a key hub in this process, but many other countries have emerged as important upstream participants as well, including advanced industrial countries. The latter play especially important roles in organizing value chains, linking stages from early design through manufacturing, and ultimately marketing.

[38]The membership of the Asia-Pacific Economic Cooperation (APEC) forum currently includes 21 economies: Australia, Brunei Darussalam, Canada, China, Chinese Taipei, Chile, Hong Kong, Indonesia, Japan, Malaysia, Mexico, New Zealand, Papua New Guinea, Peru, the Philippines, Russia, Singapore, South Korea, Thailand, the United States, and Vietnam.

Radical changes in technology are also creating new markets and forms of economic integration. The most prominent example is the rise of the digital economy, made possible by the internet and related innovations. ICTs reduce costs by automating business processes, improving information flows that enable producers to organize more complex production systems, and making many kinds of services, and especially new services emanating from the digital economy, tradeable (APEC, 2008). ICT is ubiquitous; internet services are regularly accessed in developed and developing economies. The digital revolution is fundamentally altering the way governments interact with their citizens as well as with each other (Tapscott and Agnew, 2015).

As noted above, rules for the digital economy have not kept pace with these massive changes; indeed, they are still in their infancy. The Uruguay Round was concluded well before the digital economy took shape. The more recent WTO-based Information Technology Agreement (ITA) and now the ITA2 are pertinent to the digital economy but relate to trade in physical goods, not electronic goods and services, and do not address rule-making issues (see Chapter 17).

Economic development is also making the service sector increasingly prominent. The share of services value-added in GDP has risen virtually in every country in recent years and is now greater than 50% in most developed and emerging markets; in the TPP economies, for example, the range is between 70% and 78% for the high-income economies, and 50% to 63% for all others, with the exceptions of Brunei Darussalam (31%) and Vietnam (43%).[39]

Negotiations on the TPP began in 2008, building on the earlier Trans-Pacific Strategic and Economic Partnership agreement concluded by Brunei, Chile, New Zealand, and Singapore. When signed, the TPP included 12 APEC members representing about two-fifths of world output and one-quarter of global trade, ranging widely in income from Vietnam and Peru to Japan and the United States. The 2011 TPP Leaders Statement underscored five objectives: comprehensive market access; a regional agreement that facilitates production and supply chains; cross-cutting trade issues; challenges emerging from new technologies; and a "living

[39] World Bank, http://data.worldbank.org/indicator/NV.SRV.TETC.ZS.

agreement" that brings in additional members over time (Congressional Research Service 2015). After a long and intense negotiating process, the accord itself accomplished these goals; in October 2015, an agreement was reached and a final version was signed by all 12 economies in February 2016. Many countries ratified the agreement over the following year. But the Trump Administration decided to notify its 11 partner countries of its intention to withdraw from the agreement on Monday, January 23, 2017, three days after officially taking office.

The TPP negotiations do not include China, the second largest economy in the Asia-Pacific. Chinese membership could have advanced the agreement's integration objectives given its size and role as a trading hub in East Asia, and would have made the agreement a more compelling model for future agreements. Including China might have made it more difficult to agree on rules, and, to some observers, it was better to start this unusually challenging negotiation with a less complex mix of partners. Estimates that compare the benefits of the TPP with those of larger alternatives that include China confirm the importance China's participation (Petri *et al.*, 2012).

The TPP seeks to improve market access by eliminating "high-hanging" barriers that escaped prior rounds of liberalization in the WTO. Even so, negotiators could not agree on a region-wide approach, and so left it up to each member whether to keep old FTAs in place or rewrite them, and whether to make new offers on a bilateral or multilateral basis (Barfeld 2011). Thus, new liberalization focused on links not previously covered by FTAs, such as Japan–US trade, as well as sectors neglected in existing FTAs, such as agriculture, technology, and services. It also included provisions to help align technical and sanitary and phytosanitary (SPS) standards.

The TPP attempted to circumscribe national regulations in several areas in order to promote deeper integration. For example, it liberalizes all services and trade-related investments except for sectors reserved on a negative list; attempts to keep digital trade free of barriers that apply to physical goods; prescribes national treatment in government procurement; prohibits government support for state-owned enterprises as they compete abroad; offers foreign investors national treatment both pre- and post-investment; and allows foreign investors to challenge government decisions that reduce the value of their assets through an arbitration process ("investor–state dispute settlement", or ISDS). In several of these policy areas, regulations to protect health and safety are explicitly reserved for national governments, including especially the regulation of the consumption of tobacco.

6.2. *TPP withdrawal by the Trump administration and its implications*

During the 2016 US presidential campaign, all major candidates opposed the TPP agreement. This included Hilary Clinton, who was a strong supporter of the agreement while US Secretary of State. Trade is an easy target for criticism in any US election — as well as elsewhere — but the strong negativity of this election cycle was in many ways unprecedented. Democratic contender Bernie Sanders and Republican candidate Donald Trump were the most emphatic in targeting trade as a multinational corporate grab and "threat to democracy" (Sanders) or "another disaster done and pushed by special interests" (Trump). While presenting no convincing evidence to support these arguments, the candidates used antagonistic rhetoric toward trade in general and the TPP in particular as means to extoll their populist credentials. Globalization has become a political punching bag as never before in the United States, at least in the post-war era.

Rhetoric aside, the TPP was actually a good deal for the United States. After all, the United States was clearly the lead party in the TPP negotiations and pushed relentlessly its own interests and comparative advantage industries, e.g., taking strong positions on the inclusion of services, tariff reductions in high-end manufactures, ISDS, disciplines on state-owned enterprises, IPR, and labor standards, often facing significant resistance from its partner-countries (one reason why it took over six years to negotiate). It should come as no surprise that economic studies would suggest that the United States would gain the most. Indeed, parties criticizing the TPP in the other 11 member-economies often opposed it because, in their view, it was an agreement that favored US interests disproportionately.

In addition to foregone losses in terms of economic welfare, pulling out of the TPP is costly to the United States for a number of additional reasons:

- Since Asia-Pacific economies are interested in opening up markets with or without the United States, with the rejection of the TPP, additional bilateral, sub-regional, and even regional FTAs will be formed, with the United States suffering the associated trade and investment diversion;
- In particular, this pertains to RCEP (discussed below). Not only will the United States likely suffer from trade and investment diversion in certain sectors but it would also find itself outside of discussions that may well serve to create templates for new rules governing trade, given the vacuum left by the TPP;
- The United States has extensive strategic interests in the Asia-Pacific region, which was true long before President Obama's "Asian Pivot". With its withdrawal from the TPP, the region will continue to question the commitment of the United States as a strategic partner (this is a point frequently made in the related literature, including the USITC February 2016 Report);
- The US commitment to being a leader in forging new, effective global governance of the international marketplace in the 21st century is also now being questioned, not just due to its withdrawal from the TPP but also the new Administration's insistence on renegotiating NAFTA, possibly renegotiating or even abrogating the US–South Korea FTA, and threats to withdraw from the WTO;
- The Labor Chapter of the TPP and associated bilateral side letters with Vietnam, Malaysia, and Brunei are remarkable in that they ensure that there will be no race to the bottom in terms of labor regulations and guarantee freedom of association, the right to strike, prohibition against child labor, and other measures consistent with the *ILO Declaration*. This is not the first time that the US has included a labor chapter in an FTA, but it is the first time that it is actionable.

Perhaps the greatest irony is that the biggest beneficiary from the US withdrawal from the TPP will likely be China, despite the fact that during the US 2016 Presidential campaign Donald Trump was particularly critical of China and the role that it is playing in the global system. Strangely, one of his arguments against TPP was that it would help China, even though China was not a negotiating partner and, in fact, was slated to lose the most due to trade and investment diversion and preference erosion (Petri and Plummer, 2016). In addition to not having to bear the negative effects of US participation in the TPP, the loss of stature of the United States in the region is clearly a big win for China. In fact, in terms of leadership, China now holds the upper hand in Asia, and certainly not just due to the US exit from the TPP. China has been active in providing technical assistant to many of the region's economies; its One Belt, One Road (OBOR) initiative is slated to increase Chinese economic integration with Asia and Europe significantly; and its Silk Fund is being devoted to helping spur economic development along the OBOR.

The US exit from the TPP will also increase incentives to finish the RCEP agreement by the end of 2017, a goal that has been set by China and other regional economies. The combined GDP of RCEP economies came to about $24 trillion in 2016, somewhat larger than the TPP12. [40] RCEP is the culmination of long-standing efforts to pursue Asia-centered integration, mostly based on ASEAN's efforts to promote regional economic integration (Petri and Plummer, 2014). ASEAN's "Guiding Principles and Objectives for Negotiating the Regional Comprehensive Economic Partnership" define the goal of RCEP as "a modern, comprehensive, high-quality and mutually beneficial economic partnership agreement… [to] cover trade in goods, trade in services, investment, economic and technical cooperation, intellectual property, competition, dispute settlement" (ASEAN 2012). The Principles stress the need for flexibility, and the agreement will likely include special and differential treatment for developing members.

Thus far, there have been 18 RCEP rounds, with the last one having taken place in Manila in May 2017. Discussions have focused hitherto on trade in goods, trade in services, investment, economic and technical cooperation, intellectual property, competition policy, and legal and

[40] http://dfat.gov.au/trade/agreements/rcep/pages/regional-comprehensive-economic-partnership.aspx.

institutional issues, including dispute settlement. E-commerce has also been part of the deliberations. Modalities for negotiations in various areas are still being ironed out. While sessions have been closed-door, information that can be gleaned from the websites of various RCEP participating governments suggests that some progress is being made on trade in goods and services and intellectual property, but much remains to be done by the end of 2017.

The exact results of RCEP cannot easily be predicted at this time. On one hand, most negotiating members are outward-oriented and have incentives to produce a solid, comprehensive agreement, and official statements support this goal. Still, a history of shallow agreements by key member economies (such as ASEAN's and India's bilateral agreements), the focus on flexibility, the hesitancy to include behind-the-border measures, and the inclusion of economies that have been skeptical of deep economic integration in the past does not bode well for an ambitious outcome. Instead, the results may well include a relatively comprehensive deal on market access in goods; a limited positive-list approach in services; a strong agreement on investment, albeit with numerous carve-outs; few improvements in intellectual property rules; and provisions on economic and technical cooperation as well as special and differential treatment for low-income economies. The agreement will likely avoid strong commitments on behind-the-border issues.

But in any event, a successfully completed RCEP agreement will certainly enhance the leadership credentials of China at the same time that US credibility is fading in light of its exit from the TPP. Indeed, these trends are not only in evidence at the regional level but also globally: at the 2017 Davos meetings, President Xi of China gave a strong pitch in favor of globalization, while the newly assembled Trump administration was emphasizing "America First" and an inward-looking agenda. Some scholars are even concerned that the current administration could be undoing the post-war liberal order put in place by the United States.[41]

---

[41] See, for example, Friedman, U. (2016). "How Donald Trump Could Change the World," *The Atlantic*, November 7. Available at: https://www.theatlantic.com/international/archive/2016/11/trump-election-foreign-policy/505934/ .

6.3. *Post-TPP Scenarios*

It is fair to say that the US withdrawal from the TPP took the region by surprise. The United States had led the negotiations with great enthusiasm and had negotiated hard to promote its comparative advantage industries, not only in services but also in high-tech manufactures and certain agricultural sectors and markets that had hitherto been closed to US exports. It included chapters on rules that the United States deemed were necessary for the 21st century, e.g., with respect to IPR and the digital economy, and in creating a level playing field for the global economy, e.g., with respect to state-owned enterprises and foreign investment. The TPP also included a rigorous chapter on labor protection that was actionable to ensure that there would be no "race-to-the-bottom" in terms of labor rights and also a forward-looking environmental chapter. As noted above, in the national politics of many TPP partner-countries, the agreement was criticized for being too unbalanced in favor of US interests.

In addition to being unanticipated, the US withdrawal from the TPP was greeted with great trepidation in the region, particularly in its being a first concrete manifestation of the so-called "America First" dogma and the new inward-looking protectionist bilateralism of the Trump Administration. Asia-Pacific economic cooperation, led by the United States, had been growing in importance and ambition since the creation of APEC in 1989. This cooperation had been growing *pari passu* with Asian economic cooperation, from ASEAN to the "noodle bowl" of bilateral arrangements. It is interesting to recall that the United States in the past was worried that East Asian cooperation might lead to "a line drawn down the middle of the Pacific," as the Secretary of State under George H.W. Bush, James Baker, once famously quipped, but the line apparently is now being penned by the United States itself. In any event, Asian economic integration will likely proceed apace. In fact, the US withdrawal from the region increases the incentives for key countries to push forward on Asian integration.

This will likely first be in the form of the RCEP, as discussed above. The failure of the 12-country TPP to move forward gives a strong incentive to TPP negotiating members to finalize this other Asian megaregional, for not only the inherent economic benefits but also to keep the momentum toward concerted economic liberalization going (for the reasons discussed above). It also increases the potential gains to major RCEP members that were not party to the TPP, especially China. After all, it was always clear that the TPP would be more ambitious than RCEP; now, RCEP is the only game in town.

But that may not be exactly true. The remaining 11 TPP member-states are now considering going forward with the agreement, with the exception of the necessary revision in Chapter 30 to allow for the exclusion of the United States.[42] The Abe Administration in Japan was at first skeptical about the TPP without the United States but has now decided to join in negotiations for a new TPP, with the hope that United States might join at a later date. On April 25, 2017, Prime Minister Abe appointed a Deputy Foreign Minister to be its chief negotiator for the new TPP without the United States.[43] The reasons for the change of heart no doubt derive from an interest in keeping the Asian integration process going and cementing strategic relationships, goals prominent in the aspirations of other TPP members as well.

With Japanese leadership, the likelihood of the TPP11 scenario is high. It would be an attractive option for economies in the region, particularly given that the TPP is such an advanced agreement that could also serve as a template for domestic economic reform and a model for future negotiations, including for other megaregionals such as the Free-trade Area of the Asia-Pacific. It might also continue to push the RCEP to be a more comprehensive agreement, something that Japan and other TPP economies would like to see. Moreover, should TPP11 continue to move forward, it is likely that other countries who have expressed interest in joining the original TPP would consider joining.

The Trump Administration has expressed interest in bilateral, rather than regional, agreements. The potential benefits of bilaterals are limited for a number of reasons, including the fact that tariffs on US trade with

---

[42] Chapter 30 required, among other things, that at least 6 TPP members constituting at least 85% of aggregate TPP GDP would be necessary in order for the TPP to enter into force. Given the size of the US economy, its inclusion (as well as Japan's) is necessary.

[43] http://www.japantimes.co.jp/news/2017/04/25/business/deputy-foreign-minister-named-japans-chief-tpp-negotiator/#.WQBtuIjyiUk .

Asia-Pacific partners are already relatively low, the benefits of bilateral FTAs in spurring regional production networks — an important motivation for Asia-Pacific economic integration — are limited due to the "Asian noddle bowl" problem, and a main motivation for megaregional arrangements in Asia is to avoid the inefficiencies associated with a bilateral approach. Nevertheless, the possibility of a bilateral FTA with Japan — by far the largest economy in the TPP after the US itself — has been tabled.

In short, Asia-Pacific economic integration will proceed with or without the United States. All countries would have been better off if the United States had moved forward with the TPP and continued to engage in concerted economic liberalization in the region, but the backlash against globalization and a populist agenda has put the US policy on a new course not seen since the 1930s. The editors of this book hope and expect this to be a temporary setback. Ironically, the US temporary withdrawal may create new opportunities for a more diversified megaregionalism 2.0.

Megaregionalism 2.0 building blocks — The book's message

The main message of this book is that megaregionalism has the potential to improve the welfare of the region's citizens and that of the world as a whole. These benefits derive in part from the traditional economic effects of trade and investment liberalization via lower tariff and non-tariff barriers. But in the context of the globalized 21st century economy, non-traditional aspects of integration are increasingly important in improving efficiency, productivity, and competitiveness, and "trade and innovation" links are a critical component of this process. Moreover, as with technological change and automation, structural change via trade implies that some economic agents will gain but others will lose. Hence, the key policy challenge facing governments is not merely how to write new rules for the modern economy and bring down barriers to economic integration but also to keep the pain of transition as mild and as rapid as possible and ensure that the most vulnerable in particular need to be protected. This includes well-funded policies in support of education and training, in order to foster local capabilities for management and innovation.

Globalization needs to be managed via enlightened economic and social policies rather than impeded. This book takes a fresh look at these issues, by bringing together chapters from trade economists, policymakers, and trade negotiators, as well as from experts on innovation, intellectual property rights, competition law, technical standards, labor standards, and industrial development.

The book has five parts. Below, we offer a very brief summary of the salient issues addressed in the chapters forming each part. To begin, **Part 1** focuses on economic analysis. In "The TPP: Potential Gains and Impediments" (Plummer), the economics of megaregionalism in general and the TPP in particular are evaluated in order to give context to other trade-and-innovation-related chapters of this volume. Using estimates from Petri and Plummer (2016), it shows that the TPP should generate large economic benefits throughout the region and could increase incentives to complete on-going megaregional accords, such as the RCEP. These results are echoed in, "The TPP: Promoting Waves of Innovation", by Lippoldt, who presents optimistic scenarios under which TPP might have boosted incentives for innovation, drawing on a unique data set.

Part 2 considers trade and innovation within global networks. In "Beyond Value Capture: Exploring Innovation Gains from Global Networks", Ernst explores how increasingly complex and diverse global corporate networks of production and innovation create new, yet little understood, challenges for national policies that seek to foster economic growth and prosperity through productivity-enhancing innovation. As countries differ in their capabilities to learn and innovate, they differ in their capacity to capture potential gains from integration into global networks. This analysis is followed by a Taiwanese case study on supply chain positioning and innovation by Chen Tain-Jy, which shows how a firm's position in global supply chains affects its ability and scope for innovation. The chapter reviews and evaluates the multiple innovation traps that Taiwanese firms face as a result of their deep integration into mostly modular global networks. Next, Hill focuses on trade and investment barriers in a key sector for innovation, i.e., solar and wind global value chains, and examines where value-added is located in related global production networks and what impact megaregionalism might have on trade and investment barriers. A chapter on trade secrets had to be withdrawn from the volume prior to publication.

Topics related to intellectual property, strategic patenting, and trade are addressed in **Part 3**. In "Patents and Technology Transfer through Trade and the Role of Regional Trade Agreements", Maskus explores what positive policy steps designers of regional trade agreements could take to enhance flows of technology stemming from an agreement, such as changes in patent rules and increasing flows of skilled labor. Feldman emphasizes that efforts to develop regional and global intellectual property frameworks need to focus on streamlining complexity and on the creation of new products, rather than the creation of new intellectual property games and stripped markets. "Harmony and Disharmony in Intellectual Property Law" (Chien) notes that the gap between the agreed-upon provisions within trade agreements on the one hand and the on-the-ground implementation on the other is large, due to how they are formed as well as due to problems associated with the absence of domestic policymakers at the table in responding to dynamic and evolving needs in implementation. Kahin provides powerful evidence that the digital economy has added new strategic dimensions to the use of patents. While megaregional trade agreements like the TPP may have little direct impact, they add new uncertainties related to the expanding use of ISDS by large multinationals. Corrective policies are needed to reduce the disadvantages for smaller, domestic-scale competitors.

Part 4 considers the critical role of technical standards in trade and innovation. Tassey looks at the impacts of technical standards on global trade and economic efficiency, developing a public–private asset growth model where investment decisions are based on variables that reflect (1) the complex systems nature of modern technology, and (2) the required public and private sector roles for providing the supporting technical infrastructure. Ramel and Blind examine the concentration of control over Standard Essential Patents and their economic impacts on the distribution of gains from trade for innovation. In "Patents, Standards and Borders," Contreras argues that SMEs and firms from less-developed countries are systematically disadvantaged in global standard-setting organizations (SSOs) due to a lack of a large patent portfolio and experience within SSOs. Karachalios and McCabe underscore that standards now operate in an increasingly complex environment which forces us to rethink the role of standards in global trade agreements. The chapter discusses how megaregional trade agreements could be used to promote globally open, inclusive, collaborative, and consensus-based standardization processes that would enhance cross-border trade and innovation. Wang Ping and Zheng Liang focus on China's standardization system and present a reform agenda to strengthen decentralized standard development in line with China's deep integration into international trade and global networks of production and innovation.

Finally, **Part 5** offers new perspectives on trade and innovation policy. Carlos Braga considers whether a multinational-led model of development — i.e., outward orientation with strong protection of IPRs — will become the norm for developing countries in the coming years, and what needs to be done to enable megaregional trade agreements to enhance innovation capabilities in developing countries. In "The Information Technology Agreement, Manufacturing and Innovation — China's and India's Contrasting Experiences", Ernst examines how a country's manufacturing and innovation capacity in a particular industry affects its approach to and its position in megaregional trade agreements, comparing the experiences of China and India. Helper and Kruger explore how to promote win–win outcomes in the development of global supply chains, comparing three alternative scenarios of the impact of global supply chains and trade agreements on innovation, wages, and sustainability. Lastly, Brown looks at the labor implications of the TPP and argues that, while not generally supported by organized labor, it is a progressive "game changer", particularly in light of the three TPP-side agreements negotiated with Vietnam, Malaysia, and Brunei that should dramatically elevate domestic labor rights.

Part 1

What We Know from Economic Analysis

Chapter 1

The Transpacific Partnership: Potential Gains and Impediments in a Global Context

Michael G. Plummer

The Johns Hopkins University
East-West Center
mplummer@jhu.edu

1. Introduction

Economic integration has served the world economy well over the past quarter century. The Asia-Pacific region in particular has benefited from the regional and international marketplace; at the same time that it has embraced an outward-oriented approach to economic development, the region has seen great improvements in virtually all economic indicators. For example, using $1.90 per day as a poverty threshold, the World Bank calculates that in the East Asia and Pacific region, the poverty headcount has fallen from 60.1% in 1990 to 7.2% in 2012, with absolute numbers of the poor falling from 1 billion to 147 million,[1] which is historically unprecedented. The Asian middle class has been rising rapidly; the Asian Development Bank (ADB) predicts that by 2030, about three-fourths of

[1] *Source*: http://povertydata.worldbank.org/poverty/region/EAP .

the residents in India, China, and ASEAN will be middle class, up from about one-fifth in 1990.[2] Many economies in Asia and the Pacific were able to make significant progress toward achieving the Millennium Development Goals via stable macroeconomic policies, effective public policy, and outward orientation. These factors will be no less important in meeting the ambitious Sustainable Development Goals (SDGs).

Trade has been an essential component of the successful Asian economic development experience, and regional economic integration has grown rapidly over the past quarter century. There have been a number of drivers behind this trend. First, East Asian governments have recognized the importance of the international marketplace as an engine of growth and as a weapon against poverty. The chronology, extent, and speed with which regional economies have adopted outward-oriented reforms vary, but unilateral liberalization is now a key feature of virtually all economies in the region. Second, rapid improvements in transportation infrastructure have lowered the costs of integration, and the explosion of Information and Communication Technologies (ICT) has opened up new markets and reduced the importance of borders and costs associated with geography. Third, the liberalization of barriers to economic interaction via a half century of multilateral liberalization under the General Agreement on Tariffs and Trade (GATT) paved the way for unusually rapid growth in trade, especially in developing countries (Dollar, 1992; Winters, 2004).

The share of trade in GDP varies considerably across the region, but it is high and/or rising in essentially all countries and subregions; for example, over the past 20 years until 2014, trade as a percentage of GDP rose from 73% to 102% in ASEAN, making it the most open large region in the world (World Bank, 2016). Trade is connected to growth and poverty reduction through multiple channels (Winters *et al.*, 2004). It acts as the vehicle through which countries exploit comparative advantage, improve the efficiency and productivity of domestic resources, and tap into international markets. It raises incomes, increases consumption and investment, and reduces risks associated with shocks to domestic

[2] See: http://www.adb.org/sites/default/files/publication/27726/ki2010-special-chapter.pdf and http://www.adb.org/publications/asean-prc-and-india-great-transformation .

production and markets. It enables access to greater varieties of goods, services, and intermediate inputs, and allows local firms to use or acquire improved technologies (Petri and Plummer, forthcoming). Trade creates better and higher-paying jobs, provides incentives for learning, and helps to attract Foreign Direct Investment (FDI).

More recently, a strategy of concerted liberalization via bilateral and subregional Free Trade Areas (FTAs) has been adopted by essentially all countries in the region, with most having many FTAs in place with regional and extra-regional partners. This trend has been especially important beginning at the turn of the 21st century and is gaining momentum; for example, the ASEAN Economic Community (AEC) was put in place in December 2015 (though a good deal of the project remains to be completed) and is the most comprehensive and ambitious ("deep") regional economic integration accord in the developing world. In 2000, there were 35 FTAs that were in effect in Asia; in 2016, the number rose to 140.[3]

However, a major concern regarding the FTA movement pertains to the fact that FTAs are mostly bilateral in nature, with different degrees of sectoral coverage, divergent rules and regulations, and inconsistent rules of origin associated with each arrangement. Such inefficiencies have been described as the Asian "noodle bowl" effect; while empirical work on these agreements suggests greater efficiency and welfare gains, the limited scope of bilateral arrangements constrains potential net benefits. In part in response to these shortcomings, over the past decade there has been a movement toward the consolidation of bilateral arrangements into large regional agreements, a trend that is known as "megaregionalism."

The recently signed (but not yet ratified) Trans-Pacific Partnership (TPP) agreement, which includes 12 developing and developed economies in the Asia-Pacific,[4] is the most ambitious of these megaregional arrangements, but it is not the only one: negotiations for the Regional Comprehensive Economic Partnership (RCEP), composed of 16 countries

[3] ADB, Asia Regional Integration Centre. See: https://aric.adb.org/fta, Table 1. Accessed 1 July 2016.

[4] Namely, Australia, Brunei, Canada, Chile, Japan, Malaysia, Mexico, New Zealand, Peru, Singapore, the United States, and Vietnam.

in East Asia and India,[5] were launched in November 2012, and the Free Trade Area of the Asia-Pacific (FTAAP), which could include all APEC economies and possibly others, is slated to begin negotiations in 2020. The Transatlantic Trade and Investment Partnership (TTIP) between the 28 countries of the EU and the United States is also considered part of the megaregionalism movement — though obviously with only one Asia-Pacific economy — and began negotiations in 2013. If ratified, these arrangements would all have important effects on not only the regional economy but also the world.

The goal of this chapter is to consider the economics of megaregionalism in general and the TPP in particular to provide context to other trade-and-innovation-related chapters of this volume. As is clear from the chapters, lower international barriers to trade at borders and behind them, as well as such trade-related policies as common rules and regulations, standards, intellectual property protection, investment regulations, and so forth facilitate the creation of supply chains and global production networks and, as such, promote the dissemination of information and the flows and applications of technology. As noted in Chapter 3, global innovation networks will continue to rise in importance as contributors to economic integration and progress. Since these megaregionalism agreements are also slated to be advanced, deep, and comprehensive in scope and orientation, they will no doubt have an important bearing on global innovation networks as well as production networks. The TPP is likely to be the most advanced of these agreements — except possibly the TTIP, which still has a long way to go and negotiations have currently been shelved — and its ratification will pave the way for additional member-economy applications; Indonesia, the Philippines, Thailand, Chinese Taipei, and South Korea have all expressed interest in possibly applying for a membership. Moreover, a successful conclusion to the TPP could light a fire under the RCEP and TTIP negotiations and set the stage for the FTAAP. Hence, the TPP is key.

This chapter begins with a brief review of the economics of regionalism in Section II, followed by a review of the TPP agreement itself in

---

[5] Specifically, the 10 ASEAN countries, China, South Korea, Japan, Australia, New Zealand, and India.

Section III. Next, the chapter summarizes the net aggregate economic effects of the TPP, based on Petri and Plummer (2016). Section IV endeavors to explain controversies surrounding these megaregional accords. Section V concludes.

2. Economics of FTAs: A primer

The economics of FTAs are complicated. While they remove discrimination between partner-country and domestic firms and therefore form a more efficient regional division of labor ("trade creation"), the discrimination introduces a distortion between the partner and non-partner countries, and countries within the FTA may end up purchasing imports from less efficient partner countries, which not only hurts non-partner exports but raises the price of imports ("trade diversion"). Hence, trade diversion results in a negative terms-of-trade effect, as a country potentially purchases imports from higher-cost sources. Moreover, trade diversion associated with an FTA increases the potential scope for "investment diversion," in which FDI flows to a country merely to take advantage of protected regional access or rules of origin. Investment could then flow to inefficient sectors that could dampen employment generation (e.g., if in capital-intensive sectors) and productivity growth.

Trade and investment diversion underscores why FTAs are referred to as "second best," as opposed to "first best" unilateral or multilateral liberalization, which do not introduce any discrimination across trading partners. Still, FTAs tend to be more comprehensive in terms of the coverage of goods and services, comprehensive removal of tariffs and non-tariff barriers, investment, and other aspects of economic integration (from rules governing economic interchange to competitive neutrality disciplines on state-owned enterprises) than is the case of multilateral liberalization because it is easier to negotiate complicated and politically sensitive measures among a smaller group of like-minded countries than in the context of the diverse 164-economy membership of the WTO. Hence, FTAs discriminate across *countries,* but less across *products and services* and trade-related behind-the-border barriers, implying less potential negative effects in terms of actual protection accorded to national value-added (i.e., "effective rates" of protection). The rules, regulations, and measures necessary to meet the needs of the private sector require comprehensive, deep integration, and this has hampered progress at the Doha Development Agenda negotiations of the WTO, which have been at an impasse for the past four years. In an outward-oriented, dynamic region like the Asia-Pacific, much more is possible.

An important drawback of FTAs regards restrictive rules of origin. While rules of origin in any FTA are necessary to prevent trade deflection — whereby exporters take advantage of the FTA to export through the market with the lowest external tariff — they can be inefficient in that they: (1) complicate an FTA with product-based rules of origin that can differ across FTAs, an increasing problem in the context of the proliferation of bilateral FTAs; (2) they can often require excessive documentation and can be costly, particularly for Small- and Medium-sized Enterprises (SMEs), one reason why SMEs often do not take advantage of FTAs; and (3) excessive rules of origin can lead to another form of investment diversion in order to take advantage of FTA value-added-content requirements (World Bank, 2016).

Addressing the problems associated with overlapping and complicated rules of origin is an important goal of "megaregional" FTA arrangements like the TPP and RCEP. These agreements will allow for "cumulation" of rules of origin, that is, value-added in any TPP country counts toward meeting minimum rules of origin thresholds in intra-regional trade. Cumulation is particularly important for regional production networks; the "Asian Noodle Bowl" of overlapping bilateral FTAs limits the benefits of economic cooperation to multinationals seeking to develop these networks. An example might illustrate the point. Suppose that there is a 10% tariff on thermionic tubes in all TPP markets and there is a rules of origin requirement of 40% in all US bilateral FTAs. A US multinational produces 25% of the value of a thermionic tube at home, while 55% is produced in China, 10% in Peru, and 10% in Australia, from which it is re-exported back to the US. The US has a bilateral FTA in place with both Peru and Australia, but the company will not be able to benefit from any preferential tariffs associated with either agreement because the value-added would be only 35%. To avoid the 10% tariff, the company would then face difficult decisions regarding its optimal distribution of production; for example, it may be less efficient to raise value-added in Australia, but there would be an incentive to do so (either by reducing production in China or in Peru such that it meets the 40% threshold) in order that the company could at least profit from preferences associated with the US–Australia agreement. The higher the rules of origin and existing Most Favored Nation (MFN) tariffs, the greater the incentive for value-added in a bilateral FTA partner, and the more a regional division of labor will be distorted. Cumulation can solve this problem. In the above example, the US multinational company would benefit from the TPP without changing any production.

Regionalism has other "dynamic" effects that tend to grow in importance over time. These would include, *inter alia*, economies of scale via greater market access, technology transfer, and improved management techniques through FTA-generated FDI and greater economic interaction; lower costs of trade via more efficient customs procedures and other trade facilitation measures; harmonization of product standards and professional qualifications; and structural policy change and reform. For example, an important motivation for FTAs in the Asia-Pacific region relates to the need to reduce transaction costs associated with the creation of a regional production base, which would have the effect of furthering regional integration and improving competitiveness. Fragmented trade in regional production networks relies on this type of facilitation; it can gain from economies of scale due to a larger, freer market. With greater FDI inflows, the potential for technology transfer improves.

Moreover, FTAs tend to require harmonization of various microeconomic policies and present a strong case for the adoption of "best practices" in terms of accounting standards, management techniques, adoption of common product standards, and so on. These types of "deep integration" policies are exactly the type of measures that multinational corporations seek; by reducing transactions costs and forming a more cohesive environment in which to exploit a vertical division of labor, FTAs can create a common marketplace that transcends national borders. While it is more difficult to quantify these dynamic effects compared to the "static" (i.e., price) effects of trade creation and trade diversion, they are clearly more pervasive in their salutary effects on the economy.

Regional cooperation plays a key role in attracting welfare-enhancing FDI. By bringing down barriers to regional economic integration, relaxing "behind-the-border" constraints, providing more protection for intellectual property, and increasing protection of investments, economic cooperation enhances the attractiveness of host countries in the region to FDI as well as inclusion in supply chains and global and regional production and innovation networks. Indeed, as the networks function by sending components of the production chain across many borders, the costs of any obstacles to trade are multiplied; concerted regional liberalization is especially advantageous to the private sector. The desire to promote existing regional production networks and create incentives to build new ones has been behind the drive for FTAs in general and regional FTAs in particular.

Furthermore, regional economic cooperation and integration can enhance connectivity. The Asia-Pacific region's highly interdependent production systems are making new demands on the infrastructure of regional integration. Transit/transport corridors constitute excellent examples of the advantages FTAs and other forms of regional economic cooperation offer, as by their very nature they are regional in scope. These corridors can serve to connect different areas of a single country, or they can connect subregions and regions (Wignaraja *et al.*, 2015). They are composed of roads, railways, bridges, and port access. They can be developed to increase trade within the region, provide access to landlocked countries, and create international access to trade in goods by connecting a region to ports.

In fact, transit/transport corridors are particularly important to trade and growth prospects of landlocked countries. High transport costs undermine potential gains from trade liberalization and can negate the beneficial price effects of reductions in tariffs and non-tariff barriers via regional cooperation. As such, they limit the ability of economies to take advantage of the global marketplace and participate in production networks in particular.

In addition, connectivity can be enhanced via "soft infrastructure" such as trade facilitation and via technology. Although there is no internationally agreed upon definition of "trade facilitation," the term is commonly understood as the "simplification" of practices involved in the movement of goods across borders with a specific aim to reduce trade-related transaction costs. Trade facilitation includes both border and behind-the-border practices; the former refer to such areas as logistics and customs, whereas the latter regard the specific environment in which trade transactions take place, such as domestic laws and policies, government/institutional structures, as well as telecommunications and e-business services to support border practices. Good governance is central to trade facilitation because it ensures implementation of rules, regulations, and investment in automated customs systems.

3. The TPP agreement

Negotiations on the TPP began in 2008, building on the earlier Trans-Pacific Strategic and Economic Partnership agreement concluded by Brunei, Chile, New Zealand, and Singapore, and includes 12 APEC members[6] representing almost two-fifths of world output and one-quarter of global trade, ranging widely in per capita income from Vietnam and Peru to Japan and the United States. On October 4, 2015, the 12 negotiating partners came to an agreement on the substance of the TPP, and the actual accord was signed on February 4, 2016, by all member-governments. It now awaits ratification by member countries at the time of this writing.

The TPP seeks to improve market access by eliminating "high-hanging" barriers that escaped prior rounds of liberalization in the WTO. The TPP will eliminate tariffs on over 96% of goods traded in the region (the US, for example, will drop 99% of its tariffs[7]), albeit with sometimes long implementation periods. More importantly, its 30 chapters address myriad non-tariff barriers, and new topics hitherto neglected or excluded in FTAs, and include provisions to help align technical and sanitary and phytosanitary (SPS) standards. The bulk of the TPP text updates or develops new rules for issues that have either emerged or

[6]The membership of the Asia-Pacific Economic Cooperation (APEC) forum currently includes 21 economies: Australia, Brunei Darussalam, Canada, China, Chinese Taipei, Chile, Hong Kong, Indonesia, Japan, Malaysia, Mexico, New Zealand, Papua New Guinea, Peru, the Philippines, Russia, Singapore, South Korea, Thailand, the United States, and Vietnam.

[7]See Freund *et al.* (2016).

grown dramatically in importance in recent years, including trade in services, FDI, the digital economy, competitive neutrality for state-owned enterprises, intellectual property, and regulatory coherence. The agreement is notable as much for its comprehensive approach as for specific provisions. The TPP seeks to encourage trade, investment, and development in all member countries and implicitly assumes that high-quality rules, *albeit* implemented gradually, benefit economies irrespective of their stage of development. It should create fertile ground for regional and global production and innovation networks. In sum, its architects sought to create a "21st century" agreement based as much as possible on a "gold standard" of trade policies, though obviously the agreement itself includes many compromises (as is the case with all trade agreements).

The TPP also includes a chapter specifically devoted to development, a topic often present in developing-country FTAs but not in developed-country agreements. The provisions include expressions of shared goals and principles regarding the promotion of inclusive growth and poverty reduction; an emphasis on the importance of supporting women in the workplace and as entrepreneurs; the objective of promoting education, science and technology, research, and innovation; and encouragement of joint activities to promote development.

Finally, the TPP includes *enforceable* provisions on labor and environmental issues. These chapters require members to adopt practices consistent with international conventions. With respect to labor, the TPP requires members to follow the International Labor Organisation's Declaration on Fundamental Principles and Rights at Work, which includes the right of workers to organize. This chapter will be particularly relevant in Vietnam, Malaysia, and Brunei Darussalam, with which the United States has bilateral letters outlining expectations with respect to domestic policies *vis a vis* labor. With respect to the environment, the TPP references a list of international environmental agreements. In practice, labor and environmental provisions seek to achieve a difficult balance among three goals: protecting shared values; preserving the rights of governments to pursue domestic political and social agendas; and mobilizing political support across the agreement's members.

4. Estimates of TPP effects: Petri and Plummer (2016)

Estimating the economic effects of such a complicated agreement is no mean feat. The author, together with his colleagues Peter Petri and Fan Zhai, has used a cutting-edge Computable General Equilibrium (CGE) model featuring heterogeneous firms to estimate the potential effects of the TPP, beginning as early as 2011 at the East-West Center. The details of the model, including the modeling challenges, can be found in Petri *et al.* (2012). In Petri and Plummer (2016), the authors, updated the model to include trade liberalization information from the actual TPP agreement, including in terms of tariff-reduction coverage and other aspects of the agreement. Below, we summarize the results of the model as a means of underscoring the likely effects of the agreement. More details can be found in Petri and Plummer (2016) and at www.asiapacifictrade.org.

The aggregate effects on income on selected economies and regions are summarized in Table 1. Overall, the authors estimate the net gains to global income relative to the baseline to be $494 billion, or 0.4% of the global GDP, which is more than earlier studies estimating the effects of an ambitious conclusion to the Doha Development Agenda. The TPP member-countries would, of course, reap most of the gains: $467 billion, or 1.1% of regional GDP. Benefits would be broadly shared across the TPP. The largest members — the United States and Japan — would see the largest absolute gains, but smaller economies would do best in proportional terms. Malaysia and Vietnam lead relative gains in the TPP with income increases of 7.7% and 8.2%, respectively, in 2030, relative to the baseline. Both countries would benefit from reducing distortions in their own markets and establishing stronger positions in regional supply chains through deeper integration. For example, Vietnam's textile and garment exports are projected to expand by more than 50%, benefiting from the reduction of tariffs that are now as high as 25% for some product categories in the United States (Petri and Plummer, 2016). Preferences would accelerate the shift of labor-intensive textile and garment industries from China to Vietnam.

As noted earlier, the TPP targets deeper integration and will likely generate a very different mix of benefits than earlier agreements. The simulations show that only 6% of the benefits from the TPP for its members will be due to tariff effects, while reducing NTBs on other goods,

Table 1. Petri–plummer estimates of the effects of the TPP.

	Baseline ($bill)				Change with TPP ($bill)			Percent change		
	2015	**2020**	**2025**	**2030**	**2020**	**2025**	**2030**	**2020**	**2025**	**2030**
Americas	**21,962**	**25,177**	**28,473**	**31,544**	**41**	**130**	**206**	**0.2**	**0.5**	**0.7**
Canada	1,981	2,227	2,472	2,717	8	22	37	0.4	0.9	1.3
Chile	269	329	397	463	0	2	4	0.1	0.5	0.9
Mexico	1,339	1,598	1,868	2,169	3	11	22	0.2	0.6	1.0
Peru	219	287	363	442	1	6	11	0.4	1.6	2.6
United States	18,154	20,736	23,372	25,754	29	88	132	0.1	0.4	0.5
Asia	**22,806**	**29,752**	**38,179**	**47,386**	**52**	**135**	**202**	**0.2**	**0.4**	**0.4**
Brunei	20	24	27	31	0	1	2	1.1	3.3	5.9
China	11,499	16,058	21,689	27,839	–1	–8	–19	0.0	0.0	-0.1
Hong Kong	300	358	412	461	2	4	6	0.5	1.0	1.2
India	2,210	3,086	4,197	5,487	0	–2	–5	0.0	-0.1	–0.1
Indonesia	927	1,240	1,687	2,192	0	–1	–2	0.0	-0.1	–0.1
Japan	4,214	4,462	4,693	4,924	39	91	125	0.9	1.9	2.5
Korea	1,384	1,672	1,967	2,243	–1	–5	–8	–0.1	-0.2	–0.4
Malaysia	349	444	553	675	7	28	52	1.6	5.0	7.7
Philippines	329	436	547	680	0	–1	–1	0.0	-0.1	–0.1
Singapore	320	380	437	485	2	8	19	0.5	1.9	3.8

Taiwan	511	619	707	776	0	1	1	0.1	0.1	0.2
Thailand	411	516	656	812	–1	–4	–7	–0.2	–0.6	–0.8
Vietnam	209	281	378	497	7	22	41	2.3	5.8	8.2
ASEAN NIE	124	175	228	283	0	–1	–1	–0.1	–0.2	–0.4
Oceania	**1,896**	**2,203**	**2,533**	**2,854**	**2**	**13**	**23**	**0.1**	**0.5**	**0.8**
Australia	1,704	1,986	2,292	2,590	1	8	16	0.0	0.4	0.6
New Zealand	192	217	241	264	1	4	7	0.6	1.8	2.8
Rest of World	**34,371**	**39,492**	**45,506**	**52,017**	**16**	**44**	**63**	**0.0**	**0.1**	**0.1**
Europe	17,893	19,746	21,451	23,189	12	34	48	0.1	0.2	0.2
Russia	2,244	2,462	2,903	3,371	0	1	2	0.0	0.0	0.1
ROW	14,235	17,283	21,152	25,456	3	8	12	0.0	0.0	0.0
WORLD	**81,035**	**96,623**	**114,690**	**133,801**	**111**	**321**	**494**	**0.1**	**0.3**	**0.4**
Memorandum										
TPP members	28,969	32,971	37,094	41,011	98	293	467	0.3	0.8	1.1
Non-members	52,066	63,652	77,596	92,790	13	28	27	0.0	0.0	0.0

Source: Petriand plummer (2016).

services barriers, and barriers to investment will account for 37%, 41%, and 16% of the total effects, respectively (Petri and Plummer, 2016).

At the same time, the TPP will accelerate structural transformations based on comparative advantage and scale economies. In advanced countries, these mechanisms favor traded services, advanced manufacturing, and sometimes primary products and investments. In middle- and low-income economies, they benefit manufacturing, especially in labor-intensive industries, and some primary production. Based on the heterogeneous firms' mechanisms, the model also suggests increased intra-sectoral trade, consistent with stronger regional supply chains.

The effects of the agreement on productivity take place via the assumption of firm heterogeneity; however, other likely channels of productivity increases are not modeled. For example, both increased exports and imports have been shown in the empirical literature to have important productivity spillover effects (e.g., Stone *et al.*, 2011). Moreover, to the extent that the TPP is able to enhance global production and innovation networks via lower transactions costs, greater regulatory coherence, improved treatment of foreign affiliates, and improved intellectual property protection (particularly in the area of trade secrets), the result will be greater innovation and development of new inventions and efficient production, which could have an important effect on productivity. Thus, by excluding these latter channels of technological spillovers, the model no doubt underestimates the ultimate effects of the TPP.

With respect to ramifications for non-partners, TPP simulations reveal some trade diversion and preference erosion, but these effects are small. For example, in 2030, the TPP would generate net *gains* for all non-members of $27 billion. These net results include some negative (diversion) effects; for example, in the TPP, in absolute terms, these would be larger for mid- and low-income economies (primarily China, which will be negatively affected by trade diversion), and in percentage effects they would be larger for high-income economies (South Korea, which will be affected by preference erosion since it will lose preferential treatment in the US market).

Of course, other studies have estimated the effects of the TPP as well. Most notably, at the request of Congress, the United States International Trade Commission (USITC) released its report on the effects of the TPP in May 2016. An extensive summary of the empirical TPP literature can be found in that study; in addition, it generates its own empirical estimates, via its own standard CGE model. It has an important advantage in that it focuses on disaggregated sectoral effects of the TPP, which permits it to include details from the actual TPP agreement in the estimates that is beyond the ability of the relatively aggregate Petri and Plummer (2016) approach.

Consistent with most of the rest of the TPP empirical literature, it also calculates a net positive effect on the US economy — $57 billion — but it is less than one-half of the Petri and Plummer (2016) calculations. The differences can be explained as follows: First, the USITC study does not include multilateral spillover effects from liberalization of non-tariff barriers under the TPP, whereas Petri and Plummer (2016) assume that 20% of these benefits would accrue to all countries rather than just partner countries (e.g., the empirical literature on the effects of the EU Single Market Programme generally puts the spillover effects at 20–60%). The logic is that trade facilitation improvements in TPP developing countries — for example, through improvements in customs procedures — will not be limited only to TPP partners. The same will be true of improvements in regulatory coherence, intellectual property protection, FDI, many dimensions of services trade, and so forth. Second, by employing the Melitz assumption of heterogeneity in firm productivity, Petri and Plummer (2016) are able to show "extensive margin" changes in trade (i.e., changes that derive from the entry and exit of exporting firms, rather than just changes in exports of existing firms, or the "intensive margin") along with productivity gains, something that is not possible in the USITC model. Indeed, the biggest differences between the two studies regard changes in trade: Zhai (2008) notes that excluding firm heterogeneity creates a strong downward bias regarding trade effects in CGE models, underscoring that the econometric literature suggests that the gains to trade of regional integration are considerable and far beyond the usual CGE effects. Petri and Plummer (2016) estimate that, for the US economy, US exports rise by 9%, whereas the USITC model produces a mere 1% increase in exports, an implausible result for such an advanced and comprehensive trade agreement. Another difference regards the effects on employment; as the TPP is implemented over the long-term, Petri and Plummer (2016) employ the typical CGE closure that the TPP does not affect employment.[8] However, the USITC estimates that the TPP will increase employment in the United States by 0.7%, or 128,000 jobs (positive, but miniscule given that job churn in the United States is about 55.5 million separations annually).

5. Opposition to the TPP

The TPP was signed in February 2016. Although talks were led by the United States throughout its seven years of negotiations, it is not clear whether the United States will actually ratify the agreement (it has two years to do so). If the United States does not ratify the agreement, then the TPP will be stillborn: the agreement says that six parties constituting at least 85% of the aggregate GDP of the TPP countries need to ratify the agreement for it to take effect, and as the US and Japanese economies each constitutes more than 15% to total TPP GDP, they both need to agree to the accord for it to take effect. There is nothing to keep the other TPP countries from adopting the accord themselves in a separate arrangement, but without its largest participant, its value will be much less to most markets.

The election of Donald Trump as President of the United States on 8 November, 2016, could well lead to the TPP's being stillborn, as he consistently criticized the TPP (and other trade agreements, from NAFTA to the WTO itself) during his election campaign. It is unclear whether President Trump will be as adamantly opposed to the TPP as Candidate Trump, particularly since he will be working with Republicans in Congress, who generally supports the TPP (indeed, it was originally a Republican initiative). Still, its future is in doubt.

The reasons why the TPP is struggling in the United States are many, both straightforward and complicated. But these reasons are also present in most other megaregional agreements being considered, including the TTIP, RCEP, and ultimately will be the case with the

[8]As noted by Krugman (1993), employment is determined by aggregate demand in the short run and aggregate supply in the long run. Trade models should focus on efficiency effects.

FTAAP (Petri and Plummer, 2016). Indeed, the TTIP itself has now been put on the backburner for at least a year.

First, all increases in efficiency, be it via trade or technology, lead to structural adjustment. While a trade agreement can lead to significant welfare increases for the economy as a whole, these gains are always distributed unequally, with the winners being different from the losers. Active government policies to compensate those who lose through liberalization — by, for example, facilitating reintegration of those who lose jobs via training programs, job-matching schemes, or wage insurance — can theoretically lead to better outcomes for all. However, setting up these measures and programs has proven challenging; for example, in the United States, the Trade Adjustment Assistance (TAA) program has been set up to help workers who are laid off due to trade. But the literature evaluating how effective the TAA program has been is mixed. An evaluation sponsored by the US Department of Labor found that, while workers receiving TPP benefits performed better than a control group without assistance in some ways, on the whole participation in the TAA program ultimately had a negative effect on total income. The report concludes that, if one does not include the benefits of enhanced trade on the economy as a whole, the net benefits of the program overall were negative (Social Policy Research and Mathematica, 2012).

Second, and related to the first point, critics have suggested that the TPP will exacerbate income distribution and negatively affect vulnerable groups at the lower end of the distribution. During the US 2016 Presidential Campaign, this argument formed in part the narrative of the far left on the Democratic side (Bernie Sanders) and the far right (Donald Trump) on the Republican side. Still, there is little evidence to support this argument; in fact, the literature would suggest that the opposite will likely be true. The Petri and Plummer (2016b) results and work by Lawrence and Moran (2016) conclude that the TPP will: (1) not significantly affect general employment levels; (2) raise productivity and wages for both skilled and unskilled labor; (3) displace some workers during implementation but also make it possible to compensate them generously from national gains; (4) improve the US income distribution slightly, by raising wages relative to capital returns and lowering prices particularly for low-income households; (5) prevent a "race to the bottom" in labor standards and improve labor rights in several economies; and (6) disproportionately benefit poor countries and labor within them (Petri and Plummer, 2016b).

Nevertheless, as noted above, the empirical literature underscores that structural adjustments can weigh heavily on some workers (Autor *et al.*, 2014). Petri and Plummer (2016) estimate that the number of jobs affected — in the sense of jobs lost in import-impacted industries and gained in export-impacted industries — could range from 18,900 to 160,700 jobs per year, depending on assumptions about how far structural adjustments extend beyond directly affected firms into other sectors of the economy. Lawrence and Moran (2016), using somewhat different methods, estimate a range of 23,800–169,000 jobs. These numbers increase the ongoing flow of employment changes in the US labor market, or "job churn." Given 55.5 million job changes in 2014, a typical number for non-recession years, mid-point estimates of the effects of the TPP would increase job churn by less than 0.1%.

Moreover, the labor chapter of the TPP represents the state-of-the-art in FTAs between developed and developing economies. The agreement commits parties to the ILO *Declaration on Fundamental Principles of Rights at Work*, guaranteeing freedom of association; the right to collective bargaining; and the prohibition of forced labor, child labor, and discrimination. The labor provisions are actionable, and their implementation is spelled out in side letters between the United States and Vietnam, Malaysia, and Brunei, respectively. Such provisions are understandably controversial in developing countries, but they will benefit workers in the long run and, most importantly, enable trade relations to deepen despite rising expectations for labor standards in developed markets.

6. Conclusion

In the absence of progress at the multilateral level, regional trading arrangements are emerging as second-best initiatives to reduce barriers to trade and update rules for 21st century commerce. The most advanced of the megaregional trading agreements in the Asia-Pacific region is the TPP. Like all other trade agreements, it is imperfect. But it breaks new ground, is ambitious, and is remarkably forward-looking. It has built in it the possibility of expanding in terms of coverage and membership in the future; at least six other economies have already expressed official interest in joining once the accord is ratified. We note above that the TPP should generate significant economic benefits to member-economies without much trade diversion that will negatively affect non-partner economies.

By facilitating production networks and providing greater protection of FDI, improving the protection of intellectual property, harmonizing product standards, and reducing other impediments to a regional integration, the TPP will likely have an important effect on trade in technology-intensive industries and in the diffusion of innovation in the region. As the chapters in this volume show, the transmission processes are many and complicated and, of course, trade policy is only one part of the trade and technology puzzle. But it is arguably an important part.

Acknowledgments

The author would like to thank participants at the *Megaregionalism: Trade and Innovation within Global Networks* conference held in Honolulu, Hawaii, in January 2016 for their comments. He would especially like to thank Dieter Ernst and Alissa Tafti for their substantive input.

References

Autor, D. H., D. Dorn, and G. H. Hanson (2014). "The China Syndrome: Local Labor Market Effects of Import Competition in the United States," *American Economic Review*.

Dollar, D. (1992). "Outward-Oriented Developing Economies Really Do Grow More Rapidly: Evidence from 95 LDCs, 1976–1985," *Economic Development and Cultural Change*, 40(3), pp. 523–44. Available at: http://www.jstor.org/stable/1154574?seq=1#page_scan_tab_contents.

Freund, C., T. Moran, and S. Oliver (2016). "Tariff Liberalization," Ch. 2 in *PIIE Briefing 16-1, Assessing the Transpacific Partnership, Market Access and Sectoral Issues*, Volume 1 (Peterson Institute for International Economics, Washington, D.C.), pp. 31–41. Available at: http://www.piie.com/publications/interstitial.cfm?ResearchID=2909.

Krugman, P. R. (1993). "What Do Undergrads Need to Know About Trade?" *American Economic Review,* 2 (May), pp. 23–26.

Petri, P. A. and M. G. Plummer (Forthcoming). *The TPP and Developing Economies*, World Bank Working Paper Series, Global Economic Prospects Division.

Petri, P. A. and M. G. Plummer (2016). "The Economic Effects of the Transpacific Partnership: New Estimates," Ch. 1 in *PIIE Briefing 16-1, Assessing the Transpacific Partnership, Market Access and Sectoral Issues*, Volume 1 (Peterson Institute for International Economics, Washington), pp. 6–30. Available at: http://www.piie.com/publications/interstitial.cfm?ResearchID=2909. Also, published in January 2016 as PIIE Working Paper 16-2. Available at: http://www.iie.com/publications/interstitial.cfm?ResearchID=2906.

Petri, Peter A. and Michael G. Plummer (2016b). "The Economics of the Trans-Pacific Partnership: Distributional Impact," VoxEU.org, Apri. Available at: http://voxeu.org/article/economics-tpp-winners-and-losers

Petri, P. A., M. G. Plummer, and F. Zhai (2012). *The Trans-Pacific Partnership and Asia-Pacific Integration: A Quantitative Assessment* (Peterson Institute for International Economics, Washington, D.C.).

Social Policy Research and Mathematica (2012). "The Evaluation of the Trade Adjustment Assistance Program: A Synthesis of Major Findings," Department of Labor mimeo, December. Available at: file:///C:/Users/Plummer/Downloads/TAA_Synthesis.pdf.

Wignaraja, G., P. Morgan, M. G. Plummer, and F. Zhai (2015). "Economic Implications of Deeper South Asian-Southeast Asian Integration: A CGE Approach," Asian Economic Papers.

Winters, L. A. (2004). "Trade Liberalisation and Economic Performance: An Overview," *The Economic Journal*, 114, pp. F4–F21. doi:10.1111/j.0013-0133.2004.00185.x.

WTO (2011). "World Trade Report 2011."

World Bank (2016). "Potential Macroeconomic Implications of the Trans-Pacific Partnership," Ch. 4 in *Global Economic Prospects: Spillovers and Weak Growth* (World Bank, Washington, D.C.), pp. 219–255. Available at: http://pubdocs.worldbank.org/pubdocs/publicdoc/2016/1/847071452034669879/Global-Economic-Prospects-January-2016-Implications-Trans-Pacific-Partnership-Agreement.pdf.

Chapter 2

The Trans-Pacific Partnership Agreement: Promoting Waves of Innovation

Douglas Lippoldt

HSBC Global Research

Abstract

The Trans-Pacific Partnership (TPP) is the most ambitious international trade deal in 20 years. Although it still has to be ratified, once in place, it could lead to substantial improvements in intellectual property protection and market openness. TPP would offer innovative firms the prospect of a larger addressable market and that may incentivize R&D expenditure and new product development for the expanded market. This chapter reviews the trade and innovation nexus in the context of the proposed rule-based trade reforms under the TPP. It considers empirically the recent experience with Intellectual Property Rights (IPR) reforms and the positive association of such reforms with indicators of economic and innovation performance. It reviews the changes embodied in the TPP and concerns that have been raised. It concludes that TPP offers an important economic opportunity for the TPP partner countries and highlights innovation issues to consider going forward.

1. The Trans-Pacific partnership — A trade and innovation accord[1]

Once ratified, the Trans-Pacific Partnership (TPP) will change the rules of the game, liberalizing international trade and trade-related investment across an area encompassing almost 40% of the global GDP. It will promote innovation by delivering improved alignment of standards for the recognition and protection of innovative content and processes and open markets for the associated trade and investment. This will facilitate market-driven technology development and transfer, reinforcing the dynamic effects of trade liberalization. This matters for the economic performance of innovative firms and economic development.

Leaders from the 12 TPP countries signed the agreement on February 4, 2016. The ongoing ratification process could take another year or more. The agreement requires ratification by at least six countries, representing 85% of the GDP of the region, before it can enter into force. Preliminary estimates indicate that the agreement is in the economic interests of each of the members (e.g., see Petri *et al.*, 2014; Petri and Plummer, 2016).[2] While success in the ratification process is not guaranteed, if TPP advances, we anticipate that it could deliver meaningful trade liberalization in the medium term.

This chapter takes stock of the anticipated liberalization under TPP with a focus on its implications for trade and innovation. In doing so, it provides insights relevant to business, governments, and other stakeholders. It underscores that lessons from the TPP may also have application in other proposed megaregional agreements such as the Regional Comprehensive Economic Partnership (RCEP) and the Trans-Atlantic Trade and Investment Partnership (TTIP).[3] As the relationship between

[1]Disclaimer & Disclosures: Please refer to the HSBC version of the research report, and click on the following link for important disclosures and disclaimers in the Disclosure appendix: https://www.research.hsbc.com/R/29/mhSssCni6nsF.

[2]Not all assessments of TPP are positive. A critical stream has emerged in the literature, focusing in particular on the labor market impacts of TPP in import-competing industries. For more on these views, see Rodrik (2016) or Capaldo *et al.* (2016).

[3]RCEP engages the ten ASEAN countries and six ASEAN preferential trading partners, including China and India. TTIP includes the European Union and US.

trade and innovation is complex, this chapter points to issues for further research where additional policy refinements may be feasible (Box 1).

1.1. *What is the TPP and what does it do for innovation?*

The TPP is a trade agreement among 12 countries including Australia, Brunei, Canada, Chile, Japan, Malaysia, Mexico, New Zealand, Peru,

Box 1. IPRs and megaregional agreements: An agenda for further research

Catching up with the developed countries is not an easy task, and there does not appear to be a one-size fits all solution. Cherif and Hasanov (2015) found that, in a 40-year period upto 2010, only 9 developing countries succeeded in joining the ranks of the high-income countries.[4] Clearly, there are issues where policymakers would benefit from further insights as to what works under which conditions, including with respect to megaregional trade agreements and IPR reform:

Competition and IPRs — IPRs afford rights-holders a measure of market exclusivity, but not a timeless monopoly. They remain vulnerable to competitive pressures. Patents and copyright protection expire. Trade secrets can be overturned if competitors can independently and fairly discover the insights that underpin them. A heathy competition regime can help to ensure that an appropriate balance is achieved between innovators and potential rivals. There is a growing literature from economic and legal scholars.[5] The TRIPS Agreement raises some relevant issues such as IPR abuse in the case of licensing and technology transfer, and the TPP includes a chapter on competition. But, is the appropriate balance between competition and market exclusivity achieved in the TPP and the other pending megaregional trade accords in order to drive rights-holders to continue innovating and new entrants to challenge them with alternatives?

(*Continued*)

[4]According to Cherif and Hasanov (2015), the successful economies include Cyprus, Czech Republic, Greece, Ireland, South Korea, Malta, Portugal, Slovenia, and Taiwan.
[5]For example, see US Federal Trade Commission (2004).

Box 1. (*Continued*)

IPRs and price effects — A related issue under megaregional accords concerns price effects of increased stringency of IPR standards. What can we say about price effects of IPRs in such agreements? There is some literature, for example, on pharmaceuticals with respect to the TRIPS Agreement. But, the TPP and other megaregionals go beyond the TRIPS Agreement. With stricter IPR standards and market opening under the TPP, for example, will increased international competition from alternative varieties moderate the price effects of market exclusivity for products with patent protection? What will be the price effects in countries beyond the TPP members?

Alternative business models that rely less on IPR protection — In an open, regional, and competitive landscape, firms may opt for alternative approaches to the formal IPR processes, which take time and can be costly (e.g., patent registration and litigation take years to reach conclusion). Alternatives to patents, for example, may include accelerated innovation processes and exploitation of first-mover advantages. To what extent is this actually the case? What factors drive firms' strategic decisions? How does this vary by sector? To what extent do firms compete via such alternative means and use IPR protection simply for defensive purposes (e.g., to have patents that could be offered for cross-licensing in the event of an infringement lawsuit).

IPR stringency and barriers to international trade — Regional trade agreements span increasingly diverse countries in terms of the levels of development and approaches to legal system. Does increased IPR stringency translate well across the various systems? Can the requirements under megaregional accords be equitably implemented across such diverse environments without opening opportunities for abuse or overly-constraining trade (e.g., overly strict controls could lead to delays at the border in some countries but not others)?

IPR protection and digital trade — Intangible assets are virtual and can easily be traded across borders in digital format. Are current protections adequate and can they be put in place without constraining digital trade and free flow of information? Will the existing TPP provisions be robust in the face of technological change? What would be the optimal enforcement strategy consistent with market openness objectives?

Inclusive trade, technical co-operation, and absorptive capacity — Are the support provisions of the TPP and the other megaregional accords adequate to ensure that the developing partners have a chance to benefit from the potential incentives for innovation and technology transfer?

Singapore, the US, and Vietnam. If the ratification process is successful, the agreement will be the most ambitious trade package of the last 20 years in terms of the scope and depth of the anticipated liberalization. Through improvements in the availability of legal protection for innovative ideas and practices and increased market openness to international trade among the 12 economies, the TPP agreement is expected to bolster the pace of innovation and its diffusion. (Table A1 in the Appendix presents an overview of key provisions in this regard.)

Spanning 30 chapters, the TPP rulebook will liberalize trade in industrial goods, agricultural products, services, and trade-related investment. While deeply reducing traditional trade barriers such as tariffs, the TPP goes further and addresses behind-the-border issues such as standards, regulations, and protection of intellectual property rights (IPRs). Although membership does not yet include some big economies in the region, such as China or India, the TPP members have stressed that the TPP is a living agreement open to new members and capable of addressing new and emerging issues in future.

2. IPR and the TPP

Innovation can be viewed as a new idea or practice that delivers value through its application in products or processes. But, it can be difficult for innovators to find ways to benefit from the fruits of their labor. Whether the innovation is a new film or a technology upgrade for a smartphone, a recipe for a secret sauce or a catchy logo for a new brand, it remains vulnerable to competitors.

Property rights are a fundamental element for the functioning of markets. Yet, in the case of ideas, practices, or other intangible assets, this can pose a problem (Demsetz, 1967). Being virtual, these innovative assets can be used by many users simultaneously (i.e., they are non-rivalrous) and may be relatively easy to duplicate (Maskus, 2000). Consequently, it can be difficult for innovators to appropriate a portion of the economic gains from the use of such assets or other compensation (e.g., social recognition). Weakness in the ability to reap benefits from an innovation may reduce the incentive to develop such assets. In addition, innovators may face further challenges in distributing their products internationally if the destination markets are further constrained by trade barriers or different rules of operation.

Fortunately, governments have moved to address this challenge via the global trading system. Treaties administered by the World Trade Organization (WTO) and World Intellectual Property Organization (WIPO) spell out rights and protections for innovators and subsequent rights holders. Subject to certain legal requirements,[6] an innovation may be considered as property and its owner as having IPRs. With respect to trade, the key treaty is the WTO's Agreement on Trade-Related Intellectual Property Rights (TRIPS). It defines the key types of intellectual property as including copyright and related rights, trademarks, geographical indications,[7] industrial designs, patents, layout-designs of integrated circuits, and undisclosed information (trade secrets).[8] Still, the requirements under TRIPS and related agreements are not of a uniformly high standard applicable in all markets. The strength of protection varies somewhat from market to market. Weakness in the available protection for one or more types of intellectual property in a market may make it difficult, if not impossible, for a rights-holder to do business there.

2.1. *TPP — Taking IPR protection to a higher level*

As can be seen in Table 1, changes with respect to IPRs in the TPP are incremental, building on the WTO TRIPS Agreement. In some cases, the incremental change is quite substantial. Among the various IPR reforms mandated by the TPP, most concern refinement of protection for types of

[6]These vary depending on the type of intellectual property. For example, in some cases, registration is required (e.g., for patent protection of an invention), while in others, special requirements must be met (e.g., for a trade secret, the concept must be secret, reasonable efforts must be made to keep it secret, and it must have value as a consequence of being secret).

[7]Geographical indications are names or signs that indicate the origin of a product. Examples from the TPP countries include Tequila from Mexico or Idaho potatoes from the US.

[8]The treatment of trade secrets varies by country. In some cases, they are recognised as property, in others as objects that entail rights and duties on the part of interested parties.

Table 1. Key TPP provisions to enhance IPR protection.

Issue	Action under the TPP
Insufficient transparency in national systems	Improves transparency in IPR regimes via making information available on the internet and using standardized procedures; requires authorities to provide written responses to applicants in certain cases (goes beyond TRIPS in use of internet and information requirements *vis-à-vis* the public)
Technical inadequacies and process delays	Includes provisions for ongoing technical co-operation among the national IPR agencies (TRIPS only requires availability of mutually agreed technical co-operation)
Shortness in duration of trademark terms	Offers terms of protection of no less than ten years (under TRIPS the minimum is seven years)
Patent delays may disrupt earnings potential	Requires publication of a patent application after 18 months (i.e., this enables rights-holders to collect reasonable royalties from users, if the patent is then granted) (no such requirement exists under TRIPS)
Undisclosed test data submitted to authorities may be utilized by competitors	Undisclosed test data submitted for marketing approvals will be protected for agricultural chemicals for at least ten years and pharmaceuticals for at least five to eight years (TRIPS does not specify the required term of such protection)
Copyright protection duration varies significantly, by country	Sets a minimum copyright term at not less than 70 years (in TRIPS, the minimum is 50 years)

(*Continued*)

Table 1. (*Continued*)

Issue	Action under the TPP
Weak copyright enforcement can lead to abuses	Requires possible criminal prosecution for removal of rights management information or camcording movies; requires accession to the WIPO "Internet treaties" dealing with protection against circumvention of technological measures and coverage of internet abuse; requires availability of preliminary injunctions and injunctions at border (TRIPS does not address the technology issues or the relevant WIPO treaties; and it is less specific in injunctive relief requirements.)
Copyright and trademarks: boundaries for fair use and liability are not consistent across the region	Clarifies and protects fair use of copyright and trademark materials; clarifies obligations on internet service providers with respect to third-party posting of infringing material, as well as providing limitations on their liability and safe harbors (TRIPS treats fair-use provisions as optional and is less specific, does not address internet service issues)
Trade secrets protection varies widely and is not subject to a clear, enforceable minimum standard	Requires enforceable trade secrets protection, though details may be specified drawing on a list of options set out under the TPP; requires some availability of criminal penalties (TRIPS defines undisclosed information and mandates protection, but does not specify the requirements for such protection)
Members differ in capacity for a high standards IPR regime	Tailored transition periods are established to permit each country time to prepare and adjust, as necessary (the TRIPS Agreement used a similar approach)

Source: HSBC, with reference to the TPP full text (US Trade Representative, 2015, online) and the TRIPS Agreement (WTO, 1995, online).

intellectual property that already benefit from comparatively effective minimum standards in the TRIPS Agreement.

However, the TPP goes beyond TRIPS to close a very important gap with respect to trade secrets, an area that was not effectively protected under TRIPS. The TPP specifies a minimum standard of protection that requires availability of a legal means for rights-holders to protect trade secrets and mandates provision of criminal procedures and penalties for certain acts. A further related and economically important benefit will come from institution of minimum protection for undisclosed test data submitted to authorities in satisfaction of requirements for marketing approvals for agricultural chemicals (ten years minimum) and pharmaceutical products (five to eight years minimum, depending on the nature of the drug).

Trade secrets represent a tremendous store of value for businesses. Covering confidential information in any area ranging from recipes to manufacturing techniques, and from business plans to technical specifications and beyond, many businesses utilize trade secrets as a first line of defense for their innovations. Trade secrets protection is generally available without registration. This is especially important for small businesses that may not have the resources to pursue patent registration or for businesses that have non-public innovation of a nature such that it is not covered by patent protection (e.g., specially annotated customer lists). Due to the requirement of secrecy, it is not feasible to give a precise estimate of the size of these assets, but in view of their known heavy utilization, it cannot be ruled out that there is more value in trade secrets than in patents.[9] Thus, in light of the gap in current practices between TPP in developed and developing countries with respect to trade secrets,[10] it is possible that the TPP will have a meaningful innovation and economic

---

[9] In most countries, reasonable efforts to protect secrecy constitute the main advance action required for availability of trade secrets protection. Similarly, copyright protection is automatic, and no mandatory registry is maintained (though in some countries copyright owners may voluntarily register their copyrights with customs authorities for purposes of screening for unauthorized imports).

[10] For details on specific country provisions on trade secrets, please see Lippoldt and Schultz (2014).

effect in several countries on this issue (e.g., Malaysia, Mexico, Vietnam, or Peru).[11] Being better able to protect trade secrets, businesses may be encouraged to develop more.

In other areas of IPR protection, the TPP introduces potentially important requirements that go beyond TRIPS as well. Copyright protection will see significant improvements. On the one hand, the minimum copyright protection will be increased to at least 70 years (under TRIPS, the minimum is 50 years). There also will be a possibility of criminal prosecution for removal of rights management information from products or camcording of movies for purposes of commercial abuse; protection will be extended to items in the digital environment.[12] On the other hand, the TPP clarifies areas of fair use for copyright-protected material (e.g., research[13]) or trademarks. It also offers safe harbors to internet service providers from liability for infringing items posted by third-parties, provided they have procedures and make every effort to block or remove such items once they are identified. Thus, an effort to ensure balance is embodied in the agreement.

Under the TPP, countries will need to align their patent systems with the standards in the agreement. Essentially, that means addressing weaknesses and bringing systems up to the standards of the US and Japan. For example, in some cases, countries will need to take steps to ensure patent applications are published after 18 months. In the event the patent is actually granted, this will enable rights-holders to claim reasonable royalties should another party begin to utilize the invention in the interim. Also, the TPP countries will need to ensure the availability of patent term extensions in the event of delays in review during the application process. Although the gaps in patent protection among TPP partners are more

[11] Availability of protection for trade secrets and patent rights is reviewed in Lippoldt and Schultz (2014) and Lippoldt (2011).

[12] The TPP requires accession to the WIPO internet treaties including protection against circumvention of technological measures and coverage of internet abuse.

[13] The TPP specifically mentions fair use as including criticism; comment; news reporting; teaching, scholarship, research, and other similar purposes; and facilitating access to published works for persons who are blind, visually impaired, or otherwise print-disabled.

modest than in the case of trade secrets,[14] some reforms will be needed in view of the TPP patent requirements. Countries such as Malaysia, Mexico, Peru, and Vietnam or even some developed participants, such as New Zealand, may need to reinforce their available protection.[15]

Two other areas for IPR reform include trademarks and overall procedures. For trademarks, the TPP will ensure they are granted a minimum ten-year term of protection (whereas under TRIPS it is seven years). And, with respect to procedures, the TPP aims to ensure consultation of interested parties in the formation of regulation and public availability of the procedures for registration or recognition of intellectual property and the resources available in cases of abuse.

2.2. *IPRs in the TPP — Summing up*

In most of the requirements discussed earlier, the TPP will go beyond the requirements of the WTO TRIPS Agreement (Table 1). The TPP will address gaps and misalignment in IPR provisions among the TPP members and it will open markets and promote complementary policies (e.g., transparency and consultation in trade-related regulatory policies) that may contribute to improved innovation performance and diffusion.[16] This can be part of a coherent trade policy package that encourages businesses and others to invest in intangible assets and provides the rights-holders with an improved means to engage in international, market-based technology transfer and commerce across the TPP area. As noted, the leading innovators in the TPP area have a vast stock of technological innovation that can be accessed.

As discussed in Section 3, the establishment of such an improved policy framework for IPR can be associated with improved domestic innovation performance, including in developing countries. Thus, innovators

---

[14] For example, see Park (2008) for an indication of the differences in the strength of patent protection by country.

[15] See Lippoldt and Schultz (2014).

[16] See Cavazos and Lippoldt (2010) on the economic importance of policy complements to IPR reforms.

in emerging markets are also expected to benefit. Yet, it should be kept in mind that IPR reform is not a standalone silver bullet solution. It works best where complementary policies help to support technological capacity; such policies go beyond trade to address rule of law, good governance practices, and human capital development, among other issues (e.g., see Cavazos Cepeda and Lippoldt, 2010).

3. Can changes in IPR stringency affect real-world economic performance?

What has been the actual experience with such IPR reform? Can the types of policy changes embedded in the TPP make a difference? In the years since the advent of the WTO TRIPS Agreement in 1995, countries around the world have engaged in reform of their intellectual property systems. In some cases, such as Singapore or the US, the reforms have gone beyond the minimum requirements of TRIPS either unilaterally or in the context of bilateral or regional free trade agreements. In other cases, such as Russia or Israel, countries have increased protection of intellectual property during the process of accession to WTO or OECD.[17,18] Some of these reforms touch on issues addressed in the TPP, such as the availability of protection for trade secrets or undisclosed test data.

IPR reforms have raised concerns among some stakeholders that there could be potential braking effects of IPR stringency on innovation processes. For example, they could limit the exchange of information inputs to subsequent innovative processes (Williams, 2013). In the context of TPP, innovation concerns have been raised with respect to the development of creative content (and its subsequent use) and access to affordable

[17] OECD is an international organization of primarily advanced economies, promoting trade and growth.

[18] While US legal protections for domestic intellectual property have their foundation in the US Constitution of 1787, critics sometimes note that the US and other leading economies in the 1800s did not extend intellectual property protection to non-domestic innovation. However, the environment at the time was much different. In the 21st century, the world is a more open place, and the rules-based international trade system is much more developed, providing access to innovation via market-based approaches.

drugs, among other issues. The Electronic Frontier Foundation (2016), for example, summarizes 11 concerns with the IPR aspects of TPP. These range from potentially excessive stringency in the copyright enforcement provisions (including criminal penalties in some cases) to the potential chilling effects on digital innovation and the work of hackers and tinkerers. With respect to medicines, Baker (2016) notes that heightened stringency of patent protection could, in effect, inhibit market entry of competitors and lead to more conservative government decision-making (in light of potential damage claims from IPR holders under investor–state dispute settlement provisions[19]).[20] Cumulatively, such effects could potentially weigh on innovation processes.

In response to innovation concerns, it is useful to consider empirically the recent experience around the world with increased IPR stringency. How have economies reacted? Studies by Lippoldt (2011) and Lippoldt and Schultz (2014) aimed to test the hypothetically positive association of IPR strength with economic and innovation performance in the period since 1990. This was done using a regression analysis controlling for other factors such as GDP size, endowments, and other policies. Figures 1 and 2 present the positive results with respect to patent rights and protection of trade secrets.

Patents are an important indicator of technological progress because, by definition, they must involve an inventive step, be non-obvious, and have industrial application. Not all inventions are patented, but patents represent new technological ideas that have been reviewed by experts and accepted as meeting certain minimum standards.[21] They also require the

---

[19] Capaldo *et al.* (2016) also allude in passing to the potential effects of the TPP investor–state dispute settlement mechanism in mandating changes to national legal orders in the context of enforcement of IPRs.

[20] *Médecins Sans Frontières* has been raising similar concerns about TPP. See, for example, the MSF Access Campaign Issue Brief, Trading Away Health, 2012. Available at: http://www.msfaccess.org/sites/default/files/MSF_assets/Access/Docs/Access_Briefing_TPP_ENG_2012_update.pdf

[21] In the US, patents may be re-examined. However, the quality of granted patents is such that most survive. They meet the test of novelty, inventive step, and industrial application. Yet, occasionally there are rejections. For example, such was the fate of a patent for making a crustless peanut butter and jelly sandwich in 2005 (see: http://ipfrontline.com/2007/03/reexamination-of-the-peanut-butter-and-jelly-sandwich/).

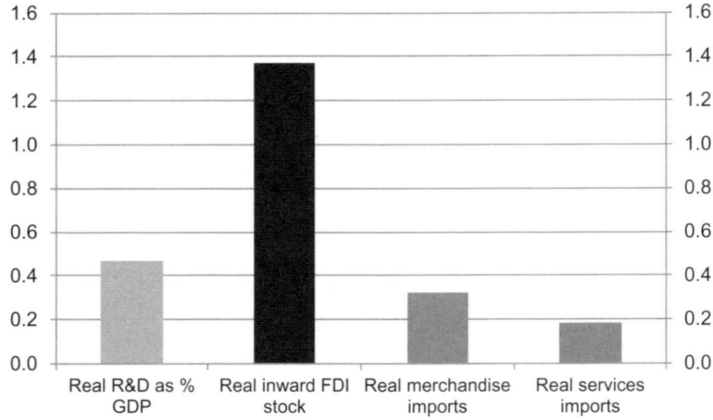

Figure 1. Economic responses in developing countries to a 1% change in the Patent Rights Index, 1990–2005.

Note: Based on a global sample of 91 developing countries. The coefficients are drawn from a regression assessment of the association of change in the index to changes in economic indicators, controlling for other factors. All coefficients shown are significant at the 95% level, except real services imports at 90%.

Source: Lippoldt (2011).

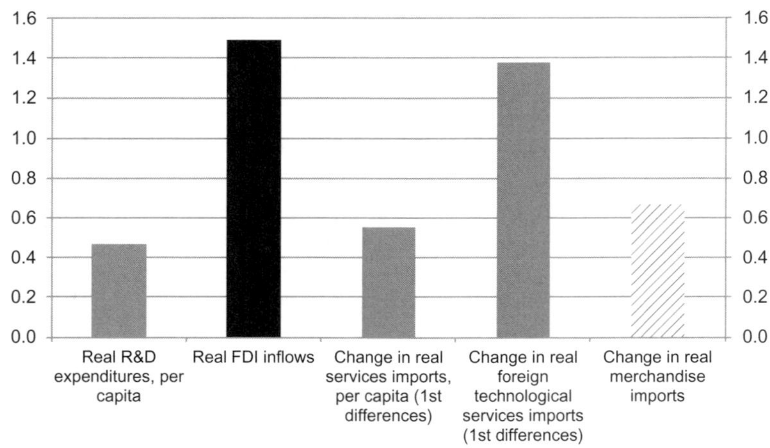

Figure 2. Economic responses to a 1% change in the Trade Secrets Protection Index, for a diverse global sample of countries, 1990–2010.

Note: Based on a global sample of 37 countries at all stages of development. The coefficients are drawn from a regression assessment of the association of change in the index to changes in economic indicators, controlling for other factors. All coefficients shown are significant at the 95% level or better. The TSPI is lagged (−1) for the R&D run.

Source: Lippoldt and Schultz (2014).

public disclosure of the basic functioning of the invention, which can lead to further innovation by others (e.g., to build upon the idea or to invent around it).

Figure 1 highlights the positive responses in key national-level indicators to the strengthening of patent rights in developing countries. Patent rights are associated with increased innovation as measured by expenditure on research and development (R&D), increased technology transfer via foreign direct investment (FDI), and increased availability of technology from abroad through increased imports of goods and services.[22] Thus, other factors being equal, TPP countries like Chile, Malaysia, or Mexico (and subsequently, Vietnam) that improved their protection of patent rights were better off for it. On average, such countries had better performance with respect to the indicators shown than would have otherwise been the case.

Figure 2 presents similar results for a global sample of developed and developing countries with respect to the strengthening of trade secrets protection.[23] As with patent rights, TPP countries that improved the protection of trade secrets, such as New Zealand, Singapore, and Japan (and to a more modest extent Peru or Vietnam), might be expected to be better off than they would have been, other factors being equal.

These studies point to a generally positive relationship between the strength of IPR protection and two sets of key economic and innovation performance indicators. While it should be kept in mind that these studies apply to certain ranges of reforms and particular time periods, they cover a range of experience that is not unlike the reforms being considered under TPP and may be indicative.[24] In many cases, the past IPR reforms covered by these data sets helped countries move from weak systems to more effective systems, with favorable results.

---

[22] In a disaggregation, not shown here, the positive responses were confirmed separately for middle-income and least-developed countries.

[23] NB, the regression approaches shown for trade secrets are not strictly comparable, as the methods were varied in order to test robustness (e.g., by lagging the dependent variable in one of the runs).

[24] However, the studies do not provide a basis to argue that ever-increasing IPR stringency will lead to similar returns.

3.1. *So, IPRs and market openness can be important … but do they matter in the TPP?*

By aligning IPR regimes and opening markets, the TPP will help to provide improved opportunities for members to tap into a pooled library of existing and future innovation. This matters because just a few countries in the world account for much of the available technology. Openness to trade and investment is essential for countries that wish to promote technology transfer and development (Keller, 2009). As of 2011, the top seven countries in the world with respect to R&D expenditure accounted for just over 75% of the global total (Table 2). Two of the three biggest were TPP countries: the US and Japan.

Yet, technology transfer alone is not sufficient for catch-up and sustainable advantage. Local technology creation and related innovation (e.g., building brand and reputation) as well as absorptive capacity also have a role to play. These dimensions, as well, may be potentially influenced by the types of reforms envisaged in the proposed megaregional trade accords. Where successful, such reform efforts can support business efforts at product-differentiation–enabling firms to reap increased margins and develop "moats" *vis-à-vis* global competition.

3.2. *Patently successful*

Why does this matter? It matters because there are huge flows of new patented innovation in the US and Japan (Fig. 3). Yet, as can be seen in Fig. 3, there are also patent flows in the other countries. With improved access to the larger TPP market, each country will gain improved access to the overall pool of patented TPP innovation and thus increase their stock of potential technologies available to be utilized locally. The availability of improved technology is an important element in boosting total factor productivity (TFP) and economic development. Among the TPP developing countries, TFP is only about two-thirds that of the US.[25]

[25] TFP data cited here are drawn from the Penn World Tables. See: http://febpwt.webhosting.rug.nl/Dmn/AggregateXs/PivotShow#

Table 2. Gross expenditure on research and development, 2011 (billions, current US$).

Global ranking	Country	R&D value	GDP	R&D as % of country GDP	Country R&D as % share of world total	Cumulative R&D % share of world total
1	United States of America*	429.1	15,517.9	2.77	31.70	31.70
2	China	205.4	7,492.4	2.74	15.17	46.88
3	Japan*	148.4	5,905.6	2.51	10.96	57.84
4	Germany	97.0	3,751.9	2.58	7.16	65.00
5	Korea	58.4	1,202.5	4.86	4.31	69.31
6	France	53.3	2,862.5	1.86	3.94	73.25
7	United Kingdom	39.2	2,592.0	1.51	2.90	76.15
8	India	36.2	1,835.8	1.97	2.67	78.82
9	Russian Federation	35.2	1,904.8	1.85	2.60	81.42
10	Brazil	27.4	2,615.2	1.05	2.03	83.45

(Continued)

Table 2. (*Continued*)

Global ranking	Country	R&D value	GDP	R&D as % of country GDP	Country R&D as % share of world total	Cumulative R&D % share of world total
Data for the other available TPP countries:						
12	Canada*	24.8	1,788.8	1.38	1.83	
20	Mexico*	8.1	1,171.2	0.69	0.60	
22	Singapore*	7.0	275.4	2.54	0.52	
26	Malaysia*	4.9	289.3	1.69	0.36	
34	New Zealand*	1.8	166.1	1.06	0.13	
For reference: World total expenditure on R&D as of 2011 = US$ 1,353.6 bn.						

Note: TPP countries are indicated with an asterisk (*). The year 2011 was selected to ensure broad country coverage. The table does not include the following TPP countries: Australia, Brunei, Chile, Peru, and Vietnam. The cumulative % share sums a country's contribution with those of all the countries above it in the table.

Source: UNESCO, World Bank, HSBC. Notes: Globally, the data set covers 79 countries in 2011.

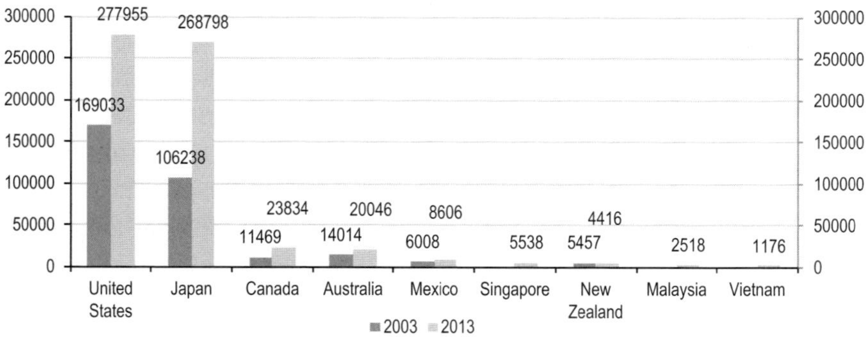

Figure 3. Annual patent grants in the TPP area.

Note: Data are not available for Chile, Brunei, and Peru and for Singapore, Malaysia, and Vietnam in 2003.

Source: Aistemos; HSBC.

While this represents an increase of a few percentage points since 1990, it is still far short of their potential.

By aligning patent systems, the smaller TPP economies will improve their opportunities for technology transfer. The alignment may encourage rights holders of US or Japanese patents to patent their technologies in the other TPP member states. This is often a first step in entering the markets with new, innovative technologies. Technology transfer may then take place via sales of products by the foreign rights-holder, by direct investment and local production by the foreign rights-holder (or as part of joint ventures with locals), or by licensing the technology to local firms.

The effects of progress in strengthening IPR regimes can be boosted by complementary policies to ensure rule of law, ease of doing business, and availability of human capital.[26] For example, IPR reform has been an integral part of Singapore's development strategy in conjunction with such complementary policies and it appears to be one factor contributing to productivity growth. TFP has grown more rapidly in Singapore than in the US: between 1990 and 2010, Singapore's TFP score grew to a point where it led the US by about 10%.

[26]Hall and Jones (1999) provide a useful discussion on the key drivers of productivity growth.

3.3. *Yes, but do firms use it?*

To answer this question, one can look at patents. Via the patent process, technological inventions are registered and publicly notified and hence available to track. Moreover, while applications can be submitted via WIPO or regional IPR authorities, the patent grants are national, and the geographic distribution can be followed using these national-level data. Thus, patents can provide some useful insights. (However, they are not necessarily representative of all types of innovation.)

Much technological innovation is concentrated in the leading R&D countries, and private sector enterprises hold a substantial share of this through their own innovation or via acquisition. Thus, it is important that the IPR system facilitates technology transfer via market-mechanisms. Firms will often register patents in anticipation of entry into a market via trade or investment, so that their innovations are protected in the new market.

Figure 4 presents the case of Toyota (the motor vehicle company based in Japan), highlighting the stock of patent grants in force for the

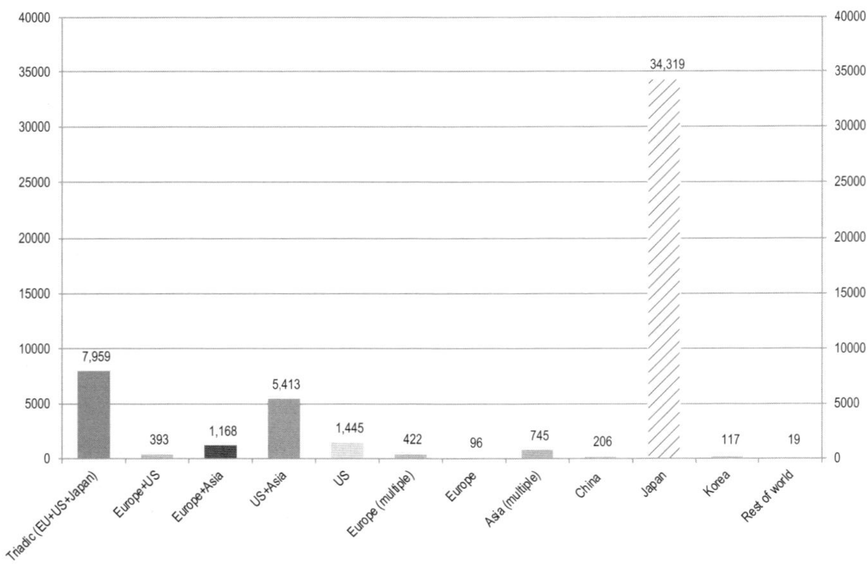

Figure 4. Patent grants for Toyota Motor Corporation (Japan).

Source: Aistemos; HSBC.

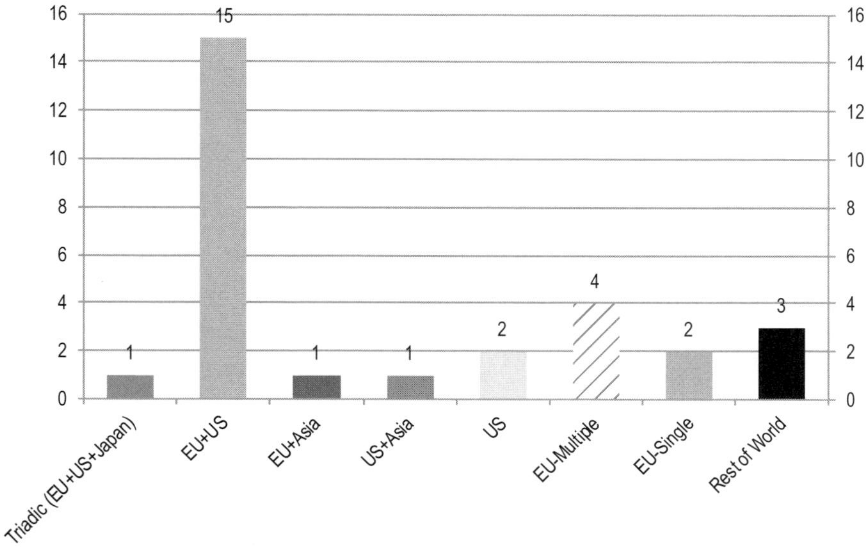

Figure 5. Patent grants for Cemex SAB de CV (Mexico).

Source: Aistemos; HSBC.

company as of 2014. Overall, the company held more than 52,000 patent grants, including some 34,000 in Japan.[27] But, it is interesting to note that the remaining 18,000 patents were widely distributed across Asia and Europe. Some of these may represent local innovation or acquisition in the countries concerned, but some will also represent international transfer of technology for trade or investment purposes. From these data, it appears that the IPR system has incentivized a significant amount of successful R&D at the firm (i.e., patents are reviewed for feasibility by experts) and also encouraged its diffusion.

Gradually, in the period since TRIPS, developing-country firms are improving their innovation performance. While there remains a significant gap with leading world innovators, it is encouraging to see that there is uptake of opportunities by some developing-country competitors as well. For example, Fig. 5 highlights the case of Cemex, a Mexican supplier of cement and other building supplies. The company has built or acquired a

[27] The company data cited in this section are from Aistemos. See: https://aistemos.com/.

portfolio of 29 patents. These grants are primarily in the US and EU member states, but there are several grants in Asia or elsewhere in the world. Thus, the company appears to be leveraging or defending its patented innovation across a number of markets.

These cases illustrate successful firm-level innovation, its formal recognition as property under the IPR system, and the leveraging of the rights for temporary market exclusivity in selected economies around the world. The accordance of these rights will facilitate the ability of these firms to sell, trade, or license the intellectual property or to exploit it directly in products or processes via trade or investment.

4. Market openness and the TPP

Seen from an innovation perspective, the TPP will support market openness in a number of areas that will complement the enhanced IPR environment. These include reductions in tariffs, market access in many services, use of clear and least-trade-restrictive regulation and standard setting, assurances for trade-related investment, and requirements for trade facilitation, among other benefits. Moreover, in most areas, the TPP provides for national treatment in order to reduce potential discrimination *vis-à-vis* the host-country stakeholders.

Table A1 in the Appendix provides more details of the key TPP reforms with potential innovation implications. Some of these changes are quite significant. For example, according to our estimates, roughly one-third of trade among TPP partners is burdened by tariffs that can be quite high in some cases (e.g., 70% for imported cars in Vietnam). About 85% of this burdened trade will benefit from reduced or eliminated tariffs upon the entry into force of the TPP. However, not all TPP liberalization will enter into force immediately upon ratification; transition periods are built in to provide time for necessary preparations and adjustment. Thus, in the case of tariffs, tariffs on an additional 5% of the burdened trade will be liberalized within seven years of the entry into force.

Market openness to trade can influence the pace of innovation in part by providing access to a broader range of inputs than would otherwise be available domestically. Imported inputs can be, in some cases, more competitive than domestic products in terms of price, quality, or functionality. On the other hand, market openness means that innovators have larger markets to sell into. This can mean opportunities to specialize and target niche markets, where economies of scale may not be adequate to justify exclusively domestic production. It can also mean opportunities for innovators to leverage their innovation by transferring technology internationally through sale, partnership, licensing, or direct investments.

4.1. *Building a moat for one's business castle?*

The TPP reform package for IPR fits well with the so-called "New-New Trade" theory developed by economists in recent years to examine differentiation among firms in the globalized economy.[28] They find that successful firms tend to be innovative, able to capitalize on market openness, and exploit potential economies of scale and specialization. This type of strategy can fuel productivity improvements and growth. Domestic firms that find it difficult to compete can potentially up their game by co-operating with innovative international firms through engagement in global value chains.[29] At the same time, firms that fail to provide competitive value propositions risk being pushed aside. Most firms do not trade internationally and their lack of exposure to external opportunities and competition can result in their falling behind and becoming vulnerable to disruption going forward.

Product differentiation through innovation in technology, content, design, and branding can help firms to compete internationally, with myriad examples ranging from Apple to Uber, Shiseido to Billabong, and CNN to Sony. Relevant innovation processes may focus on incremental technological change in serving existing clients and neglected market segments or displacing existing technologies and business models, among other possibilities.[30] Often businesses will strive to secure a measure of market exclusivity by protecting their intangible assets using some form

[28] For example, see Melitz and Trefler (2012).

[29] For example, see Keller (2009).

[30] There is extensive literature on this topic. See, for example, Christensen *et al.* (2015) and Christensen (2003).

of intellectual property protection. As noted earlier, this can help them leverage the fruits of their investments in innovation, enabling them to appropriate a portion of the societal gains from their innovation.

It should be noted that the reinforcement of IPRs can be compatible with open innovation models. In some cases, innovators choose to invoke intellectual property protection to permit use while preventing users from claiming exclusive rights. This is the approach, for example, of the Creative Commons initiative that provides a framework for protecting creations and offering non-exclusive licenses that provide access for others to the content.[31] Also, availability of IPR protection does not preclude the use of competitive strategies that forego recourse to IPR protection. For example, exploitation of rapid innovation and first-mover advantage in entering markets may provide enough differentiation to enable a firm to outcompete rivals (e.g., see Boldrin and Levine, 2007). In some cases, firms may make some of their innovations freely available (e.g., through open-source software) and then build their business models on provision of complementary proprietary products and services.

5. Conclusions

The TPP offers a tantalizing prospect of liberalization in two areas that can catalyze increased innovation: intellectual property protection and market openness. Steps toward liberal reform in both areas have been undertaken at the global, regional, and national level. Yet, under the main international legal instrument promoting IPR protection in global trade, the WTO TRIPS Agreement, there remain some important gaps and uncertainties for market participants. The TPP will strive to address some of these gaps and uncertainties, while delivering substantial improvements in market openness among its members.

A larger market can provide opportunities for bigger returns as well as exposure to more competition and can thereby incentivize R&D expenditure as businesses strive to produce new products for the expanded market. By removing barriers and aligning regulatory regimes and standards, the TPP can expedite the transit of goods and services and also create

[31] For more information, see: www.creativecommons.org.

opportunities to specialize and serve small markets that were previously fragmented but now have sufficient scale to be attractive (e.g., to develop pharmaceutical products for a rare disease).

Market opening can also permit firms to source new and competitive inputs from a larger pool of potential suppliers. This can stimulate innovation as businesses will have greater freedom to combine diverse inputs with their own innovations to create new products. In some cases, liberalization of trade and investment may permit the development of new collaborative approaches. Thus, trade liberalization can affect both the production process and distribution of the products. Together with improved market openness, the IPR solutions embodied in the TPP should offer businesses improved opportunities to leverage investments in innovation.

Will the TPP generate waves of innovation? Its provisions appear conducive to creation and diffusion of innovation in the TPP area. This may lead to further dynamic gains through improved productivity and new approaches to doing business. But, also, to the extent that the lessons from the TPP are picked up and incorporated into other regional accords, the TPP may serve as a template for further reforms. If that is the case, we may see subsequent expansion of the opportunities for innovation, perhaps even going global (e.g., Baldwin, 2014). Thus, the economic gains from reform could then compound.

Finally, in view of the complexity of economic and innovation processes, it will be important to monitor developments once TPP implementation begins. There may be opportunities to refine the policy framework further with respect to the changes embodied in TPP. More broadly, there remain open questions in some areas of the relationship between trade and innovation, and the implications of increased stringency in protection of IPRs (see Box 1). Examples include finding the appropriate balance between protection of IPRs and a healthy competition policy, identification of effective policies to build technological capacities in developing countries, and elaboration of strategies for IPR protection while promoting digital trade and the free data flows. Examination of such areas may provide additional insights for use in the further refinement of trade and innovation policies in the TPP area and beyond.

References

Baker, B. K. (2016). "Trans-Pacific Partnership Provisions in Intellectual Property, Transparency, and Investment Chapters Threaten Access to Medicines in the US and Elsewhere," *PLoS Medicine*, 13(3), p. e1001970. doi:10.1371/journal.pmed.1001970.

Baldwin, R. (2014). "Mulilateralising 21st Century Regionalism," Paper Presented at *OECD Global Forum on Trade: Reconciling Regionalism and Multilateralism in a Post-Bali World.*

Boldrin, M. and D. K. Levine (2007). *Against Intellectual Property.* UCLA Department of Economics.

Capaldo, J., A. Izurieta, and J. K. Sundaram (2016). "Trading Down: Unemployment, Inequality and Other Risks of the Trans-Pacific Partnership Agreement." Working Paper No. 16-01, Global Development and Environment Institute, Tufts University, January.

Cavazos Cepeda, R. H. and D. C. Lippoldt (2010). "The Strengthening of IPR Protection: Policy Complements," *The WIPO Journal*, 2(1), pp. 99–122.

Cherif, R. and F. Hasanov (2015). "The Leap of the Tiger: How Malaysia Can Escape the Middle-Income Trap," IMF Working Paper WP/15/131, June.

Christensen, C. M. (2003). *The Innovator's Dilemma* (Collins).

Christensen, C. M., M. E. Raynor, and R. McDonald (2015). "What is Disruptive Innovation?" *Harvard Business Review*, December.

Electronic Frontier Foundation (2016). "What Is the Trans-Pacific Partnership Agreement (TPP)?" Available at: https://www.eff.org/issues/tpp. Accessed on 26 August 2016.

Hall, R. and C. Jones (1999). "Why Do Some Countries Produce So Much More Output per Worker Than Others?" *Quarterly Journal of Economics.*

Keller, W. (2009). "International Trade, Foreign Direct Investment, and Technology Spillovers." NBER Working Paper 15442, October.

Lippoldt, D. (2011). *Patent Rights, Developing Countries and the Economic Influence of the Multilateral Trading System* (Sciences Po, Paris).

Lippoldt, D. C. and M. F. Schultz (2014). "Uncovering Trade Secrets — An Empirical Assessment of Economic Implications of Protection for Undisclosed Data." OECD Trade Policy Papers No. 167, OECD Publishing, Paris. doi: http://dx.doi.org/10.1787/5jxzl5w3j3s6-en.

Melitz, M. J. and D. Trefler (2012). "Gains from Trade When Firms Matter." *The Journal of Economic Perspectives*, 26(2), pp. 91–118.

Maskus, K. (2000). *Intellectual Property Rights in the Global Economy.* Institute for International Economics.

Park, W. G. (2008). "International Patent Protection: 1960–2005," *Research Policy*, 37, pp. 761–766.

Petri, P. A., M. G. Plummer, and F. Zhai (2014). "The TPP, China and the FTAAP: The Case for Convergence." In *New Directions in Asia-Pacific Economic Integration*, G. Tang and P. A. Petri (eds.) (East-West Center, Honolulu, HW, USA). Available at: http://ssrn.com/abstract=2438725.

Petri, P. A. and M. G. Plummer (2016). "The Economic Effects of the Trans-Pacific Partnership: New Estimates," Working Paper 16-2, Peterson Institute for International Economics, January.

Rodrik, D. (2016). "The Trade Agreements Dilemma: Comment," *LiveMint*, February 15.

Romer, P. M. (1990). "Endogenous Technological Change," *The Journal of Political Economy*, 98(5), pp. S71–S102.

Solow, R. M. (1956). "A Contribution to the Theory of Economic Growth," *Quarterly Journal of Economics,* 70, pp. 65–94.

Tsai, C., F. Samb, M. Ganslandt, and D. Lippoldt (2011). *Enhancing Market Openness, Intellectual Property Rights, and Compliance Through Regulatory Reform in Chile* (OECD).

US Federal Trade Commission (2004). "To Promote Innovation: The Proper Balance of Competition and Patent Law and Policy," *Berkeley Technology Law Journal*, 19(3), pp. 861–883.

USTR (2015). "TPP Full Text," Office of the US Trade Representative. Available at: https://ustr.gov/trade-agreements/free-trade-agreements/trans-pacific-partnership/tpp-full-text.

Williams, H. (2013). "Intellectual Property Rights and Innovation: Evidence from the Human Genome," *Journal of Political Economy,* 121(1), pp. 1–27.

Appendix

Table A1. An illustrative overview of key TPP provisions for innovation and its diffusion.

Intellectual property rights (IPRs)	Trade in services	Financial services	Market access for goods	Trade facilitation	Technical barriers to trade	Investment
National treatment	National treatment and most-favored nation treatment	National treatment and most-favored nation treatment for TPP providers of financial services	National treatment	Customs laws, regulations & procedures to be published online & in English; contact points	Technical standards & recommendations are to be the least trade restrictive measures, as feasible	National treatment and most-favored nation treatment, in like circumstances, for TPP investors
Transparency in IPR regime via information on Internet, standardized procedures; technical co-operation among TPP members	No quantitative restrictions on supply of services or number of suppliers	No restrictions on numbers or value of financial service providers, assets or transactions	Eliminates most customs duties on qualifying goods from the TPP region (with transition periods and exceptions notably in sensitive agricultural products)	Release of goods within 48 hours, though customs determination may be still pending	Mutual recognition of TPP conformity assessment bodies; tests/ certification will be valid for the TPP area	Minimum standards of treatment with access to police protection & justice system; expropriation only for public purpose, with due process & prompt, adequate, transferrable compensation

Covers all main types, e.g.: TM: offers terms > 10 years; patents: publication after 18 months; protects undisclosed test data: for ag chemicals 10 years, pharma 5–8 years; copyright: life of author + 70 years; possible criminal prosecution for removal of rights management information or camcording	Transparency provisions, notification and consultation on new requirements; system shall be fair and follow reasonable manner of operation	No requirement to establish locally, but to permit oversight, may require registration or authorization for cross-border trade	Requires accession to WTO's Information Technology Agreement (duty-free treatment for selected electronics products)	Availability of advance rulings, within 150 days of request and valid for at least 3 years	New rules to be published in advance; stakeholders and interested parties will have opportunities to participate in setting of the rules	TPP investors will have right to pursue neutral, international arbitration in event of an alleged violation by a TPP member of commitments under the chapter, subject to transparency, time limits, protection against frivolous cases
Requires availability of enforceable trade secrets protection, including criminal penalties in some cases	Retains right to regulate, license, establish credentials; some non-conforming measures remain	Specifically permits cross-border transfer of information for data processing in ordinary course of doing business	Prohibits import licensing that requires export performance or use of domestic distributors	TPP preferences are subject to rules of origin	Annexes protect proprietary information; permit reasonable regulation of ICT or food and beverage products	Ban on performance requirements, including local content, technology transfer, or localization, exports

(Continued)

Table A1. (*Continued*)

Intellectual property rights (IPRs)	Trade in services	Financial services	Market access for goods	Trade facilitation	Technical barriers to trade	Investment
Requires accession to the WIPO internet treaties, including protection against circumvention of technological measures, coverage of internet abuse.	No requirement to establish locally; free flow of funds transferred in relation to services (exceptions apply)	Minimum standard of treatment under the agreement, but also some non-conforming measures accepted	Requires publication of the rules, procedures, rates; transparency in export licensing procedures and measures concerning biotechnology trade	Express shipments shall be expedited, with duty-free treatment if below the national threshold in weight or value	Requires due process & scientific basis for cosmetics, medical devices & pharmaceutical regs.	Freedom to transfer funds, though subject to possible limitations to manage volatile capital flows
Availability of preliminary injunctions, injunctions at border; protection of fair use of copyright & TM materials; ISP liability safe harbors	Liberalization applies to all services except for those specifically exempted		Elimination of agricultural export subsidies for TPP trade and limits on any agricultural export restrictions to 6 months with due notification to partners	Co-operation among customs authorities	Members must ensure a reasonable implementation period for new rules	TPP investors are to be able to appoint senior managers without regard to nationality
Transition periods: established for each country	No new non-conforming measures allowed		Ability to trade remanufactured goods without discriminatory restrictions			Right to regulate in public interest, e.g., for public health, safety, financial stability, the environment

Note: National treatment means subject to the same rules as nationals of the country concerned; ISP = internet service provider.

Source: USTR (2015); HSBC.

Part 2

Innovation and Trade within Global Networks

Chapter 3

Beyond Value Capture — Exploring Innovation Gains from Global Networks

Dieter Ernst

East-West Center, Honolulu
Centre for International Governance Innovation, Waterloo, Ontario, Canada

Abstract

Increasingly complex and diverse global corporate production networks (GPNs) and global innovation networks (GINs) are transforming international trade, production, and innovation, creating new, yet little understood, challenges for national policies that seek to foster economic growth and prosperity through productivity-enhancing innovation. As countries are integrated into these global networks (GNs), they are forced to rethink a broad array of policies. A fundamental tension needs to be addressed between global knowledge sourcing, which requires trade liberalization, and domestic capability development, which requires supporting industry and innovation policies. To reap innovation gains from GNs, countries need to address simultaneously two challenges. From the "outside," national policies have to deal with a multitude of bilateral and regional trade agreements (RTAs) and attempts to establish a new architecture of megaregional trade agreements, like the Trans-Pacific

Partnership (TPP), that seek to create and harmonize new international benchmark standards for national policies, rules and regulations, for example, for intellectual property, technical standardization, government procurement, taxation, and competition policy. National policies need to take into account these changes in the governance of the international economy (as demonstrated throughout the book). This chapter focuses on a second set of challenges that originate from the "inside": countries differ in their capabilities to learn and innovate, and hence they differ in their capacity to capture potential gains from integration into GPNs and GINs. The chapter outlines a conceptual framework and describes the research needed to trace the dissemination (or lack thereof) of innovation resources and capabilities through knowledge-sharing within those networks that span developed and developing worlds. The analysis focuses on the experience of China and other leading Asian electronics manufacturing exporting countries. All of these countries are deeply integrated into international trade and foreign direct investment (FDI) through GPNs, and in some cases, through GINs. The author argues that research on GNs needs to move beyond value capture. Instead more effort should be invested in exploring whether and how integration into these networks might foster or erode a country's absorptive capacity and firm-level innovation capabilities. The chapter reviews what is known about driving forces and characteristics of GPNs and GINs and highlights the increasing diversity and complexity of these networks in the electronics industry. To establish whether GN integration might foster or erode the host country's absorptive capacity and firm-level innovation capabilities, three scenarios are considered: (i) the "Gains from Trade" effect; (ii) the "Domestic Disintegration" effect; (iii) the "Innovation Trap" effect; and (iv) the "Limits to Modularity" effect.

1. Trade and innovation within global networks

Increasingly complex and diverse global corporate production networks (GPNs) and global innovation networks (GINs) transform international trade, production, and innovation, creating new, yet little understood, challenges for national policies that seek to foster economic growth and prosperity through productivity-enhancing innovation. These networks integrate dispersed production, engineering, product development, and research across geographic borders.

The spread of GNs has outgrown the capacity of multilateral World Trade Organization (WTO) agreements. Megaregional trade agreements signal the beginning of big changes in international commercial governance, moving beyond tariff reduction to the harmonization of rules, regulations, and policies. This trend has far-reaching implications for national trade policies and for the link between trade and innovation. For instance, deep regional trade agreements (RTAs) and the provisions of the Trans-Pacific Partnership (TPP) Agreement, released on November 5, 2015 and signed on February 4, 2016,[1] allow enforcement of contracts and intellectual property rights (IPRs) across borders. These mega-regionals exceed commitments covered by WTO (WTO "plus") in areas such as free movement of foreign direct investment (FDI), technical barriers to trade, liberalization of services, dispute settlement, customs cooperation, and government procurement.

As countries are integrated into GNs (some more than others), they are forced to rethink a broad array of policies, well beyond the realm of international trade. In order to foster innovation within GNs, new ideas are required on how to adjust policies on education, research, and innovation, especially climate change, environment, and green technology; widespread but neglected diseases; the development, use, and protection of intellectual property; technical standards and certification needed to enhance technology transfer and knowledge diffusion; the treatment of workers and investors; and cybersecurity. Equally important are policies to reduce harmful side effects of "wild innovations" in the financial sector that have completely lost sight of fundamental rules that are necessary to balance profits with due process and risk analysis.[2]

Some basic definitions are in order to study the gains from innovation that countries and companies might capture via participation in GNs.

---

[1] For the full text of the original TPP agreement, see: https://ustr.gov/trade-agreements/free-tradeagreements/trans-pacific-partnership/tpp-full-text. While the Trump administration has withdrawn from TPP in early 2017, the remaining 11 members are exploring how to retain TPP without the US under a new name: the Comprehensive and Progressive Agreement for Trans-Pacific Partnership, or CPTPP.

[2] See Lazonick (2016).

Innovations convert ideas, inventions, and discoveries into new products, services, processes, and business models. Radical breakthrough discoveries and inventions through scientific research are only the tip of the iceberg.[3] It is important to emphasize that innovation is more than research and product development; that users must perceive an advantage to pay for the innovation; and that entrepreneurs are not just founders of Internet start-ups, but vary in terms of size, business model, and organization.[4] Critical are innovations in generic industrial technologies that allow for new ways of manufacturing existing products (e.g., with new materials or by using three-dimensional (3D) printing) as well as for manufacturing new products derived from new ideas, discoveries, and inventions (e.g., implanted sensors). Without large-scale advanced manufacturing facilities, which can draw on an integrated domestic industrial value chain, even the most sophisticated research and development (R&D) capabilities are of little use to generate innovations.[5]

To reap innovation gains from GNs, countries need to address simultaneously two challenges. From the "outside," national policies have to deal with a multitude of RTAs and attempts to establish a new architecture of megaregional trade agreements, such as the TPP, that seek to create and harmonize new international benchmark standards for national policies, rules, and regulations, for instance, for intellectual property, technical standardization, government procurement, taxation, and competition policy.

This chapter, however, focuses on a second set of challenges that originate from the "inside": countries differ in their capabilities to learn and innovate, and hence they differ in their capacity to capture potential gains from integration into GPNs and GINs. A huge gap in innovation capacity continues to separate emerging and developing countries from the US, the main European countries, and Japan.[6] For the technology followers, this raises important challenges.

---

[3]Ernst (2009b).

[4]This broad definition is in line with Peter Drucker's classic statement: "The test of an innovation, after all, lies not in its novelty, its scientific content, or its cleverness. It lies in its success in the marketplace" (Drucker, 1985, p. viii).

[5]As demonstrated in Helper *et al.* (2012).

[6]See Cornell University, INSEAD, and WIPO (2015).

This chapter outlines a conceptual framework and describes the research needed to trace the dissemination (or lack thereof) of innovation resources and capabilities through knowledge-sharing within GNs that span developed and developing worlds. The analysis focuses on the experience of China and other leading Asian electronics manufacturing exporting countries. All of these countries are deeply integrated into international trade and FDI through GPNs, and in some cases, through GINs.[7]

Section 2 argues that research on GNs needs to move beyond value capture and explore whether and how integration into these networks might foster or erode a country's absorptive capacity and firm-level innovation capabilities. Section 3 reviews what is known about driving forces and characteristics of GPNs and GINs and highlights the increasing diversity and complexity of these networks in the electronics industry. To establish whether GN integration might foster or erode the host country's absorptive capacity and firm-level innovation capabilities, Sec. 4 considers the following scenarios: (i) the "Gains from Trade" effect; (ii) the "Domestic Disintegration" effect; (iii) the "Innovation Trap" effect; and (iv) the "Limits to Modularity" effect. The chapter concludes with implications for policy and research in Sec. 5.

2. Moving beyond value capture

Research on GNs has mostly concentrated on *trade in value-added* and the *capture of value* across different locations. The joint OECD-WTO Trade in Value-Added (TiVA) database, for instance, measures the foreign

[7] For these Asian economies, international trade now approaches 100% of GDP on average, ranging from 42% in quasi-continental China to 78% in Korea, 136% in Malaysia, and 140% in Taiwan. (Merchandise trade as a share of GDP is the sum of merchandise exports and imports divided by the value of GDP, all in current US dollars. See: http://data.worldbank.org/indicator/TG.VAL.TOTL.GD.ZS) Developing Asia has captured the largest share of global FDI inflows, growing from $401 billion in 2012 to $465 billion in 2014, with China ($129 billion) and Hong Kong ($103 billion) ahead of the US ($92 billion) and the UK ($72 billion). At the same time, Asia's share in FDI outflows has increased from $299 billion in 2012 to $432 billion in 2014, accounting for almost one-third of global FDI outflows, with Hong Kong ($143 billion) and China ($116 billion) now being the second and third largest home economies, after the US (with $337 billion). See UNCTAD (2015a).

content share of exports as a proxy for GN integration, while using a growing domestic content share of exports as an indirect signal of "an increasing ability to upgrade within the value chain".[8] In a similar vein, research at the World Bank explores policies that might help to capture value from GN integration for developing countries.[9] A recent report by the International Labour Organization (ILO) seeks to identify ways to capture value for employment,[10] while UNCTAD attempts to assign "labor and profit components of the value-added …[to] … the locations where participating firms are headquartered (for profit accounting) and have located their factories (for labour accounting)".[11]

At the center of much of the research on value capture is the assumption that the proliferation of GNs will continue unabated, as an increasing number of countries participate in international trade and investment. Implicit in this assumption is that the current slowdown of international trade (discussed below) is cyclical rather than structural, and that GNs will soon return to their role as the engine of international trade.

A second assumption is that, as GNs proliferate, "[i]t is necessary to value imports as well as exports, to reduce time delays as well as tariffs, and to look 'behind the border' at regulatory measures as well as 'at the border' measures".[12] This has important implications for policy. "In a world characterized by geographical fragmentation of production, policies aimed at supporting production of specific industries by protecting them from foreign competition seem even less likely than in the past to reach the intended target of sustaining activity in the protected industries".[13]

The first assumption is contentious. The year 2016 marks the eighth consecutive year in which annual trade growth has fallen below 3% and the fifth year in which world trade will have grown at roughly the same

[8] See: http://www.oecd.org/sti/ind/measuringtradeinvalue-addedanoecd-wtojointinitiative.htm; and http://www.oecd.org/sti/ind/tiva/CN_2015_China.pdf.

[9] Taglioni and Winkler (2014).

[10] ILO (2015).

[11] UNCTAD (2015b). However, this report concedes "challenging" data problems as teardown reports, which physically disassemble and examine component parts of a product, are available for the IT industry but not for most other sectors.

[12] OECD, WTO, World Bank (2014, p. 7).

[13] *Ibid*. p. 33.

rate as world GDP (at market exchange rates), rather than twice as fast, as was previously the case.[14] Such a long, uninterrupted spell of slow trade growth is unprecedented. There is evidence "that the slower pace of expansion of global supply chains is an important determinant of the trade slowdown".[15] And the most recent *World Bank June 2016 Global Economic Prospects* concludes that the persistent trade slow-down is structural: "Declining commodity prices; China's shift towards a slower, more sustainable, growth path; and soft activity in advanced economies appear to have been mutually reinforcing drivers for weaker merchandise trade growth".[16] Further support for a slowdown of GPNs comes from research by the Canadian Trade Commissioner Service, which highlights a disproportionally large fall in the exports of intermediate goods as a main contributing factor.[17]

The second assumption brings back into the analysis insights from Endogenous Growth Theory, which posits that "trade restrictions prevent new types of goods and new types of productive activities being introduced from abroad",[18] and thus may stifle productivity and technology diffusion. Paul Romer's important proposition, however, needs to be balanced with the "infant industry" argument, formalized by Greenwald and Stiglitz.[19] At the same time, innovation theory tells us that strengthening national innovation capabilities improves a country's ability to engage in

[14] See: https://www.wto.org/english/news_e/pres16_e/pr768_e.htm.

[15] Constantinescu *et al.* (2015): page 26. See also World Bank (2015, p. 173), which finds that the trade slowdown may be explained by "shifts in the structure of value chains, in particular between China and the US, with a higher proportion of the value of final goods being added domestically — that is, with less border crossing for intermediate goods."

[16] See: http://pubdocs.worldbank.org/en/842861463605615468/Global-Economic-Prospects-June-2016-Divergences-and-risks.pdf , p. 16.

[17] The Canadian Trade Commissioner Service (2015) finds that, "while total exports rose US$47.5billion in 2012, intermediate exports fell US$ 116.8billion … [T]his decline could … indicate firms' withdrawal from participating in GVCs" (p. 2).

[18] Romer (1994a).

[19] Greenwald and Stiglitz (2006). Philippe Aghion summarizes the argument as follows: For developing countries with a nascent industrial sector, "full trade liberalization will make it very costly for domestic industrial sectors to invest in learning by doing … since domestic costs are initially higher than foreign costs … [and] … the social benefits from learning by doing are not fully internalized." See: Aghion (2014).

and benefit from the international trading system".[20] This gives rise to a central proposition of this chapter: *there is a fundamental tension between global knowledge sourcing, which requires trade liberalization, and domestic capability development, which requires supporting industry and innovation policies.*

It is important to highlight a third assumption underlying the "value capture" research agenda which may no longer hold today. For Paul Romer, for instance, "the most important policy questions about growth … [are] …: In a developing country…, what are the best institutional arrangements for gaining access to knowledge that already exists in the rest of the world? In a country like the US, what are the best institutional arrangements for encouraging the production and use of new technology?"[21] This statement seems to assume that the division of innovation capacity between the US and developing countries remains unchanged, and that the gains from trade for innovation will largely remain uneven.

Whether or not this clear-cut division of innovation capacity still holds requires empirical research. However, innovation has not been at the center of trade in value-added research. It is time to move beyond value capture and to conduct empirical research on the gains for innovation from GN integration and on the domestic capabilities that might help to capture these gains.[22]

3. Driving forces and the increasing diversity and complexity of GNs

Global production networks began to proliferate during the 1970s, signaling a shift from *partial* to *systemic globalization.*[23] With a "partial globalization" strategy, multinational companies (MNCs) invest in a loose patchwork of stand-alone affiliates, joint ventures, and suppliers

[20] See Carlos Primo Braga's MCTI Think Piece, p. 3.

[21] Romer (1994b).

[22] For related ideas, see the chapters by Susan Helper and Tim Krueger; Keith Maskus; and Chen TainJy. A case study on the tension between global knowledge sourcing and innovation policy can be found in: Ernst (2015).

[23] Ernst (1997).

that are scattered across the globe and which co-exist without much interaction and coordination across functions and locations. "Partial globalization" implies that the MNC cannot reap the full benefits of international specialization. With "systemic globalization," a lead firm (a "network flagship") proceeds to integrate its dispersed operations ("intra-firm") and inter-firm relationships worldwide, both across functions and locations. Integration enables the flagship to internalize and combine resources and capabilities without running into the constraints of excessive centralization.

A more recent development is the emergence of GINs, driven by the relentless vertical disintegration of engineering, product development, and research. A gradual opening of corporate innovations systems disperses R&D across firm boundaries and national borders. Take Intel, whose R&D activities are expanding most rapidly in India and China. A closer look at Intel Bangalore shows substantial progress in the portfolio of innovative capabilities, but it also reveals a persistently hierarchical division of labor.[24] Bangalore's strength continues to lie primarily in the so-called back-end design tasks, which includes re-verifying the design, chip layout, running manufacturability tests, and performing product engineering.

Overall, GINs have dispersed innovative capabilities to new players, but this dispersion remains highly concentrated. For instance, with respect to chip design, instead of the established global centers of excellence (such as Silicon Valley), there are now a handful of rapidly expanding new clusters emerging in Asia in places like Hsinchu, Taipei, and Tainan (in Taiwan); Shanghai, Suzhou, Hangzhou, Beijing, Shenzhen, and Xi'an (in China); Seoul, Incheon, and Daedok Science Town (in Korea); Bangalore, Noida, Chennai, Hyderabad, Mumbai, Pune, and Ahmedabad (in India); Penang and Kuala Lumpur (in Malaysia); and Singapore.

The electronics industry has been in the vanguard of this transformation. Today, however, GPNs and GINs are driving production and trade in myriad goods and services sectors, from clothing, food processing, motor vehicles, and pharmaceuticals, to accounting, finance, and legal services.

---

[24] Ernst (2009a).

3.1. *Global production networks*

Trade economists have explored the importance of changes in the organization of international production as a determinant of trade patterns.[25] Their work demonstrates that (i) production is increasingly "fragmented," with parts of the production process being scattered across a number of countries, hence increasing the share of trade in parts and components; and (ii) countries and regions which have been able to become a part of GPNs are the ones that have industrialized the fastest.

Building on this work, the author has developed a broader concept that emphasizes three essential characteristics[26]: (i) *asymmetry*: lead firms ("flagships") dominate control over network resources and decision-making; (ii) *knowledge diffusion*: the sharing of knowledge is the necessary glue that enables these networks to grow; and (iii) *scope*: GPNs encompass all stages of the value chain, not just production. Over time, the focus of outsourcing is shifting from assembly-type manufacturing to knowledge-intensive support services, such as supply chain management, engineering services, and new product introduction. Outsourcing may also include design and product development. This indicates that GPNs also differ from traditional forms of sub-contracting — much denser interaction between design and production and other stages of the value chain require substantially more intense exchange of information and knowledge. Network flagships increasingly rely on the skills and knowledge of specialized suppliers to enhance their core competencies.

In short, GPNs are a major organizational innovation that enables "network flagships" to reap the combined advantages of outsourcing ("vertical disintegration") and integration. Outsourcing allows for the separation of labor-intensive processes (that move to low-cost locations) from capital- and knowledge-intensive processes and their dispersion across firm boundaries and national borders. Integration of the dispersed production, supplier and customer, and knowledge bases is necessary to reduce the high costs and risks of coordinating cross-border exchanges of products, people, information, and knowledge.

---

[25] For instance, Feenstra (1998) and Jones and Kierzskowski (2000).
[26] Ernst (1997). See also Ernst (2002, 2005a).

Asymmetry is a fundamental characteristic of this process. Network flagships dominate and define network organization and strategy. Control over and coordination of network resources and decision-making enables the flagship to directly affect the growth, the strategic direction, and network position of lower-end participants (e.g., specialized suppliers and sub-contractors).[27]

GPNs cover both intra-firm and inter-firm transactions and forms of coordination: a GPN links together the flagship's own subsidiaries, affiliates, and joint ventures with its sub-contractors, suppliers, service providers, as well as partners in strategic alliances. A network flagship like Apple or Intel breaks down the value chain into a variety of discrete functions (by product or by production process) and locates them wherever they can be carried out most effectively, where they can improve the flagship's access to resources and capabilities, and where they are needed to penetrate important growth markets.[28]

A great variety of governance structures is possible. GPNs range from loose linkages that are formed to implement a particular project and that are dissolved after the project is finished (so-called "virtual enterprises") to highly formalized networks ("extended enterprises") with clearly defined rules, common business processes, and shared information infrastructures. Formalized networks do not require common ownership — these arrangements may or may not involve control of equity stakes. What really matters for innovations is whether relations between the parties in the GPN are collaborative or arm's-length.[29]

Arbitrage of differences in factor costs across locations has been a fundamental motivation behind the construction of those networks. In the garments industry, for instance, local suppliers are under constant pressure to reduce costs and speed-up delivery. In collaborative networks, however,

[27] The complexity of these global networks is mind-boggling. According to Peter Marsh, the *Financial Times*' manufacturing editor, "…[e]very day 30 m tones of materials valued at roughly $80 billion are shifted around the world in the process of creating some 1 billion types of finished products." See Marsh (2010). For a detailed case study of the multilayered global production networks in Asia, see Ernst (2004).

[28] Apple, for instance, relies heavily on Foxconn's gigantic production complexes in Shenzhen, Chengdu, and other major Chinese cities.

[29] Helper (1991).

local suppliers are expected to constantly upgrade their capabilities. If they fail to upgrade, these suppliers will perish.[30] Flagships seek to gain quick access to skills and capabilities at lower-cost overseas locations that complement the flagships' core competencies. As a result, two-way knowledge-sharing among network participants has gained in importance. As Ernst and Kim[31] demonstrate, network flagships transfer both explicit and tacit knowledge to local suppliers through formal and informal mechanisms. This is necessary to upgrade the local suppliers' technical and managerial skills so that they can meet the flagships' specifications. Once a network supplier successfully upgrades local capabilities, this creates an incentive for flagships to transfer more sophisticated knowledge, including engineering, product, and process development. This process, however, is not automatic: for upgrading to run its course, supportive industrial and innovation policies are required to foster the absorptive capacity of local suppliers.

3.2. *What distinguishes global innovation networks?*

Since the turn of the century, outsourcing has included product development, design, and research, giving rise to GINs. There is an important element of continuity: GINs emerge as a natural extension of GPNs and hence share most of their characteristics. As explained below, GINs are prone to systemic pressures to internationalize innovation that distinguishes them from GPNs.[32]

Good data on the international dimension of innovation remain scarce, in particular, on the importance of GINs and the motivations for establishing these networks. One such attempt is a database, developed at the East-West Center, which draws on findings of questionnaire-based case studies of GINs in the integrated circuit design industry.[33] The questionnaire surveys and case studies provide basic information (some

[30] Ernst and Kim (2002).
[31] *Ibid.*
[32] See the analysis below on enablers and driving forces. For another attempt to capture such differences, see Cooke (2013).
[33] As described in Ernst (2009b).

quantitative, but mostly qualitative) on the location and type of activity of offshore R&D labs, date of establishment, the size and composition of the workforce, and the educational background and work experience of senior managers.[34] In addition, the database contains information on the rationale for establishing a GIN rather than following a traditional integrated R&D model; the scope and stability of these networks and their governance; management practices to implement network coordination; approaches to learning and knowledge sharing; and the types of innovation and innovative capabilities at different network locations and their interaction.[35]

Conducting such in-depth and standardized case studies requires deep industry knowledge. Collecting these data is a time-consuming and costly affair — an important drawback from the perspective of research funding agencies. Nevertheless, this type of industry-specific, micro-level "detective work" is necessary to move research on innovation gains from GNs beyond the fairly general statements typical of much of the value capture research mentioned at the beginning of this chapter.

Since space is limited, two examples must suffice to illustrate the depth of findings of such case study research on GINs in China's chip design industry. As for enabling factors, the interviews emphasize (in this order): modular design which enables vertical specialization (i.e., the slicing and dicing of the innovation value chain); the role of information and

[34] To address this research agenda, data were collected for a sample of almost 150 companies in the electronics industry. Companies differ in terms of size, ownership, business model, and nationality, ranging from large global brand leaders from the US, Asia, and Europe to specialized suppliers of technology, core components, and product development services. Also collected are profiles of mini-GINs for small trans-Pacific start-up companies of foreign-born engineers from Taiwan, China, and India that are headquartered in Silicon Valley. These start-ups had to commit to conducting product development and research work in Asia in order to receive venture capital funding. The database is representative, capturing important protagonists that shape the dynamics of GINs in the electronics industry and providing important qualitative insights into the nature of these global transformations.

[35] Questionnaires were distributed to engineers involved in research and product development and were customized for China-based chip design companies (both international and domestic) and for buyers of those chips (the so-called Original Equipment Manufacturers (OEMs), both international and domestic).

communications technology in enhancing cross-border coordination of diverse tasks; as well as trade and investment liberalization through multilateral trade agreements within GATT and WTO, and, more recently, RTAs, including megaregional trade agreements.

With respect to driving forces, the interviews highlight their systemic nature — labor cost differentials matter, but they are only one among a package of competitive pressures. Global companies expect China's integration into their GINs (either through local affiliates in China or through outsourcing to independent local suppliers) to…:

- enable them to increase the return-on-investment on R&D, despite the rising cost, complexity, and uncertainty of R&D;
- facilitate the penetration of high-growth emerging markets in compensation for the slow demand growth in core OECD countries;
- accelerate speed to market in line with shorter product life cycles;
- gain access to lower-cost pools of knowledge workers;
- tap into the resources and innovative capabilities of new competitors and emerging new innovation hubs;
- provide them with opportunities for so-called "regulatory arbitrage," i.e., the exploitation of differences in IPR regimes; incentives; tax laws (especially for transfer pricing); and in regulations on finance, the environment, and health.

3.3. *Increasing diversity and complexity of global innovation networks*

An important recent development is the *increasing diversity and complexity* of these knowledge-sharing network arrangements. Global innovation networks (GINs) now involve multiple firms and organizations that differ substantially in size, business model, market power, and nationality of ownership, giving rise to a variety of networking strategies and network architectures (Table 1).

The flagship companies that control key resources and core technologies, and hence shape the hierarchical intra-firm and inter-firm networks, are still overwhelmingly from the US, the EU, and Japan. However, there are also now network flagships from emerging economies, especially from

Table 1.

Asia, which construct their own GINs. Huawei, China's leading telecommunications equipment vendor, and the second largest vendor worldwide, provides an example of a Chinese GIN that illustrates the considerable organizational complexity of such networks. Huawei's own GIN now includes, in addition to at least eight R&D centers in China, five major overseas R&D centers in the US, and at least ten R&D centers in Europe.[36]

In addition, international public–private R&D consortia are no longer exclusively originating from the US, the EU, and Japan. Asian countries are also quite active now in global sourcing through such cross-border public–private partnerships. Taiwan's Industrial Technology Research Institute (ITRI) provides a telling example of such global knowledge sourcing from the erstwhile periphery.[37]

[36] For details, see Ernst (2009b).

[37] Within Europe, ITRI's global knowledge network concentrates on Germany, the Netherlands, and France, where it covers a broad array of science disciplines and technologies. By contrast, ITRI's presence in Russia is heavily focused on the country's leading research institutes for advanced mathematics and physical sciences. It is also noteworthy that ITRI has a much larger and widely diversified presence in the US, both

An important recent transformation are splintered GINs with diverse network flagships which increasingly complement the erstwhile dominant hierarchical networks. This indicates that vertical specialization within GINs continues unabated. Different types of splintered GINs are emerging, and network flagships now might include core component suppliers (Intel, Microsoft; ARM; Qualcomm) which control technology platforms; mega-contractors (e.g., Taiwan's Foxconn) which can co-shape strategic direction and provide integrated solutions; global mega-distributors of semiconductors (e.g., Arrow Electronics; Avnet) which can provide integrated solutions; and suppliers of new generic technology platforms like IBM, which seeks to capture industry-specific data and knowledge from major users like leading US healthcare providers to enable its Artificial Intelligence "Watson" platform to solve real-world business problems.[38]

Arguably, the most important manifestation of rising network complexity is the convergence of Information and Communication Technology (ICT) infrastructure for the Internet, wireless and mobile communications, and cloud computing that culminates in "The Internet of Everything." While the vision of an "Internet of Everything" certainly exaggerates what will be possible over the next decade, concepts like "Connected Manufacturing" highlight how global manufacturers are implementing "... bidirectional information-sharing through the global manufacturing value chain — from research and development (R&D) to the customer and back; from suppliers to plants to sales-channel partners, and conversely".[39]

4. Capturing innovation gains from global networks — alternative scenarios

Economic theory still has a long way to go to catch up with the new world of increasingly complex GINs. As indicated, research has mostly

---

with leading universities and with global industry leaders. Finally, ITRI's knowledge network closely interacts with private GINs established by leading Taiwanese companies. For instance, Taiwan's TSMC (the world's largest provider of chip-manufacturing services) has a strong presence in UC Berkeley and at Stanford University, with a heavy focus on leading-edge IC development for advanced computing.

[38] Waters (2016).

[39] Hartman *et al.* (2006).

concentrated on *trade in value-added* and the *capture of value* across different locations. This is important, but it tells us little about the distribution of innovation gains from GN integration and about ways to improve a country's innovation capacity.

To establish whether GN integration might foster or erode the host country's absorptive capacity and firm-level innovation capabilities, the remainder of the chapter considers the following scenarios of contingent impacts: (i) the "gains from trade" effect[40]; (ii) the "domestic disintegration" effect[41]; (iii) the "innovation trap" effect[42]; and (iv) the "limits to modularity" effect.[43]

4.1. *The "Gains from trade" effect*

The first scenario considers how GN integration might unlock new sources of industrial innovation through technology diffusion and global technology sourcing. Research on gains from trade for innovation typically focuses on trade liberalization through the reduction of tariffs and non-tariff barriers to trade, such as quotas, rules of origin, embargoes, sanctions, levies, and other restrictions, as well as "trade facilitation," such as improved customs procedures. By lowering import prices, improving market access for exporters, and enhancing competition, trade liberalization is expected to strengthen a country's innovation capacity through three specific channels: (i) imports, FDI, and technology licensing; (ii) learning-by-exporting, which would expose the region to both foreign technology and intangible knowledge as a source of product and process innovation; and (iii) competition, which may reduce monopoly rents from innovation and create pressure to increase productivity.[44]

Important theoretical contributions are Paul Romer's analysis of the welfare costs of trade restrictions and their constraining effects on innovation, described earlier in this chapter; Robert Feenstra's analysis of Integration of Trade and Disintegration of Production in the Global

[40] As formalized in Romer (1994a).
[41] Rodrik (2015).
[42] See chapter 4.
[43] Drawing on Ernst (2005b) and Chesbrough (2003).
[44] Kiriyama (2012) and Onodera (2008).

Economy,[45] and Lee Branstetter's work on the role of FDI as a channel of knowledge spillovers.[46] These arguments have been popularized into widely disseminated policy prescriptions by Richard Baldwin's proposition that GN integration provides "the 21st century's fast lane to industrial development".[47] Baldwin, however, also proposes restrictions on trade (such as enhanced TRIPS-Plus IP protection and constraints on national technical standardization in line with the WTO Agreement on Technical Barriers to Trade[48]) so as to incentivize multinational participation.

Research on Asia's global innovation network hubs in the electronics industry finds ample opportunities for knowledge diffusion and learning through GN integration. The research shows that foreign R&D centers can act as important catalysts for accelerated learning and capability development. Interviews with foreign affiliates of global corporations as well as with independent Asian network suppliers indicate that integration into global innovation networks can improve access to state-of-the-art innovation management practices, tools, ideas, and opportunities for innovation.[49]

---

[45] Feenstra (1998, 2008).

[46] Branstetter (2006).

[47] Baldwin (2014). See also Baldwin (2013).

[48] Unfortunately, the WTO's concept of standardization is from a bygone era. There is little reason to believe that the people behind the TBT Agreement anticipated the requirements, imposed by the Internet and resulting complex networks and technology systems, for the development and use of consensus international standards. In addition, the TBT Agreement seems to apply only to governments and a few recognized standard development organizations (SDOs), ignoring the reality of ubiquitous, global standards that do not fall under these categories. In its current form, the TBT Agreement thus provides an imperfect policy tool for addressing the challenges of international trade for standardization.

[49] For instance, Chang *et al.* (2006) find that exposure to state-of-the-art innovation management practices of global R&D operations can improve innovation management in Taiwan firms and force them to be "more innovative." Chen (2006) shows that the R&D intensity of foreign-owned affiliates in Taiwan's manufacturing industry has increased from 1.5% in 2002 to 1.9% in 2003. Chen argues that foreign-owned subsidiaries with high export intensity and which rely on Taiwanese original equipment manufacturing/original design manufacturing suppliers "may need to devote more effort to R&D in order to effectively interact with their local suppliers." In turn, this requires that domestic R&D reaches a critical threshold so that it can "serve as a complement to, rather than a substitute for, the R&D activities of foreign affiliates."

Competition in the electronics industry combines intense price competition with the speed of new product introduction, with the result that product life cycles become shorter and shorter. Only those companies that succeed in bringing new products to the relevant markets ahead of their competitors will thrive. Of critical importance for competitive success is that a firm can build specialized capabilities quicker and at a lower cost than its competitors.[50] For many electronics companies, competing for scarce global talent thus has become a major strategic concern. Global sourcing for knowledge workers now is as important as global manufacturing and supply chain strategies. The goal is to diversify and optimize a company's human capital portfolio through aggressive recruitment, especially in emerging Asia's lower-cost-labor markets. Over time, global firms realize that, in order to retain these knowledge workers, it is necessary to transfer exciting projects to new locations in Asia that provide opportunities for learning and knowledge sharing.

All of this implies that innovation systems of global corporations are being opened to outsiders, at least in a few select areas. This has substantially reduced entry barriers and opened up new opportunities for network suppliers to engage in global knowledge sourcing. Research on Chinese IC design firms provides supporting evidence.[51] According to the CEO of China's leading IC design company, Spreadtrum, "the availability of IC design tools, semiconductor fab services, and open-source smartphone software [Android] allows Chinese firms to circumvent their weak spots and develop their strengths in hardware, IC design, and integration".[52] Deep integration into the global semiconductor value chain thus enables Chinese firms to globally source technology and capabilities on a scale never thought possible before. In addition, as the global semiconductor industry critically depends on China's huge and rapidly growing market, this enhances China's bargaining power in negotiations on global technology sourcing.[53]

---

[50] Kogut and Zander (1993).

[51] Ernst (2015).

[52] Interview with Dr. Leo Li, CEO of Spreadtrum, conducted on June 22, 2012, as reported in Ernst (2015, p. 12).

[53] This also explains China's success in the recently concluded expansion of the Information Technology Agreement (ITA-2). See chapter 17.

In short, participation in GPNs and GINs might provide powerful mechanisms for global technology sourcing and learning for suppliers who are latecomers to these networks. In principle, integration into GNs thus could enable developing countries to overcome "barriers to exporting by accommodating specialization in narrow business functions and niche activities and … [to] … limit dependence on the degree of industrial development and broader skills set in the country".[54]

However, it is unclear at this stage whether and how the current international trade slow-down will affect the "gains from trade" effect. While evidence is still accumulating, it may be advisable for GN participants to prepare for a Plan B to compensate for slower GN-driven trade.

4.2. *The "Domestic Disintegration" effect*

This second scenario considers how GN integration might erode absorptive capacity and innovation capabilities. In fact, an important caveat needs to be added to the first scenario: The gains from trade through GN integration are contingent — or a country's capacity to capture those gains depends — "on the structure of specialization and the level of development".[55]

In fact, there are concerns that integration into GINs may be a poisoned chalice. It is feared that, apart from a few prestige projects that might provide limited short-term benefits, R&D by global corporations may not provide the means for upgrading the host country's industry to higher value-added and more knowledge-intensive activities.

Foreign R&D centers often intensify competition for the limited domestic talent pool, leaving domestic companies on the sidelines. Inward R&D by global industry leaders may also give rise to a reverse "boomerang effect," providing global firms with precious insights into business models and technologies developed by domestic firms. Furthermore, foreign R&D centers typically show limited interest in sharing knowledge

---

[54] OECD and World Bank (2015).

[55] Kowalski *et al.* (2015).

with domestic firms and R&D labs. In addition, as global competition is centered increasingly on the development of superior knowledge, "intellectual property" (the commercial embodiment of knowledge) will become more and more intensely guarded.[56]

A recent report for the UK's Overseas Development Institute (ODI) argues that GN integration might erode absorptive capacity and innovation capabilities "because of disconnect to the rest of the economy".[57] In a similar vein, Taglioni and Winkler (2014, p. 6; see Footnote 11) observe that "foreign investors do not actively pursue — and sometimes resist — such integration for several reasons ranging from economic constraints to technological and quality gaps with domestic suppliers to shortages in specialized workers and skills."

But domestic players also may be, in part, responsible for causing the "domestic disintegration" effect. An important under-researched topic is the role that domestic universities and research institutes may play in fostering or hindering domestic knowledge diffusion. For instance, Chinese companies have been complaining that some leading Chinese research universities are more interested in working with MNCs than local companies, "but we don't know whether this is a general phenomenon".[58] Similar concerns were raised in a recent study of the barriers to upgrading India's electronics industry, and in research for the OECD Innovation Policy study on Malaysia.[59] Korea and Taiwan had to struggle with similar problems right from an early stage.[60] While domestic knowledge exchange has improved in both countries, the MNC cooperation bias still seems to continue. Given the advanced level of development that both countries have reached today, this may have also positive effects: close links with global technology leaders now may provide pathways to domestic innovation.

---

[56] Chen (2004).

[57] Rodrik (2015). The coexistence of a country's global integration through trade and FDI and domestic disintegration has been analyzed quite some time ago for Latin America by Sunkel (1978).

[58] Prof. XUE Lan, Tsinghua University Lan, e-mail to the author, December 1, 2015.

[59] Ernst (2014a, 2014b).

[60] Ernst (1994, 2010).

4.3. *The "Innovation Trap" effect*

The third scenario explores how the position of GN suppliers of intermediate goods might constrain their ability and scope for innovation. Taiwan provides an example of how participation in GPNs and GINs may impede rather than foster the innovation capacity of Taiwanese firms. Network integration has facilitated the catching-up of Taiwanese firms as fast followers, especially in the electronics industry. However, as explained below, deep network integration has now become a mixed blessing, unless appropriate policies are put in place to develop domestic capabilities for disruptive and low-cost innovation both at the firm level and across the industry. As for R&D in the electronics industry, until recently, its focus has been on incremental innovation. There is a growing recognition that Taiwanese firms must develop new products and services in order to avoid diminishing returns from network integration.[61]

Taiwanese firms and policy-makers are still groping in the dark on what precisely that strategy requires, for the time being content with adopting a pragmatic trial-and-error approach until they find something that works. However, a new approach is gradually taking shape which combines market-led innovation and public policy coordination of multiple layers of private and public innovation stakeholders.

Due to its pragmatism and openness to new forms of public policy and private–public partnerships, Taiwan's innovation policy may, in fact, shed new light on the opportunities and challenges for capturing innovation gains from GN integration. In Chapter 4 in this volume, Chen Tain-Jy raises important and largely under-researched questions for the study of trade and innovation: How does a firm's position in GNs affect their ability and scope of innovation? And what are the specific challenges that fast followers like Taiwanese firms are facing when they seek to upgrade their innovation capabilities within these GNs?

Chen's chapter introduces multiple innovation traps to explain why Taiwanese GN suppliers of intermediate goods are stuck in a position of "passive innovators" — they innovate in collaboration with end-users that are typically brand marketers of final goods and services and hence

---

[61] Ernst (2013).

"remain weak in making market-oriented innovations in the sense of putting together different technologies to independently create final products for consumers".[62]

Research on the "innovation gap" effect provides important insights for the study of innovation gains from GNs. It shows that a company's position in a GN affects its access to knowledge and its ability to appropriate innovation rents. Equally important, however, is the analysis of strategic responses by Taiwanese firms to escape the innovation gap. This indicates that effects of GN integration are "neither inherently productive nor undesirable"[63] — support policies are necessary to strengthen a country's capacity to capture gains from GN integration.

4.4. *The "Limits to Modularity" effect*

The fourth and final scenario examines how GN suppliers might be handicapped by the "limits to modularity" that characterize many of the GNs, in particular, the demanding coordination requirements and constraints to interface standardization, which might constrain knowledge diffusion.

This pushes the analysis down to a more fundamental level. In fact, recent research has raised doubts that participation in GNs will automatically enhance the innovation capacity of GN participants. For instance, Chesbrough's dynamic theory of modularity demonstrates that, if a firm fails to adjust its organization and innovation management to the requirements of the new architecture, it risks being caught in a "modularity trap." Quite often this reflects fundamental conflicts of interest that separate, for instance, a global system player from its modular suppliers of manufacturing and design services. The dilemma facing a system player is that the more system technology he gives away to his suppliers, the better and cheaper products might become. But, at the same time, he may experience a substantial loss in the control that he can exercise over his suppliers.

A study on the limits to modularity in chip design finds that "…[i]t is …difficult to sustain the assumption, implicit in much of the modularity

[62] See Chapter 4, p. xx.

[63] As argued in Chapter 18 by Susan Helper and Tim Krueger in this volume.

literature, that modularity is the stable end state of industry evolution, and that this is true across industries and technologies. While modular design has acted as a powerful catalyst for changes in business organization and industry structure, limits to modularity are aplenty, and constrain the convergence of technical, organizational and market modularity".[64]

To somewhat oversimplify the point to gain clarity, let us focus on interface standardization — a fundamental prerequisite for effective GNs. Research has identified multiple constraints to such interface standardization, which might constrain knowledge diffusion. A surprising feature of modular systems is their considerable rigidity. Once deployed, interface standards are difficult to adjust. When performance gains from a given design architecture approach a limit, it becomes necessary to establish a new architecture. But a defining characteristic of modular systems is that any transition to a new generation of design architecture requires fundamental changes in system components, which consequently will break down established interface standards.[65]

Chip design provides an important example of the tight limits to interface standardization. Based on standard interfaces and design rules, the division of labor used to be reasonably simple during much of the 1990s. The resulting separation of chip design from fabrication has been one of the favorite examples of modularization proponents. Engineers designed chips and handed the definition to the mask makers, who then sent the masks to the wafer manufacturers (the silicon foundries). And (most of the time, at least) the result of having this modular division of labor was a chip that could be manufactured at an acceptable yield.

However, this easy phase of modularization of the semiconductor industry has vanished permanently. As process technology has dramatically increased in complexity, intense interactions are required across all stages of the semiconductor value chain, and it is no longer possible to work with entrenched standard interfaces and design rules. All participants in the semiconductor industry know that they need to find a way to organize collective and integrated solutions through effective interoperability standards. But implementing such standards is a tortuous process, due to conflicting

[64] Ernst (2005b).
[65] Chesbrough (2003).

interests of different stakeholders. Furthermore, technological uncertainty makes consensual implementation of interoperability standards extremely difficult, as does the fact that the industry is now vertically specialized.[66]

In short, limits to modularity provide powerful arguments for skepticism that participation in modular GNs will *automatically* enhance the innovation capacity of GN participants. While modularity, in principle, facilitates independent innovation within components, this requires that local suppliers master the necessary capabilities. Thus, an important insight of the above research is that the deeper a company is integrated into GNs, the more important are policies to strengthen local networks.

Public policies thus are required in order to enhance the capacity of companies within a country to reap the hidden potential gains for innovation from GN integration. The World Bank argues that "GVC participation is a necessary but not a sufficient condition for development. While GVCs open doors, they are not magical. Most of the work still has to be done at home with domestic pro-investment, pro-skills, pro-jobs, and pro-growth reforms." (Taglioni and Winkler, 2014, p. 3). And the afore-mentioned ODI report concludes that such policies need to be "…pragmatic, opportunistic, often "unorthodox" and promotes domestic manufacturing industries — such as protection of home market, subsidisation of exports, managed currencies, local-content rules, development banking, special investment zones...."[67]

Note however that, as Susan Helper and Tim Krueger observe in Chapter 18: "[M]any trade agreements forbid items on this list, especially local content rules." (p. xx). For instance, the WTO Agreement on Trade-Related Investment Measures (TRIMs) prohibits local-content requirements because they might "restrict and distort trade".[68] And the TPP, in its Chapter 9 on investment, establishes investment rules "requiring non-discriminatory investment policies and protections that assure

[66] Recently, however, attempts to avoid being trapped by prematurely frozen design parameters have led to new approaches to improve the flexibility of system-on-chip (SoC) design, for instance, through reconfigurable processors. But it remains to be seen how viable these new approaches will be.

[67] Rodrik (2015).

[68] See: https://www.wto.org/english/tratop_e/invest_e/invest_info_e.htm

basic rule of law protections, while protecting the ability of Parties' governments to achieve legitimate public policy objectives." Chapter 9 specifically prohibits "'performance requirements' such as local content or technology localization requirements".[69]

Case study research is required to establish the real effects of local content rules and other support policies recommended in the aforementioned ODI report.[70]

5. Conclusions and implications for policy and research

The proliferation of increasingly complex and diverse GPNs and GINs raises important but unresolved challenges for policies that seek to foster economic growth, competitiveness, and prosperity through productivity-enhancing innovation. Most research on GNs focuses on trade in value-added and value capture, but offers only vague insights into innovation gains that might result from integration into these networks.

Based on the evidence presented in this chapter, the author argues that it is time to move beyond value capture research. Instead, more efforts should be invested in exploring head-on whether and how GN integration might foster or erode a host country's absorptive capacity and firm-level innovation capabilities. Case study research has shed some light on the drivers, the geographic extension, and the diverse manifestation of these networks. Collecting these data requires deep industry knowledge. And it is a time-consuming and costly affair, which explains why research has been limited to a handful of industries.

This chapter highlights an important blank spot in our knowledge: How are networks that span developed and developing worlds distributing

---

[69] See: https://ustr.gov/about-us/policy-offices/press-office/press-releases/2015/october/summary-trans-pacific-partnership

[70] In line with the infant industry argument, Aghion (2014, p. 498) argues that "adequately targeted sectoral intervention, e.g. to more skill-intensive or to more competitive sectors, can enhance growth," especially for developing countries that need to overcome latecomer disadvantages. For a negative assessment, see Hufbauer *et al.* (2013).

innovation gains among countries at different stages of development and among their firms? The chapter summarizes what is known by introducing four alternative scenarios. These scenarios are useful as signposts for further research. In essence, they highlight a fundamental tension between *global knowledge sourcing*, which requires trade liberalization, and *domestic capability development*, which requires supporting industry and innovation policies.

However, we need more specific and fine-grained information in order to provide sound policy advice on what type of domestic capabilities and institutions are needed to capture the innovation gains from GNs. This is true in particular for the major challenge "from the outside" — namely, emerging RTAs and attempts to establish a new architecture of megaregional trade agreements, such as the TPP, that seek to create and harmonize new international benchmark standards for national policies, rules, and regulations, for instance, in intellectual property, technical standardization, government procurement, taxation, and competition policy. We need in-depth empirical research to assess the role that trade agreements such as RTAs and TPP could play in enhancing or constraining the innovation gains from GNs.

The following questions deserve priority attention in further research:

1. How should one measure the innovation gains that a country or a firm might capture from integration into GNs? And which combination of those measures should we use to measure long-term developments?

   (i) By type of innovation (product; process; marketing; organizational);
   (ii) By complexity (incremental; architectural; modular; radical[71]);
   (iii) Firms introducing products new to the market, in manufacturing and services;

   (iv) Firms introducing products new to the market, by firm size;
   (v) Input indicators (R&D; quality of innovation infrastructure, including standardization and certification);

[71] See Appendix I: Innovation Matrix.

(vi) Measures of industrial upgrading[72] ("firm-level upgrading" from low-end to higher-end products and value chain stages; "industry-level linkages" with support industries, universities, and research institutes). Policy challenge: How to enable firm-level and industry-level upgrading to interact in a mutually reinforcing way ("virtuous circle")?

(vii) How important are three other forms of "industrial upgrading" to guide policies for upgrading a country's position in GNs[73]?

- inter-industry upgrading proceeding from low-value-added industries (e.g., light industries) to higher-value-added industries (e.g., heavy and higher-tech industries);
- inter-factor upgrading proceeding from endowed assets (i.e., natural resources and unskilled labor) to created assets (physical capital, skilled labor, social capital);
- upgrading of demand within a hierarchy of consumption, proceeding from necessities to conveniences to luxury goods.

2. What indicators are needed to measure transfer of technology and knowledge sharing among different participants in GNs (flagships; first-tier suppliers; lower-tier suppliers) and across different sectors?
3. What type of case studies are needed to examine firms that differ in size, nationality, and business model and to examine the following questions: How does a firm's position in GPNs and GINs affect its ability and scope of innovation? And how does a firm's innovation capacity shape its position in these networks?
4. What policies might be best suited to upgrade a country's position in GNs, in which combination, and in what order?

   - Should such policies focus primarily on the demand for innovations or on the supply of innovation capabilities?
   - Should policies focus primarily on attracting FDI? Or is support for industrial and innovation policies needed?

---

[72] Ernst (2014c).

[73] For the concept of "industrial upgrading," see Ernst (2007).

- How important is liberalization of business services as a potential catalyst for host country upgrading through innovation?
- How important is *flexible policy implementation*, i.e. a more bottom-up and interactive approach to industrial policy ("smart specialization"[74]) in which the private sector discovers and produces information about new activities, and the government provides incentives for the search to happen and for the effective implementation?

5. How important and feasible is effective harmonization of international standards, especially for interoperability standards and certification? In light of the persistent diversity of national standards systems and policies, will the international standards system remain in a "multispeed mode" for the foreseeable future? And if so, how might this affect the scope for knowledge sharing within GNs?
6. What research is needed to improve our knowledge about the impact of "standard-essential patents" (SEPs) for the distribution of innovation gains within GNs?
7. Will GN integration enhance or constrain the space for domestic innovation and industrial support policies? For developing countries that seek to expand their participation in GNs, will they need to transition much faster than in the past in favor of strong IPR protection and FDI-led development? And how might this affect their domestic capability development?
8. What indicators are needed to assess whether megaregional trade agreements are likely to foster knowledge sharing and the dissemination of innovation gains from GN integration?
9. What research is needed to establish the role (if any) of domestic content requirements (DCRs)? Will prohibition of DCRs increase the dynamism of host-country clusters by reducing their insulation from outside ideas and competitors (as argued by Paul Romer)? Or will such prohibition constrain the development of dynamic, spillover-rich clusters in host-country economies?
10. Finally, what indicators are needed to determine whether the growth of GNs requires stronger and more harmonized IPR regulations, as

[74] Foray (2015).

they might raise the certainty of investment, reduce contracting costs, and permit technologies to be deployed more efficiently across borders within these networks? Conversely, what research might help to assess how relevant are concerns by developing countries that stronger and more harmonized IPR regulations might restrict technology transfer and knowledge absorption within GNs?

Acknowledgments

This chapter draws on a paper presented at the agenda-setting workshop, funded by the National Science Foundation, *Mega-Regionalism — New Challenges for Trade and Innovation* (MCTI), at the East-West Center, Honolulu, on January 20 and 21, 2016. I am grateful for comments and suggestions received from Shigeyuki Abe, Center for Contemporary Asian Studies, Doshisha University, Tokyo; CHEN Tain-Jy, National Taiwan University, Taipei; Susan Helper, Case Western Reserve University, Cleveland; Derek L. Hill, National Science Foundation and the Federal Renewable Energy Laboratory, Arlington; Brian Kahin, University of Washington, Seattle; Konstantinos Karachalios, IEEE-Standards Association, Piscataway; Bill Lazonick, University of Massachusetts at Lowell and The Academic-Industry Research Network; Kate Linton, U.S. International Trade Commission, Washington, D.C.; Doug Lippoldt, HSBC, London; Keith E. Maskus, University of Colorado, Boulder; Michael Plummer, The Johns Hopkins University, SAIS, Bologna; WANG Ping, China National Institute for Standardization (CNIS), Beijing; Mark Wu, Harvard Law School, Cambridge; and XUE Lan, Dean, School of Public Policy and Management, Tsinghua University, Beijing.

References

Aghion, P. (2014). "Afterword: Rethinking Industrial Policy." In *Creating A Learning Society. A New Approach to Growth, Development, and Social Progress*, Stiglitz, J. E. and B. C. Greenwald (eds.) (Columbia University Press, New York).

Baldwin, R. (2013). "Trade and Industrialization after Globalisation's Second Unbundling: How Building and Joining a Supply Chain are Different and

Why It Matters." In *Globalization in an Age of Crisis: Multilateral Economic Cooperation in the Twenty-First Century*, Feenstra, R. and A. Taylor (eds.) (University of Chicago Press, Chicago).

Baldwin, R. (2014). "Multilateralising 21st Century Regionalism," *Global Forum on Trade Reconciling Regionalism and Multilateralism in a Post-Bali World*, OECD, February 11–12, p. 39.

Branstetter, L. (2006). "Is Foreign Direct Investment a Channel of Knowledge Spillovers: Evidence from Japan's FDI in the United States," *Journal of International Economics*, 68, pp. 325–344.

Canadian Trade Commissioner Service (2015). "Is Value Chain-Driven Trade On The Wane?" October 20.

Chang, Y-C., C.-T. Shih, and Y.-L. Wei (2006). "Hooking up Local and National Innovation Networks: A Study of Foreign-Affiliated R&D Centers in Taiwan." Paper presented at *Greater China's Innovative Capacities: Progress and Challenges*, May 20–21, Beijing, China.

Chen, S-H. (2006). "The National Innovation System and Global R&D Strategies: The Case of Taiwan," Paper presented at *R&D Interplay in Northeast Asia: Global Corporate Strategy and Host Countries' National Innovation System*, February 24, Seoul, South Korea, pp. 15–16.

Chen, T-j. (2004). "The Challenges of the Knowledge-Based Economy." In *The New Knowledge Economy of Taiwan*, Chen, T-j., and J. S. Lee (eds.) (Edward Elgar, Cheltenham), pp. 1–18.

Chesbrough, H. W. (2003). "Towards a Dynamics of Modularity. A Cyclical Model of Technical Advance." In *The Business of Systems Integration*, A. Prencipe *et al.* (eds.) (Oxford University Press, Oxford).

Constantinescu, C., A. Mattoo, and M. Ruta (2015). "The Global Trade Slowdown: Cyclical or Structural?" IMF Working Paper WP/15/6, International Monetary Fund, Washington, D.C., p. 26.

Cooke, P. (2013). "Global Production Networks and Global Innovation Networks: Stability versus Growth," *European Planning Studies*, 21(7), pp. 1081–1094.

Cornell University, INSEAD, and WIPO (eds.) (2015). *Global Innovation Index 2015* (Geneva, Ithaca and Fontainebleau).

Drucker (1985). Innovation and Entrepreneurship. Practics and Principles, New York: Harper & Row.

Ernst, D. (1994). *What are the Limits to the Korean Model? The Korean Electronics Industry Under Pressure*, A BRIE Research Monograph (The Berkeley Roundtable on the International Economy, University of California at Berkeley, Berkeley, CA, USA).

Ernst, D. (1997). "From Partial to Systemic Globalisation. International Production Networks in the Electronics Industry." Report to Sloan Foundation, published as: *The Data Storage Industry Globalisation Project Report 97-02* (Graduate School of International Relations and Pacific Studies, University of California at San Diego, CA, USA).

Ernst, D. (2002). "Global Production Networks and the Changing Geography of Innovation Systems. Implications for Developing Countries," *Economics of Innovation and New Technologies*, XI (6), pp. 497–523.

Ernst, D. (2004). "Global Production Networks in East Asia's Electronics Industry and Upgrading Perspectives in Malaysia." In *Global Production Networking and Technological Change in East Asia*, Y. Shahid *et al.* (eds.) (The World Bank and Oxford University Press, Washington, D.C.).

Ernst, D. (2005a). "The New Mobility of Knowledge: Digital Information Systems and Global Flagship Networks." In *Digital Formations. IT and New Architectures in the Global Realm*, R. Latham and S. Sassen (eds.) (Princeton University Press for the US Social Science Research Council, Princeton and Oxford).

Ernst, D. (2005b). "Limits to Modularity — Reflections on Recent Developments in Chip Design," *Industry and Innovation*, 12, pp. 303–335.

Ernst, D. (2007). "Beyond the Global Factory Model — Innovative Capabilities for Upgrading China's IT Industry," *International Journal of Technology and Globalization*, 3(4), pp. 437–459.

Ernst, D. (2009a). "What do we know about Global Innovation Networks?" In *A New Geography of Knowledge in the Electronics Industry? Asia's Role in Global Innovation Networks*, Ch. 3, Policy Studies No. 54, August (East-West Center, Honolulu, USA).

Ernst, D. (2009b). *A New Geography of Knowledge in the Electronics Industry? Asia's Role in Global Innovation Networks*, Ch. 2, Policy Studies No. 54, August (East-West Center, Honolulu, USA).

Ernst, D. (2010). "Upgrading Through Innovation in a Small Network Economy: Insights from Taiwan's IT Industry," *Economics of Innovation and New Technology,* 19(4), pp. 295–324.

Ernst, D. (2014a). *Upgrading India's Electronics Industry — Regulatory Reform and Industrial Policy. An East-West Center Special Study* (East-West Center, Honolulu, HI, USA) 90 pp.

Ernst, D. (2014b). *The Role of Innovation for Malaysia's Economic Development.* Background study for the OECD Innovation Policy Study on Malaysia, OECD, Paris.

Ernst (2014c). "Trade and Innovation in Global Networks — Regional Policy Implications." In *Can Policy Follow the Dynamics of Global Innovation Platforms?* (The International Innovation Policy Network of the Six Countries Programme, Delft), pp. 5–6.

Ernst, D. (2015). *From Catching Up to Forging Ahead: China's Policies for Semiconductors* (East-West Center Special Study), September, p. 74.

Ernst, D. (2016). "The Information Technology Agreement, Manufacturing and Innovation — China's and India's Contrasting Experiences." Available at: http://www.eastwestcenter.org/sites/default/files/filemanager/pubs/pdfs/6-1Ernst-20160925.pdf. Accessed 11 July 2017.

Ernst, D. and Linsu Kim (2002). Global Production Networks, Knowledge Diffusion and Local Capability Formation, Research Policy, 31, no.8/9, special issue in honor of Richard Nelson and Sidney Winter.

Feenstra, R. (1998). "Integration of Trade and Disintegration of Production in the Global Economy," *Journal of Economic Perspectives*, 12(4), pp. 31–50.

Feenstra, R. (2008). "Offshoring in the Global Economy," Ohlin Lectures, Stockholm School of Economics, September.

Foray, D. (2015). *Smart Specialization. Opportunities and Challenges for Regional Innovation Policy* (Routledge, London and New York).

Greenwald, B. C. and J. E. Stiglitz (2006). "Helping Infant Economies Grow: Foundations of Trade Policies for Developing Countries," *AEA Papers and Proceedings*, May, pp. 141–146.

Hartman, C., R. Kuppens, and D. Schlesinger (2006). *Connected Manufacturing.* Available at: http://www.cisco.com/web/CA/pdf/Cisco_Connected_Manufacturing.pdf. Accessed 11 July 2017.

Helper, S. (1991). "Strategy and Irreversibility in Supplier Relations: The Case of the U.S. Automobile Industry," *Business History Review*, 65 (SplIss. 04), pp. 781–824.

Helper, S., T. Krueger, and H. Wial (2012). *Why Does Manufacturing Matter? Which Manufacturing Matters?* (Brookings Institution, Washington, D.C.).

ILO (2015). *Value Chain Development for Decent Work* (ILO, Geneva).

Jones, R. and H. Kierzskowski (2000). "A framework for fragmentation." In *Fragmentation and International Trade*, S. Arndt and H. Kierzkowski (eds.) (Oxford University Press, Oxford).

Kiriyama, N. (2012). "Trade and Innovation: Synthesis Report," OECD Trade Policy Papers No.135, OECD, Paris.

Kowalski, P. *et al.* (2015). "Participation of Developing Countries in Global value Chains. Implications for Trade and Trade-related Policies," OECD Trade Policy Papers No. 179, OECD, Paris, p. 36.

Kogut, B. and U. Zander (1993). "Knowledge of the Firm and the Evolutionary Theory of the Multinational Corporation," *Journal of International Business Studies*, 24(4), p. 625.

Lazonick, B. "How Stock Buybacks Make Americans Vulnerable to Globalization." AIN Working Paper #16-0301. Available at: http://www.theairnet.org/v3/backbone/uploads/2016/03/Lazonick.BuybacksAndGlobalization_AIR-WP16-0301.pdf. Accessed 11 July 2017.

Marsh, P. (2010). "Marvel of the World Brings Both Benefit and Risk," *Financial Times,* June 11, p. 7.

OECD, WTO, World Bank (2014). *Global Value Chains: Challenges, Opportunities, and Implications for* Policy. Report submitted to G20 Trade Ministers Meeting, 19 July, Sydney, Australia, p. 7.

OECD and World Bank (2015). "Inclusive Value Chains. Policy Options in Trade and Complementary Areas for GVC Integration by SMEs and Low-income Developing Countries." Report prepared for G20 Trade Ministers Meeting, Istanbul, Turkey, October, 6, p. 14.

Onodera, O. (2008). *Trade and Innovation: A Synthesis Paper*, OECD Trade Policy Working Paper No.72, August 7.

Rodrik, D. (2015). *Past, Present, and Likely Future of Structural Transformation*, Presentation at ODI, 23 June. Available at: http://www.odi.org/events/4213-economic-transformation-growth-dani-rodrik . Accessed 11 July 2017.

Romer, P. (1994a). "New Goods and the Welfare Costs of Trade Restrictions," *The Journal of Development Economics*, 43, p. 6.

Romer, P. (1994b). "The Origins of Endogenous Growth," *Journal of Economic Perspectives*, 8(1), p. 21.

Sunkel, O. (1978). "Transnationalization and its National Consequences." In *Transnational Capitalism and National Development — New Perspectives on Dependence*, Villamil J. J. (ed.) (Harvester Press, Hassocks, Sussex), pp. 67–94.

Taglioni, D. and D. Winkler (2014). *Making Global Value Chains Work for Development* (The World Bank, Washington, D.C.).

UNCTAD (2015a). "Global Investment Trends." In *World Investment Report*, Ch. I, Figs. 11.2 and 11.3; Ch. II "Regional Investments Trends." (United Nations, New York and Geneva).

UNCTAD (2015b). *Tracing the Value Added in GVCs: Product-Level Case Studies in China* (UNCTAD, New York and Geneva), p. 25.

Waters, R. (2016). "FT Big Read. Artifical Intelligence," *Financial Times*, January 6, p. 5.

World Bank (2015). "What Lies Behind the Global Trade Slowdown?" In *Global Economic Prospects*, Ch. 4, January, (World Bank, Washington, D.C.).

Appendix

Innovation Matrix, By Complexity

Innovations differ with regard to opportunities and barriers to learning; they also differ in the capabilities that a firm needs to implement a particular type of innovation. It is useful to distinguish between *incremental, modular, architectural,* and *radical* innovations (Ernst, 2009, drawing on Henderson and Clark, 1990). The boundaries between these four types of innovation are fluid. For instance, incremental and radical innovations are about the extent of changes caused by innovation, while modular and architectural innovations are about where the change is happening. They could therefore overlap.

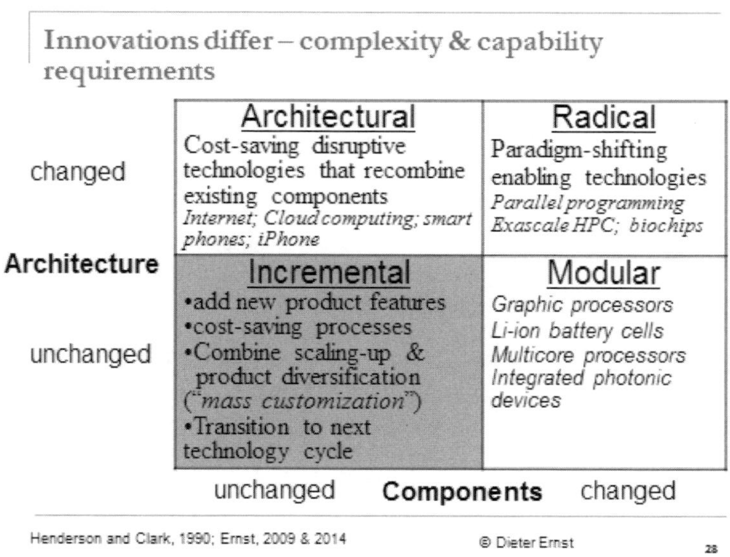

Chapter 4

Supply Chain Positioning and Innovation: Taiwan's Challenge*

Tain-Jy Chen

Department of Economics
National Taiwan University, Taipei, Taiwan

1. Innovation traps of an intermediate goods producer

We argue that supply chain positioning affects the ability and scope of innovation. This is so because positions on the supply chains affect the access to knowledge and the ability to appropriate the value coming out of innovation. Knowledge determines the capability of innovation, and appropriation power determines the incentive of innovation. Most Taiwanese firms position themselves as suppliers of intermediate goods. They innovate in collaboration with end-users that are typically brand marketers of final goods and services. Their innovations manifest several characteristics that are closely associated with their position on the global value chains:

---

*This chapter was originally presented as a think-piece for the *Mega-Regionalism — New Challenges for Trade and Innovation Workshop* held at the East-West Center, Honolulu, Hawaii.

1.1. *Technology trap*

Without direct contacts with consumers, the innovation of Taiwanese firms is mainly driven by technology rather than market demand. They depend on end-users to provide market information to chart the course of innovation. This locks them in a position as market followers. Taiwanese firms have rarely been the first to introduce a truly innovative product to the market, unless they do so in collaboration with their brand partners who define the products. The inability to define products is the Achilles heel of Taiwanese firms when they attempt to become brand marketers, as exemplified by one of the more successful Taiwanese brands, HTC. They can perform reasonably well when innovation is technology-driven rather than market-driven, for example, in the semiconductor industry where Taiwanese foundry service providers have followed the "Moore Law" by moving the processing technologies forward generation after generation. But even in this industry, Taiwanese firms cannot put their technologies to use without the cooperation of their clients, who bear the costs of new product experiments and the risk of technological failures. Cooperation is possible only if Taiwanese firms possess frontier technologies that are worth the costs and risks of experiments. Before they secured that position, co-operation with end-users in innovation was rather difficult, and Taiwanese firms had to compete by price. Price competition led to imitation rather than innovation (Aghion *et al.*, 2001). The experience suggests that, for technological followers in the intermediate goods industry, there is little room for innovation even if they wish to do so. If Moore's Law stops functioning one day, it is quite possible that Taiwan's partnership with end-users like Apple will also crumble. In contrast to the foundry service providers in the semiconductor industry, the Taiwanese IC designers seem to have more difficulties in forming alliances with the world's leading users of IC chips because market dominates technology in driving IC design innovation. They have succeeded, to some extent, in the case of smartphones, where a modular type of design architecture like Android is available to allow sub-system innovations. But Taiwanese IC designers have been almost invisible in other areas, such as automobiles or consumer electronics, where knowledge about the market is critical to innovation. This may be called a "technology trap".

1.2. *OEM trap*

As an intermediate goods supplier, the value of innovation cannot be realized without the help of end-users: In most cases, innovation is achieved in collaboration with the brand marketers, who define the products and bear the responsibility of marketing the products. While collaborative innovation reduces the risks of innovation and increases the efficiency of innovation investment, it also limits the scope of innovation. Moreover, because the value of intermediate goods innovation can only be realized by brand marketers that hold complementary assets to the said innovations, only a small portion of the value of innovations is accrued to intermediate goods innovators (Dedrick *et al.*, 2009). To enhance the value of intermediate goods innovation, it is imperative for innovators to seek adoption by other brand marketers. When succeeding, it leads to a crowded market and a quick decline in the price of final goods. To cope with the price decline, the brand marketers have a tendency to source similar intermediate goods from alternative suppliers or intermediate goods based on alternative technologies. Therefore, economic rent derived from intermediate goods innovation is only transitory, making it difficult for Taiwanese intermediate goods producers to engage in long-term R&D. They are therefore trapped in a short-term collaborative R&D cycle and unable to make autonomous innovations to enhance their position on the supply chain or the value of their relational assets. Therefore, it is hard to expect that disruptive innovation will come out of this kind of collaborative R&D. This may be called an “intermediate goods trap” or “OEM trap”.

1.3. *Modular trap*

Small Taiwanese firms specialize in a narrow range of technologies. Their scope of innovation is confined to product and process innovations based on technological improvements. There are few cases of business model or market discovery innovations. They are more successful in a modular system as opposed to an integrated one. By definition, in a modular system, innovation can be conducted independent of actions taken in the other sub-systems (Baldwin and Clark, 2000). They can also draw on some “public goods” in a modular system to achieve innovations.

This reduces the costs of innovation and increases the success rate. However, a big drawback of the modular innovation is that innovation is not possible until the system is well structured. In other words, as a modular innovator, it is impossible to be the first mover in a new industry. An apparent example is Taiwanese firms' struggle to break into the newly emerged Internet of Things (IoT) industry. They could not find an entry point because the architectures of smart homes, smart cities, smart factories, and the like, are yet to be determined. Taiwanese firms lack the capacity to make system innovation, even for a sub-system. Therefore their ability to innovate is to no avail before the system architecture is clearly defined, and they cannot innovate beyond the boundary of a sub-system. We may call this a "modular trap". The modular trap applies to Taiwan's most important Information and Communications Technology (ICT) industry, including computers and smartphones. In this industry, their ability to innovate depends on the innovative capacity of the system architects like Windows and Android. When the system architectures stop innovating, the innovations of Taiwanese firms also stall.

2. Recent developments

In recent years, as the pressure for innovation mounts, some attempts have been made by Taiwanese firms to escape from the aforementioned traps. As a result of such attempts, their role of innovation in the global supply chains seems to have changed to some extent. By reviewing these attempts, we can see the progress and limitations of intermediate goods innovation. The recent experience indicates that there is hope for Taiwanese firms to escape from the OEM and modular traps, but they remain confined in the technology trap.

2.1. *Innovation enabler*

Instead of offering innovative products, some firms offer intermediate goods to provide an impetus for innovations by end-users. In other words, they enable innovations. This is best exemplified by material suppliers. By providing new materials, they allow the downstream users to make innovations based on such materials. Functional textile is a case in point, which is produced by applying new technologies in the traditional textile-manufacturing process, like weaving, knitting, crocheting, knotting, or pressing, to generate some specific qualities of textile such as anti-bacterial, moisture management, flame-retardant, or water-repellent effects. The objective of material innovation is clearly defined by functional requirements in sportswear or fire-fighting outfits, and it is to be achieved through technologies. This falls within the capacity perimeter of Taiwanese firms. In fact, a series of innovations in functional textiles in the recent years have rejuvenated Taiwan's declining textile industry, allowing it to capture a major share in high-end sportswear. The innovators do not have to worry about the details of market trends, as long as they understand there is a demand for healthy and green living. They can make innovation without any knowledge of fashion, which is to be shaped by apparel producers. However, to market the new materials to clothing designers, they have to change the conventional way of conducting business. In the past, Taiwanese textile suppliers competed for orders from apparel makers after clothes were designed and materials designated. Now Taiwanese textile suppliers have to persuade the apparel makers to adopt their materials to make new cloth designs. With new functions contained in the new materials, they enable apparel makers to innovate products that manifest the newly available qualities. The new materials are generic in the sense that they can be used by all apparel makers and can be fashioned by different makers as distinctive products in the market. By doing so, Taiwanese firms remain an intermediate goods supplier, and their names may not show up in the final products (some do in an Intel-inside method). Nevertheless, they have escaped from the "OEM trap" in the sense that they are no longer instructed by the clients to make innovations. Instead, they innovate on their own initiatives and the clients follow. New materials create a new value to fashion brands, the owners of which can add further value with their own innovations. Brand competition no longer diminishes the value of intermediate goods innovation.

Another example is the light-emitting diode (LED). Advancements in LED technologies were contributed mostly by Japanese and American companies, but Taiwanese firms were pivotal in reducing the production costs and improving light-emitting efficiency that popularized the new materials. Cost was the cornerstone for the diffusion of LED lighting.

Taiwanese firms specialized in processing technologies, including the processing of epitaxial wafers into chips and the assembly of chips into LEDs, or further into modules for ready adoption in designing lights. They work with different end-users in various industries, and in various stages of production, to offer LED materials for innovation. LED has been applied in computers, mobile phones, automobiles, traffic lights, TVs, household and commercial lighting, etc. Again, the contributions of Taiwanese firms are essentially functional, in terms of brightness and texture (colors) of the lights emitted, and luminous efficiency, which are to be achieved by technologies. Innovation in the downstream industries by brand marketers expands the scope of LED applications, and Taiwan has since become the largest supplier of LED materials in the world.

2.2. *Co-innovation*

Although they do not initiate the innovations, they are invited to be an innovation partner from the very beginning. This happens often to components suppliers who possess the frontier technologies in certain areas, often niche areas, and the technologies are not mature enough to make industrial applications so apparent. They are invited by the brand marketers to engage in product innovations, making use of the untested technologies. Co-innovation differs from the traditional case of using mature technologies for innovation, where the technology owner plays no role in innovation. It also differs from modular innovation where innovation is conducted independently of other innovators and is conducted only after the system architecture is in place. Co-innovation is a joint effort by partners in search for a new product design rather than an independent act under a modular system. By definition, co-innovation entails a certain degree of knowledge exchange and integration. Knowledge sharing is apparent and the partners may also jointly invest in the exploration for new knowledge. Co-innovation is a step away from the "modular trap," and it becomes possible because Taiwanese firms possess frontier technologies, albeit only in niche areas, to be exploited for new product development. With this move, the partnership in innovation becomes more equal-footed than in the past, and more value may be created out of the joint innovation.

For example, when Apple developed its first iPhone in 2007, it collaborated with a Taiwanese company, TPK, in creating the finger-impacted, touch-screen panels, which are the pivotal components of smartphones. TPK was initially a component supplier in the PC industry, specializing in computer monitors. After failing in the monitor business, TPK shifted to production of touch-screen monitors used in automatic teller machines (ATM) and point-of-sales (POS) devices. The traditional touch screens were built on resistive technologies. From around 2000, TPK started to venture into the capacitive technologies of touch screens, which allow multipoint inputs and enable instant responses to finger-applied pressures. With the only but untested technology that seemed to fit Apple's design of the newly conceived iPhone, TPK was solicited by Apple in 2004 to be a collaborative partner in the invention of the touch-screen panel. It took almost three years for the collaborative R&D to yield concrete results, but the results were revolutionary as they set the standard for smartphones. Until now, TPK remains a major supplier of touch screens for iPhones, despite Apple's repeated attempts to seek alternative technologies and substitute suppliers. TPK is one of the few Taiwanese suppliers that dare to reject Apple's request to sign an exclusive supply contract, suggesting the status of equal partnership.

2.3. *Branding attempts*

Some Taiwanese contract manufacturers attempted to transform themselves into brand marketers. A few brands have been established, but their position in the world market remains fragile. Although still young in the history of brand marketing, it has become clear that innovation is indispensable for brand success. However, a success formula of innovation is yet to be discovered. According to an assessment by Inter-brand, a consulting company, the leading five brands of Taiwan (based on brand value of 2015) are Asus (computer), Trend Micro (software), Want-Want (food), Acer (computer), and Giant (bicycle). For the first time in the history of this survey, two non-ICT companies made the top-five list. The core competence of Taiwanese ICT brands appears to be in technology. It is also interesting to note that the leading ICT brands, Asus and Acer, and HTC as well (which was ranked sixth in 2015), all have their major market in

Europe rather than the emerging markets in Asia-Pacific or Latin America. This suggests that markets in advanced countries still drive the directions of ICT innovation, and the emerging markets, such as China, may only offer a chance for product differentiation. In comparison, the brands in the food (Want Want) and bicycle (Giant) industries are not necessarily technology based. Progress of brand building in the non-ICT industries gives some hope that Taiwanese firms may eventually succeed in breaking out of their technology trap, although the evidence is relatively weak compared to the case of escaping from OEM and modular traps.

3. Opportunities and challenges of China

China has risen to become a major player in the global supply chains as well as an important market for global companies. In recent years, China has also become a major player in global innovation, as indicated by the number of patents, academic papers, and major technological developments in some specific areas. The rise of China presents opportunities and challenges to global companies in general, and to Taiwanese companies in particular. It may be expected that the emergence of the Chinese market provides a chance for Taiwanese firms to overcome their Achilles heel in forging market connections. Unfortunately, the experience so far generally failed to substantiate such an expectation. However, the rise of China has offered a new opportunity for developing alternative technological paths, and for accumulating strategic resources for innovation. Both seem to reinforce the basic pattern of technology-based innovation of Taiwanese firms rather than provide an impetus for market-based innovation. Two explanations are plausible for this result: First, the Chinese market is not easily accessible. Second, the Chinese market is not sophisticated enough to drive innovations that fall within the capability scope of Taiwanese firms.

3.1. *Shanzhai innovation*

Market drives innovation, but the Chinese market drives a special kind of innovation. Chinese consumers have a strong desire for frontier products and services but refuse to pay high prices for them. This gave rise to the

Shanzhai industry, which offered products with comparable functions of the world's leading brands at significantly lower prices. Lower prices were achieved through low manufacturing costs in China, and Taiwanese firms contributed to this end by offering a technological platform that enables such kind of manufacturing. Shanzhai is not exactly an imitation, because imitation will not produce lower-cost products as many frontier products (e.g., iPhones) are already manufactured in China to achieve low costs. Shanzhai is an imitation in appearance, but a complete transformation of the way the product is designed and manufactured. Shanzhai products are most noticeable in the ICT industry, such as DVD and MP3/MP4 players and mobile phones, where Taiwanese IC design houses provided IC chips to enable easy design and production of related products. Taiwanese firms played a dominant role in offering technology platforms which had never existed before, although they did not physically manufacture or market the products.

For example, Mediatek, a Taiwanese firm specializing in mobile phone processors, provides processors that allow small-scale Chinese independent design houses to add desirable functions of their own designs without technological hurdles. Low technological barriers, together with the industry cluster which has been created by multinational firms through export platforms in Shenzhen, quickly produced a Shanzhai industry that took over the low-end segment of the Chinese consumer market. For example, in the mobile phone industry, the Shanzhai products almost drove out all foreign brands in the end of feature phones era until the smartphones came to replace them. Mediatek was crowned as "King of the Shanzhai industry."

However, to the extent that Shanzhai products mimic leading products in the advanced countries, the contributions of Taiwanese firms are still in the field of technological innovation instead of product innovation. Their dominant role in the Shanzhai industry did allow them better access to the market information, but they have yet to demonstrate their ability to make use of this information for genuine product innovations. There is no doubt that the technological capabilities of Taiwanese firms have enhanced because of the Shanzhai experience, especially in the area of technology integration. It remains to be seen whether they can compete beyond product mimicking. With the advancement of technologies by Chinese firms, the Shanzhai smartphones have all but disappeared; Mediatek is competing now with Qualcomm, Spreadtrum, and other firms for orders from local brands such as Xiaomi and Lenovo. It is not clear whether the Shanzhai experience gave Taiwanese firms like Mediatek a competitive edge.

3.2. *Geographical dispersion of innovation*

It is well noted that innovation has been more dispersed today than in the past (Ernst, 2005). The reason for this occurrence appears to be the increasing availability of R&D resources in the emerging markets rather than the increasing importance of emerging markets. At least from the perspective of Taiwanese firms, this has been the case. Both China and India offer rich human resources for high-level R&D, and multinational firms have rushed to take advantage of these newly found gold mines of human resources. Globalization and the advancement of information technologies have made knowledge sharing and R&D coordination across national borders an easier task. Taiwanese firms are also active in exploring these new resources for innovation activities. Some firms do it through direct employment of skilled workers, while some work with universities. The availability of offshore resources allows Taiwanese firms to diversify their innovation activities and increase the efficiency of research projects. Hon-Hai, for example, has been one of the leading patent producers in China in recent years through its collaboration with Tsinghua University. In 2000–2012, Hon-Hai received 4,912 patents granted by the US Patent Office in the G (physics) and H (electronics) categories, accounting for 19.6% of all China-originated patents in these areas. Many multinational firms were equally active in conducting R&D in China, but Hon-Hai was well ahead of its competitors like Microsoft (1,256 patents in the same period and categories) and IBM (707 patents) (Chen, 2015, p. 10).

The collaboration may not change the basic pattern of innovation of Taiwanese firms, but it enlarges the scope of innovation, thereby increasing the opportunity of Taiwanese firms to cooperate with multinational firms, a manifestation of "alliance capitalism". The essence of alliance capitalism is to continuously renew the firm-specific resources over time to create new opportunities for cooperation. With the renewal of firm-specific resources, the alliance partners may also change, together with the complementary resources to be linked for synergy. New partnership drives the evolution of Taiwanese firms into new business territories. Although Taiwanese firms are rarely the pathbreakers of a new industry, they are eager to embark on the new industry alongside their partners. They used to be quick followers once a new industry has emerged. This position is now taken over by a new league of countries, including China. In order to secure a new position in the global supply chain, the Taiwanese firms need to move into a new industry at the same time as the pathbreakers, even if they only play an auxiliary role in production. They have to be prepared when the horizon for a new industry arises, and patents in possession are their admission tickets to the new horizon. To this end, human resources in China are valuable.

3.3. *Escaping imitation*

Emerging markets, such as China, provide new resources for innovation, but they also present new challenges, notably the threat of imitation. Imitation reduces the potential value of innovation and therefore undermines the incentives for innovation. Lack of protection for intellectual properties encourages imitation. In some countries, including China, imitation is even cherished as socially valuable. To escape from imitation, Taiwanese firms hide their innovations inside the products of multinational firms, which have more means to protect themselves against imitation. For example, Largan Precision, a Taiwanese company which offers the most advanced camera lens for smartphones, is relatively unknown and rarely targeted by imitators. Its technologies cannot be easily tracked down through reverse engineering because the performance of the camera is a joint result of hardware technologies of Largan and software technologies of smartphone makers. The combination of hardware and software technologies entails tacit knowledge, which is hard to identify. The components are sold only to high-priced brands so as to exclude the potential imitators from accessing the relevant knowledge, therefore solidifying Largan's alliance with major brands like Apple and Samsung.

Despite the fact that China has become an important market for their products, Taiwanese firms' alliance with multinational firms can be strengthened if they have the kind of components to distance themselves from the imitators. Their alliance partners welcome and encourage innovations of Taiwanese firms that allow them to escape imitation. The supply chains of multinational firms are essentially segmented from the supply chains that serve Chinese local firms, as all are conscious of the threat of imitation. However, imitation does take place. Imitation has been more successful in the area where there is a demand for outdated and low-end products (Salazar, 2008). For example, the demand for sub-quality liquid crystal display (LCD) panels by local industries, including TV receivers, provided resources to develop the local LCD industry. When imitation of intermediate goods was successful in China, it also allowed final goods producers, such as TV-receiver makers, to compete with multinational firms, which in turn, will seek new technologies to escape imitation. This provided an impetus for innovation by Taiwanese intermediate goods providers, which eventually developed the 4K technologies in LCD panels. The threat presented by successful final goods imitators, like Xiaomi, Huawei, and Lenovo, is more damaging as they undermine the traditional alliance between Taiwanese intermediate goods producers and multinational firms. Chinese final goods producers have a tendency to seek vertical integration as the Chinese government encourages nationalization of key components through its "indigenous innovation" policy (Ernst, 2011). Therefore the Chinese final goods producers present a more serious challenge to Taiwanese firms than the intermediate goods producers.

4. Future perspectives

In short, in order to escape from the three traps of innovation as an intermediate goods producer, Taiwanese firms attempt to move up and down the supply chains. When moving up the supply chains, they innovate to make new materials or new components to enable the downstream users to make further innovations. The purpose of innovation is to create new functions in their products, and innovation is mainly achieved through technological breakthroughs, which are one step up the technological ladder from their previous task of technological improvement.

When moving down the supply chains, they perform some kind of integration in collaboration with their partners. In most cases, the kind of integration is also technology oriented. They put together technological resources to create new products in consortium with their partners. Co-innovation is more difficult than innovation-enabling because it entails knowledge sharing and the creation of new knowledge through joint efforts. Nevertheless, co-innovation seems to present more opportunities for Taiwanese firms, which embrace alliance capitalism, to enhance the value of innovation, and with it, more incentives for innovation will also be operational.

Both movements are anchored by technological progress, and the strategy will be successful only if the technological frontiers have been reached. The challenge is daunting, but the road to success is clearly mapped. Some Taiwanese firms have actually succeeded in making the moves, demonstrating the feasibility of this approach. In contrast, Taiwanese firms remain weak in creating market-oriented innovations in the sense of putting together different technologies to independently create final products for consumers.

In other words, Taiwanese firms remain intermediate goods producers, but they have tried to be proactive rather than passive innovators today. A proactive innovator engages in innovation on its own initiative rather than taking instructions from the clients to engage in "contracted" innovation. It requires some new capabilities to do so, which includes the ability to see the market value of new technologies and the ability to obtain such technologies.

It remains to be answered why Taiwanese firms are unable to become final goods innovators. The attempts to do so have recorded limited success, if not a total failure. They apparently lack the capabilities to combine market knowledge with technologies in order to design the products that meet the consumer demand. It is not clear what exactly these capabilities entail and how they can be acquired. However, like the technology learning that Taiwanese firms have experienced for at least half a century, it must be a slow and cumulative process. Considering that Taiwanese firms began to engage in market learning only recently, it may take some time before we see concrete results. Although final goods innovation commands more market value than intermediate goods innovation, it is also a more risky adventure.

References

Aghion, P., C. Harris, P. Howitt, and J. Vickers (2001). "Competition, Imitation, and Growth with Step-by-Step Innovation," *Review of Economic Studies*, 68, pp. 467–492.

Baldwin, C. and K. Clark (2000). *Design Rules: The Power of Modularity* (MIT Press, Cambridge MA, USA).

Chen, J. (2015). *The Study of the Knowledge Spillover Pattern in China through Patents in USPTO*, Master's Thesis, Department of Economics, National Taiwan University, Taipei.

Dedrick, J., K. Kraemer, and G. Linden (2009). "Who Profits from Innovation in Global Value Chains? A Study of iPod and Notebook PCs," *Industrial and Corporate Change*, 19(1), pp. 81–116.

Ernst, D. (2005)."Complexity and Internationalization of Innovation — Why is Chip Design Moving to Asia?" *International Journal of Innovation Management*, 9(1), pp. 47–73.

Ernst, D. (2011). "Indigenous Innovation and Globalization: The Challenge for China's Standardization Strategy." East West Center Report, Hawaii.

Salazar, A. (2008). "Outdated but Profitable: China's Approach for Science and Technology in the Countryside and the Future of the Spark Program," *Analisis*, 11(32).

Chapter 5

Trade and Investment Barriers in Solar and Wind Global Production Networks*

Derek L. Hill

National Science Foundation
National Renewable Energy Laboratory

1. Introduction

Commercial investment in clean energy has increased significantly over the last decade, driven by public incentives, rapidly falling costs of renewable energy, and removal of trade and investment barriers. The bulk of the investment has been in solar and wind technologies. Over the last decade, investors have poured in $1.1 trillion and $1.0 trillion, respectively, into solar and wind technologies, accounting for 80% of investment in all clean energy technologies (Figure 1).

Innovation in solar and wind technologies appears to have intensified over the last several years. The costs of solar and wind energy have fallen very rapidly, making these technologies competitive with conventional

---

*This chapter was originally presented at "Mega-Regionalism — New Challenges for Trade and Innovation", an agenda-setting workshop, co-funded by the National Science Foundation and the East-West Center, held at Honolulu, on January 20–21, 2016.

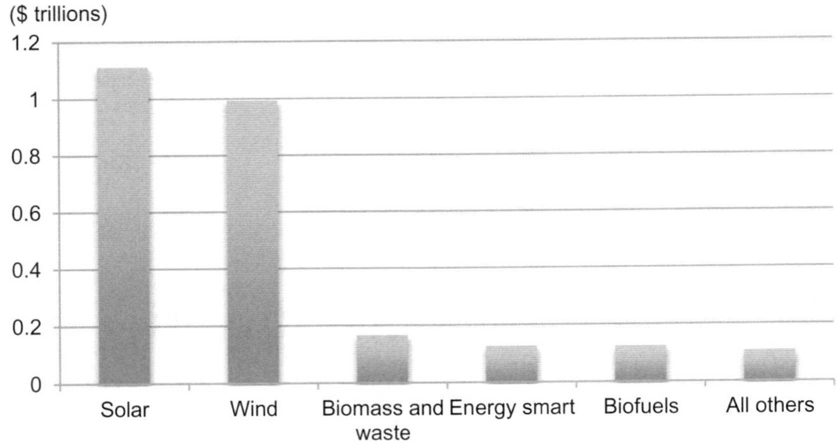

Figure 1. Cumulative investment in clean energy technologies (2006–2016). ($ trillions)

Note: Investment includes asset finance, public markets, re-invested equity, small-scale solar, and venture capital and private equity.

Source: Bloomberg New Energy Finance.

fuels in some countries. Solar power is receiving a significant share of clean energy venture capital and private equity investment. Patenting in solar and wind technologies has grown rapidly.

The burgeoning investment in solar and wind energy sectors, technological breakthroughs, and relatively free trade and investment have led to a huge increase in the production and rise of global production networks (GPNs) in both these industries.

GPNs have played a major role in the declining costs of solar and wind and the huge expansion in production. China has become the largest producer and global supplier of low-cost solar photovoltaics (PVs) that has led to a boom in the installation of solar power around the world. Solar and wind energies accounted for 43% of renewable energy and 18% of all energy sources of global power generation added between 2006 and 2015.[1]

[1]See: Bloomberg New Energy Finance. Available at: https://www.bnef.com/core/market-size?tab=Generation.

Since 2008, a number of developed and developing countries have enacted green industrial policies — local content requirements, domestic incentive measures, technical barriers to trade, and trade remedies — that encourage or protect domestic solar panel and wind-turbine manufacturers, particularly through local content requirements.[2] Although they may benefit domestic manufacturing industries, many of these green industrial policies have negatively impacted trade and investment in solar and wind GPNs. Solar and wind GPNs require free flows of trade and investment and the ability to globally source goods and services at the lowest market price to remain competitive and innovative. The proliferation of green industrial policies has prompted an escalating number of trade disputes in the WTO and imposition of tariffs.

The scaling up of renewable energy is an essential element to help mitigate climate change. Solar and wind are by far the most rapidly growing renewable industries. The IEA estimates cumulative investment of $53 trillion in renewables will be needed by 2035 to help meet the target of limiting the globe's average temperature from rising beyond 2°C.[3] However, the IEA estimates that the projected cumulative investment through 2035 will be $48 trillion based on today's policies.

It is urgent that nations change policies and remove barriers that are hampering investment in solar and wind energies to erase the projected $5 trillion shortfall in clean energy investment through 2035. The OECD has found that green industrial policies on solar and wind GPNs are likely to significantly hamper international investment inflows, primarily due to local content requirements that are attached to feed-in-tariffs.[4]

This chapter will examine the adverse effects of trade and investment barriers on solar and wind GPNs. It will explore and recommend further research on alternative measures to encourage trade and investment and help domestic producers in solar and wind. Section 2 will describe the rise of global production networks in solar and wind energy manufacturing.

---

[2] OECD (2015), p. 10.

[3] International Energy Agency (2014), p. 14.

[4] OECD (2015), p. 60. The OECD conducted an econometric analysis that determined that local content requirements attached to feed-in-tariffs have a significant and negative effect on international investment in solar and wind energy sectors.

Section 3 will discuss the emergence of green industrial policies to encourage and foster domestic manufacturing in solar and wind. Section 4 will discuss the research agenda to better understand the positive and harmful impacts of green industrial policies on solar and wind GPNs, and development of alternative policies to encourage domestic manufacturing in these technologies.

2. Global production networks in solar photovoltaic and wind energy sectors

International trade, Foreign Direct Investment (FDI), and the relocation of manufacturing to China and other countries have fueled the rapid growth and global integration of solar PV and wind GPNs. These networks allow firms to lower their production costs by sourcing activities such as equipment manufacturing based on countries' comparative advantages in costs, skills, materials, and other location-specific activities.[5] The sourcing of activities, particularly manufacturing, to other countries has made solar PV and wind energy sectors to increasingly rely on intermediate imported inputs.

Global investment[6] in the solar energy sector has increased rapidly over the last decade, from $19 billion in 2006 to $110 billion in 2016.[7] International trade has been the primary driver of globally integrating the solar PV energy sector. Domestic and foreign investment in the solar energy sector has also soared in the US, Japan, and China over the last decade. Greenfield FDI has also been a major driver of global integration, accounting for about one-third of total asset finance in the solar industry in 2012.[8]

The rapid expansion of the solar sector, the globalization of production, cost reductions in silicon and equipment, and technological advances

---

[5] OECD (2015), p. 31.

[6] Bloomberg New Energy Finance defines new investment consisting of asset finance, public markets (new equity), venture capital/private equity, re-invested equity, small distributed capacity, corporate R&D, and government R&D.

[7] See: Bloomberg New Energy Finance, https://about.bnef.com/.

[8] OECD (2015), p. 33.

have driven an exponential rise in the generation capacity of solar energy. Between 2006 and 2016, the solar sector added 317 GW in generation capacity to reach 324 GW in 2016.[9]

Global investment has also grown rapidly in the wind sector over the last decade, rising steadily from $39 billion in 2006 to $111 billion in 2016.[10] In the wind energy sector, manufacturing of turbines is increasingly globalized and concentrated. Wind turbine manufacturers often invest in plants located close to emerging and large consumer markets (Brazil, India, the US) because of high transport costs and competition.

Greenfield FDI has been the major driver of global integration of the wind energy sector, especially in the EU and US. China is an exception, with growth stemming mainly from domestic investment. The top wind energy technology manufacturers have been increasing their greenfield FDI investment in emerging and developing markets such as China.

Wind energy has become the second largest clean energy technology after large-scale hydropower, as measured by the total installed capacity. (Nuclear energy is not included as a clean energy source in Bloomberg New Energy Finance's coverage of clean energy investment and generation capacity.) Over the last decade, the generation capacity of wind energy has grown rapidly from 69 GW to 480 GW, a cumulative gain of 410 GW.[11]

3. Domestic "green energy" policies to support solar and wind technology manufacturing industries

The emergence of solar and wind energy sectors as successful and rapidly growing industries has prompted several OECD and developing countries to enact green industrial policies to generate growth and employment in their local solar and wind energy manufacturing industries. Four factors — local content requirements, domestic incentive measures, technical barriers to trade, and trade remedies — have had the most impact on harming trade and investment.

---

[9] See: Bloomberg New Energy Finance, https://about.bnef.com/.

[10] *Ibid.*

[11] *Ibid.*

3.1. *Local content requirements*

Local content requirements (LCRs) typically require developers and investors to source a specific share of manufactured components or equipment locally in order to be eligible to bid on projects or receive financial support or incentives. LCRs have been implemented in 6 OECD countries — Canada, France, Greece, Italy, Spain, and the US — and 20 developing countries, including Brazil, China, Russia, South Africa, and Turkey.[12]

Governments have used LCRs in several ways to support their local industries. The most common approach is using LCRs as eligibility criteria to qualify for the full Feed-in Tariff (FiT)[13] program or for a possible variable part of the FiT and premiums provided on top of the base FiT.[14] The LCR is determined based on renewable energy components and equipment and varies considerably across countries. Fifteen countries have used LCRs for qualifying for FITs. For example, China included a 70% LCR in bidding for wind power plants between 2004 and 2009.[15] Several countries have also used LCRs as eligibility criteria to qualify for direct financial transfers such as public subsidized loans and loan guarantees.

3.2. *Technical barriers to trade*

Technical barriers to trade are the second type of barriers that have primarily affected the wind energy industry. They consist of technical regulations; domestic voluntary product standards applied by the importing country that are inconsistent with international standards; and certification and conformity assessment procedures. These measures have primarily

[12]OECD (2015), pp. 52–54.

[13]An FiT is a policy mechanism designed to accelerate investment in renewable energy technologies. It achieves this by offering long-term contracts to renewable energy producers, typically based on the cost of generation of each technology. FiTs often include "tariff degression," a mechanism according to which the price (or tariff) ratchets down over time. This is done in order to track and encourage technological cost reductions.

[14]OECD (2015), p. 50.

[15]*Ibid.*, p. 60.

affected the wind energy industry in several countries, notably China, India, and South Africa.[16]

3.2.1. *Domestic incentive measures*

Governments have provided generous support to domestic PV and wind-turbine manufacturers, primarily through tax reductions and FiTs.[17] There are two main categories:

- Demand-side (or demand pull) policies that affect consumption, such as FiTs, renewable portfolio standards, and carbon taxes; and,
- Supply-side (or technology push) policies that stimulate production, including R&D, direct investment, and government subsidies.

3.2.2. *Preferential access to financing*

Emerging economies have frequently used preferential access to financing to support technology development and deployment in solar PV and wind-turbine equipment.[18] These policies include subsidized loans, loan guarantees, and direct financial transfers. Emerging economies have also provided access to financing for state-owned enterprises (SOEs), particularly in countries with large state-owned banks, notably Brazil, China, and the Ukraine.

Developed countries have also provided preferential access to financing through direct financial support or tax credits to support their domestic solar PV and wind energy industries.[19] Many of these programs were enacted as part of stimulus packages following the Great Recession in 2008. For example, the US provided about $14 billion in loan guarantees to domestic solar projects as part of the American Recovery and Reinvestment Act in 2009.[20]

---

[16] *Ibid.*, p. 105.
[17] *Ibid.*, p. 90.
[18] *Ibid.*, p. 92.
[19] *Ibid.*, p. 93.
[20] *Ibid.*, p. 93.

3.2.3. *Trade remedies for solar and wind industries*

The OECD and emerging economies have used trade remedies for the solar and wind sectors since 2010. (Trade remedies allow governments to take remedial action against imports that are causing material injury to a domestic industry.) Countries have initiated trade remedies to protect domestic manufacturers against unfair trade practices and alleged distortions (dumping and actionable subsidies). Governments have initiated 15 anti-dumping investigations and countervailing duty investigations in the solar and wind sectors.[21] In addition, governments have imposed nine duty measures and seven countervailing duty measures on solar PV and wind energy products. The OECD countries have used trade remedies to attempt to create a level playing field because they believe their domestic industries face unfair competition. Developing economies have also more recently used trade remedies. Most trade investigations now originate in the top solar and wind energy markets: China, the EU, and the US.

4. Impact of the Trans-Pacific Partnership on green industrial policies

The Trans-Pacific Partnership (TPP) has two provisions that could potentially impact green industrial policies: the prohibition of local content requirements and consistent rules of origin for inputs and materials used in regional supply chains within the TPP.[22] However, the impact of these two TPP provisions is likely to be minimal because key countries that have had green industrial policies, most notably Brazil, China, EU member countries, and India, are not members of the TPP. In addition, the US, a key player in solar and wind GPNs and trade disputes over green industrial policies, has formally withdrawn from the TPP. Performance requirements, including LCRs, are not permitted under the TPP. Local content requirements in solar and wind GPNs have arguably been the most contentious issue of green industrial policies, with numerous trade disputes and researchers voicing concern that LCRs may significantly

[21] *Ibid.*, p. 98.

[22] Office of the United States Trade Representative (2015).

hamper investment in solar and wind GPNs. In addition, the TPP has agreed on a single set of rules of origin to promote regional supply chains and help ensure that TPP countries, rather than non-participants, are the primary beneficiaries of the TPP agreement. To encourage regional supply chains within the TPP, the agreement provides for "accumulation" so that inputs and materials from one TPP country are accorded the same rules of origin as other TPP countries that supply inputs and materials for regional supply chains within the TPP.[23] The TPP approach acknowledges that many goods, including solar PV and wind turbines, are produced in GPNs and that encouraging regional supply chains within the TPP could minimize disputes over rules of origin among the TPP countries.

5. Implications for new research agenda

Research conducted by the OECD, other organizations, and researchers indicates that green industrial policies, defined as initiatives to build capacity in domestic solar PV and wind-turbine manufacturing industries, are restricting or distorting trade and investment in solar PV and wind GPNs. Maximizing growth of trade and investment to promote the growth of solar and wind power production is critical for the effort to mitigate climate change and maximize the economic benefits for all countries involved in the global production of solar and wind energy products and equipment.

The research agenda, going forward, consists of five main elements:

- Inform and educate policy-makers about the value-added and jobs generated at all stages of solar and wind GPNs, especially downstream activities;
- Conduct more comprehensive and timely monitoring of green industrial policies by the international community;
- Conduct more research on value-added and jobs captured by developing countries in solar and wind GPNs;

[23] *Ibid.*

- Perform more research needed to quantify the impacts of green industrial policies on trade, investment, and innovation, especially in developing countries; and,
- Conduct research that is needed to evaluate proposed policies and to develop new policies that build domestic manufacturing capacity but do not restrict trade and investment.

The remainder of this section will provide further explanation of each element of the research agenda.

5.1. *Inform and educate policymakers about the value-added and jobs generated in downstream activities of solar and wind GPNs*

The rise of GPNs in the solar and wind energy sectors means that a country's domestic production and exports rely on an increasing share of intermediate goods imported from other countries. In the solar PV sector, foreign companies account for an increasing share of value-added for solar PV module manufacturing. In the US, 70% of the value-added of PV modules is imported (largely from China) compared to 27% of the value of the total systems imported from abroad.[24] The EU has similar shares for the foreign content of PV modules and the entire PV system.

Although the manufacture of solar PV modules generates jobs and value-addition, especially in emerging markets, downstream activities, including construction, installation, system integration, operation, maintenance, and sales, are employment-intensive. Manufacturing activities in developed countries typically account for 18–24% of total employment in the solar GPNs.[25] Several studies estimate at least 50% of solar PV jobs and value-added are captured by downstream activities. In the US solar sector, downstream activities employ 70% of the

[24] OECD (2015), p. 36.
[25] *Ibid.*, p. 37.

total jobs, while cells and module manufacturing account for 20% and 10% of jobs, respectively.[26]

5.2. *Establish a more robust monitoring of national trade, investment, and other policies that affect domestic solar and wind power industries*

The OECD and other institutions have done substantial research and analysis to monitor the various green industrial policies that are affecting solar and wind GPNs. (See list of key studies following references.) The OECD has performed a thorough and timely analysis of these policies, analyzed the effects of some of these policies on trade and investment, and cited research that also examined the effects of these policies on trade and investment. Although the OECD and IEA have done substantial reporting efforts on domestic incentive measures, the international community has not adopted guidelines for calculating all types of energy subsidies and harmonizing reporting standards.

A more comprehensive and robust effort is needed to monitor national policies in a comprehensive and timely manner. A more concerted effort is needed to monitor ongoing developments and changes in green industrial policies that affect solar and wind GPNs. This monitoring could be done by several organizations, such as the UN working with the OECD and IEA.

5.3. *Conduct more research on value-added and jobs captured by developing countries in solar and wind GPNs*

Research needs to be conducted on the value-added and jobs generated in all activities of solar and wind GPNs in developing countries. China captures value-added and jobs as the world's largest manufacturer of solar PV modules and other components. Other developing countries are likely to capture far less value-added and jobs in their domestic solar manufacturing

[26] *Ibid.*, p. 39.

industries because they cannot compete with the low cost and huge scale of China's solar components manufacturing sector. The situation is similar in wind-turbine manufacturing; several countries have become major producers and exporters of wind turbines.

Evaluate and develop new policies to replace LCRs. In addition, well-designed LCRs may work in limited circumstances but more research needs to be done

Several research studies have concluded that LCRs have mixed or negative impacts on international investment, trade, employment, and welfare of solar and wind GPNs in the implementing country. (See list of research studies following references.) The OECD has conducted several econometric studies to analyze the effect of LCRs on international investment in tapping solar and wind energies. The OECD's most recent econometric analysis found that FiT policies play a significant role in attracting investment flows in solar and wind sectors.[27] However, LCRs have had a detrimental effect on global international investment and hamper the effectiveness of FiT policies that are coupled with LCRs to destination countries. In addition, the OECD surveyed global investors in clean energy in 2014 and found that investors identified LCRs as the main impediment to international investment.

A separate OECD study on the impact of LCRs on international trade show that LCRs that cause a fall in intermediate imports reduce imports in both the implementing country's main trade partners and globally.[28] Although intended to develop an entire domestic industry, the more likely outcome is that LCRs help develop a domestic intermediate industry at the cost of a loss in downstream activities and final goods.

LCRs have benefits, particularly in countries that have large markets and carefully design their LCR programs.[29] LCRs may attract international and domestic investment in the implementing country or regional hub if the LCRs are backed by large bids and aggressive development

[27] *Ibid.*, pp. 67–78.
[28] *Ibid.*, p. 61.
[29] *Ibid.*, p. 65.

policies. LCRs may also attract technology transfer. LCRs may lead to an increase of local job creation in the short run, which may expand public support for renewable energy and developing ambitious renewable energy policies and targets.

Research should evaluate policies recommended by the OECD to replace LCRs that build domestic capability without harming trade and investment:

- Well-targeted R&D support in solar and wind energy sectors has the potential to stimulate innovation in solar and wind energy manufacturing, build local manufacturing capacity, and encourage technology transfer without favoring domestic firms;
- Training programs and promotion measures can improve the technological skills of midstream manufacturers, build local capability of downstream firms, and encourage innovation across the entire value chain; and,
- Demand-side instruments, including well-designed FiTs, auctions, tax incentives, or carbon pricing instruments, can encourage wider deployment of solar and wind energy or more cost-effective carbon pricing instruments.

Additional policies to replace LCRs should be developed. A "toolbox" of policies that can be tailored to meet each country's role in the solar PV and wind GPNs could be useful. A second area of research is to examine the feasibility and develop the design of LCRs for countries that have large markets that would have minimal or no impact on trade and investment.

5.4. *Quantify the impacts of technical restrictions on the wind sector*

Technical restrictions can drive up transaction and financing costs and act as a *de facto* market-access barrier.[30] They can reduce transparency for

---

[30] *Ibid.*, p. 107.

foreign project developers and investors. They may have a negative effect on imports. Foreign firms may not be willing to provide downstream activities. Although these measures are often used for protecting human health, ensuring safety, or protecting the environment, they can distort international trade and investment and help protect the domestic industry.

The OECD and European Wind Energy Association carried out a study of the impact of technical restrictions on international trade and investment of wind energy in Brazil, Canada, China, India, and South Africa. The study found that technical standards are likely to negatively affect trade and investment in these countries.

Research needs to better quantify how technical standards on the domestic wind industries of Brazil, Canada, China, India, South Africa, and other countries affect local and global trade and investment. Are these technical standards a minor or major hindrance to wind GPNs?

5.5. *Conduct further research on the impacts of preferential access to financing*

Most demand- and supply-side policies indirectly affect trade and investment by changing relative prices domestically and abroad by affecting investment, production, and consumption decisions.[31] Policies that can discriminate between domestic and international investors include:

- Preferential access to financing, including direct financial transfers such as grants, equity injections, and subsidized loans that favor domestic manufacturers or are associated with a local content requirement;
- Export subsidies;
- Preferential tax incentives for producers or investors; and
- Government procurement of below-market-price inputs.

Preferential access to finance may lead to inefficiencies and overcapacity in solar PV and wind-turbine manufacturing by reducing the

[31] *Ibid.*, p. 95.

incentive for domestic manufacturers to reduce costs through innovation or rationalization.[32] For example, China's substantial support to its domestic solar PV manufacturers through measures such as below-market-price financing likely boosted China's solar PV exports at a cost of harming manufacturers in Europe and the US, and leading to global overcapacity in the sector.

Export subsidies consist of direct grants or concessional loans. They distort international trade by lowering the cost of exported solar PV and wind-turbine inputs that otherwise would be non-competitive by domestic manufacturers.

Emerging economies provide access to financing for SOEs, which can hinder market access and investment of foreign and domestic independent power producers in solar and wind energy. This issue is particularly important in countries with large state-owned banks, notably Brazil, China, and the Ukraine

Future research should examine how the various policies for preferential access to financing impact relative prices domestically abroad. It also might consider if and to what degree they affect trade and investment.

6. Research measures to limit use of trade remedies or develop alternative policies

Although trade remedies can help offset foreign incentives that negatively affect domestic industries, several studies have concluded that trade remedies have negative impacts on solar PV and wind energy GPNs.[33] Trade remedies have been particularly harmful on trade in solar products. The direct impact of trade remedies is the reduction of trade volume. In addition, remedies are likely to increase investment risk and cost of inputs in downstream activities in solar and wind GPNs, particularly in the former. Therefore, trade remedies may hinder competitiveness and investment in solar PV and wind energy sectors.[34]

---

[32] OECD (2015), p. 92.

[33] See*Ibid.*, pp.: 100–104 for a discussion on impact of trade remedies.

[34] *Ibid.*, p. 102.

Higher inputs costs may hamper the continued development of specialization, skills, and innovation which are essential for solar and wind GPNs to remain competitive and innovate. Increases in solar PV prices may raise electricity prices, which would reduce the cost-competitiveness of solar PV versus electricity generated from fossil fuels. The investor perception of risk and uncertainty on investment decisions has increased in reaction to trade disputes because of the unpredictability of duties. In addition, renewable energy importers have faced the possibility that the trade remedies could be posed retroactively, which increases uncertainty.

Governments have several policy options to limit use of trade remedies, according to the Swedish National Board of Trade:

- Impose duties at a level lower than the margin of dumping but adequate to remove injury (known as the *lesser duty rule*);
- Reduce the scope of trade remedies to specific product or import value; and
- Reduce the time that the duty is imposed.

A second option is to align or even replace anti-dumping rules with competition rules, aimed at targeting truly harmful anti-competition behavior.[35] Under this approach, higher thresholds would be set for companies that have a dominant market share (e.g., 40% or greater) compared with anti-dumping rules that would be triggered if the exporter has a market share of 1%. Competition rules may also lead to stronger international rules on the trade and investment effects of SOEs.

Future research should identify the most effective policies to limit or replace trade measures. This research could be used to develop consensus among producing countries and in the WTO to agree on a single, consistent approach. In addition, more research is needed on competition rules that could supplement or replace trade remedies. What would be the appropriate threshold for companies that have a dominant market share? What competition rules could have an impact on the policies of SOEs that distort trade and investment?

---

[35] OECD (2015), p. 103.

6.1. *Research needed on the impact of green industrial policies on innovation in solar and wind GPNs*

Innovation has been rapid in the solar and wind energy industries over the last decade, with deep declines in prices, technological development, increases in patenting activity, and substantial early-stage financing in solar technology. Innovation is vital for the solar and wind energy industries to continue their rapid growth and technological progress and to sustain the GPNs in each of these sectors. It does not appear that much research has been done on the impact of green industrial policies on innovation. An important priority is to conduct in-depth research on the effect of such policies on innovation.

7. Conclusion

The solar and wind energy industries have expanded rapidly over the last decade. Costs have fallen far more rapidly than most observers could have predicted a few years ago. Solar and wind energy sources have now become competitive with fossil fuels in many countries. The globalization of production in these two industries has been a major factor in their success.

However, a number of countries have enacted green industrial policies to foster development of their solar and wind power manufacturing industries. Many of these policies lead to reduced investment and trade domestically and globally in solar and wind GPNs. These barriers are inhibiting the investment needed to further build renewable energy capacity that is needed to help mitigate climate change.

What should be the next steps to implement the proposed research agenda on green industrial policies and solar PV and wind GPNs?

Organizations such as the OECD, UN, and IEA should hold meetings with senior officials in energy, finance, and development from countries involved in solar PV and wind GPNs to emphasize that downstream activities may generate as much or more value-added and jobs than upstream activities, such as manufacturing.

A stronger and more concerted monitoring of green industrial policies could be implemented by several organizations working together, such as the UN, OECD, and IEA. These organizations could share and divide the work of monitoring green industrial policies.

The research agenda could be implemented through creating a multi-disciplinary group of researchers with expertise in economics, innovation, development, and renewable energy. An international organization could provide resources and facilities for the research group. The research group would disseminate its work through the sponsoring international organization; other organizations such as the UN, OECD, and IEA; and peer-reviewed journals.

Questions for Discussion

- Are there other topics on fostering growth of solar and wind GPNs that merit research?
- Is there research or insight from GPNs in other industries on facilitating trade, investment, and innovation that can be applied to solar and wind GPNs?
- Will the TPP have any impact on trade and investment barriers in solar and wind GPNs?
- How should innovation be measured in solar PV and wind GPNs?
- Are patenting and other forms of intellectual property having an adverse or beneficial impact on solar and wind GPNs?

References

Bloomberg New Energy Finance, "New Investment in Clean Energy and Generation Capacity databases." Available at: http://about.bnef.com/.

International Energy Agency (2014). "World Energy Investment Outlook." Available at: https://www.iea.org/publications/freepublications/publication/WEIO2014.pdf.

OECD (2015). "Overcoming Barriers to International Investment in Clean Energy." Available at: http://www.oecd-ilibrary.org/environment/overcoming-barriers-to-international-investment-in-clean-energy_9789264227064-en.

Office of the United States Trade Representative (2015). "Summary of the Trans-Pacific Partnership Agreement." October. Available at: https://ustr.gov/about-us/policy-offices/press-office/press-releases/2015/october/summary-trans-pacific-partnership.

UNCTAD (2014). "Trade Remedies: Targeting the Renewable Energy Sector." Available at: http://unctad.org/en/pages/PublicationWebflyer.aspx?publicationid=1084.

Selected Research of Impact of Green Industrial Policies

Böhringer, C., and V. Aleexeeva-Talebi (2013). "Unilateral Climate Policy and Competitiveness: Economic Implications of Differential Emission Pricing." *The World Economy*, 36(2). Available at: http://onlinelibrary.wiley.com/doi/10.1111/j.1467-9701.2012.01470.x/abstract.

Cárdenas Rodríguez, M. *et al.* (2014). "Inducing Private Finance for Renewable Energy Projects: Evidence from Micro-Data." OECD Environment Working Papers, No. 67, OECD Publishing, Paris. Available at: http://dx.doi.org/10.1787/5jxvg0k6thr1-en

Cimino, C., and G. Hufbauer (2014). "Trade Remedies Targeting the Renewable Energy Sector." Green Economy and Trade. Ad hoc Expert Group 2: Trade Remedies in Green Sectors: The Case of Renewables, UNCTAD, Geneva. Available at: http://unctad.org/meetings/en/SessionalDocuments/ditc_ted_03042014Petersen_Institute.pdf.

Haščič, I. *et al.* (2015). "Public Interventions and Private Climate Finance Flows: Empirical Evidence from Renewable Energy Financing." OECD Environment Working Papers No. 80, OECD Publishing. Available at: http://dx.doi.org/10.1787/5js6b1r9lfd4-en

Hufbauer, G. *et al.* (2013). "Local Content Requirements: A Global Problem." September, Policy Analyses in International Economics, Peterson Institute for International Economics, Washington, D.C.

Kasteng, J. (2014). "Trade Remedies on Renewables: Intermediate and Long-Term Solutions." Presentation at UNCTAD Clean Energy and Trade Ad hoc Expert Group 2 on Trade Remedies in Green Sectors: The Case of Renewables, Geneva, 3–4 April.

Kuntze, J., and T. Moerenhout (2012). "Local Content Requirements and the Renewable Energy Industry — A Good Match?" 12 September. Available at: https://ssrn.com/abstract=2188607; http://dx.doi.org/10.2139/ssrn.2188607.

OECD (2015). "Overcoming Barriers to International Investment in Clean Energy." Available at: http://www.oecd-ilibrary.org/environment/overcoming-barriers-to-international-investment-in-clean-energy_9789264227064-en.

Rivers, N., and R. Wigle (2012). "Domestic Content Requirements and Renewable Energy Legislation." Available at: https://papers.ssrn.com/sol3/papers.cfm?abstract_id=2129808.

Wu, M., and S. Salzman (2014). "The Next Generation of Trade and Environment Conflicts: The Rise of Green Industrial Policy." *Northwestern University Law Review*, 108(2).

Part 3

Intellectual Property, Strategic Patenting, and Trade

Chapter 6

Patents and Technology Transfer through Trade and the Role of Regional Trade Agreements*

Keith Maskus

Department of Economics
University of Colorado Boulder, Economics Building,
256 UCB #212, Boulder, CO 80309 USA

1. Introduction

Regional Trade Agreements (RTAs) can have powerful impacts on technology transfer (TT), primarily through their effects on trade in high-technology goods and services, Foreign Direct Investment (FDI), and licensing, all of which are key channels of information diffusion. In that context, RTAs and megaregionals such as the TPP embody considerable promise for direct expansion of TT, if primarily within the area of the agreements. Moreover, recent econometric evidence largely finds that each of these flows responds positively to strengthened patent laws

*This chapter was originally a think-piece prepared for *MCTI Conference* held at East-West Center Honolulu, Hawaii in January 2016.

and other intellectual property rights (IPR), though this conclusion must be conditioned on a variety of factors (Maskus, 2012). These basic observations suggest that there could be an important complementarity between the formation of trade agreements and their IPR standards, a critical question that has not yet attracted much investigation.

In addition, it stands to reason that countries joining RTAs are likely to observe growing within-agreement patent registrations. In that context, patent applications in themselves can facilitate the international diffusion of technological information. Patents are published and may be read by firms seeking to learn new technologies and invent around them, or simply to imitate them if associated patents are not in force in a particular location. Eaton and Kortum (1996), for example, found that OECD countries have derived significant productivity growth from importing knowledge through patents. Moreover, acquiring access to patent rights is a direct form of TT, whether through within-firm or arm's-length licensing. Multinational firms are increasingly loath to license complex technologies without some reasonable legal assurance that the accompanying IPR can be protected and will bear some return (Maskus, 2012). Finally, the availability of enforceable patent rights and other IPR can support the development of international technology markets and reduce contracting costs (Arora *et al.*, 2001; Yang and Maskus, 2001).

At the same time, patents embody market power and may be deployed in ways that diminish access to technologies, whether through restrictive licensing practices or unaffordable rights fees. Further, multinational firms may take steps to reduce the ability of local firms to imitate their processes, reducing effective transfers. In response, some governments, especially in emerging and developing economies, erect policies aimed at forcing faster TT or offsetting such market power. Whether such policies are effective or simply deter inward technology flows is context-specific and not well understood (Hoekman *et al.*, 2005).

Patents and other IPR are even more controversial in their potential impacts on access to technologies that address important quasi-public goods questions, such as affordable medicines, green technologies to combat climate change, and agricultural inputs (Maskus, 2012). To the extent that patents sustain long delays in generic competition, they may sharply raise the prices of new medicines, for example. There are complex questions surrounding the roles patents play in these and other areas, with little systematic evidence overall. There is virtually no evidence about whether RTAs may facilitate or diminish access and TT in this regard, yet another shortcoming in available research.

These basic observations point to the importance of thinking about how international strategies might be devised to increase TT on affordable terms in key sectors going forward, with some emphasis on the potential roles RTAs could play. This is the point of departure for the current chapter, building on some recent contributions (Barton and Maskus, 2006; Maskus and Saggi, 2013; Maskus and Okediji, 2010). An opportunity for such reflection arises in the recently completed negotiations of the Trans-Pacific Partnership (TPP) and continuing intergovernmental and non-governmental organizations' (NGOs) work on climate change and access to medicines. In brief, the chapter argues that RTAs could play a positive role if they are supplemented with additional financing commitments, technical assistance, and knowledge sharing. In this context, they could be useful complements to larger global initiatives, including an agreement on access to basic science and technology.

Section 2 outlines the subject matter, a brief history, and the important challenges it raises. Section 3 discusses the importance of undertaking additional research to understand the role RTAs may play in conjunction with IPR to enhance or deter technology diffusion. Section 4 describes the potential policy approaches that could help facilitate future growth in technology diffusion. Section 5 provides the conclusion.

2. Background and history

The idea that intellectual property protection could increase TT flows and, ultimately, enhance local development prospects is enshrined in the TRIPS agreement at the WTO. Specifically, the objectives of TRIPS, set out in Article 7, simply are stated as:

> The protection and enforcement of intellectual property rights should contribute to the promotion of technological innovation and to the transfer and dissemination of technology, to the mutual advantage of producers and users of technological knowledge and in a manner

conducive to social and economic welfare, and to a balance of rights and obligations.

How, exactly, TRIPS could achieve these goals has been the subject of extensive debate (Maskus, 2012). On the one hand, stronger and more harmonized IPR regulations might raise the certainty of investment, reduce contracting costs, and permit technologies to be deployed more efficiently across borders within production and research networks. On the other hand, elevated patents, plant protection, and digital copyrights might diminish access and raise costs as noted earlier. Both could be true, depending on market and technological circumstances. Clearly, this is a first-order and difficult empirical question, as described in Sec. 3.

Whatever the facts, the underlying TRIPS claims have been expanded by many RTAs, especially those involving the US and the EU as developed-country partners, by incorporating increasingly complex IPR chapters. As a brief and approximate overview, focusing on US agreements, consider the following history.

- The US–Israel Free Trade Agreement (FTA) (1985) comprised one paragraph mentioning national treatment and Most Favored Nation (MFN) status.
- The US–Jordan FTA (2001, at the time considered the "gold standard" for IPR) consists of five pages, adding some TRIPS-Plus features in patent standards, pharmaceuticals, test data protection, digital copyrights, and anti-circumvention.
- The US–Chile FTA (2004) regularized test data periods and added patents for new plant varieties.
- The US–Australia FTA (2005) added further pharmaceuticals protection, including a "linkage" rule by which patent holders must be notified of potential generic entry. It also reduced the scope of exceptions and limitations in copyrights.
- The US–Korea FTA (2012) put in further limits on copyrights exceptions, added patents for new uses, removed pre-grant patent opposition, added detailed rules on the rights and obligations of internet service providers, and set out extensive enforcement obligations.
- The TPP (pending) added test data protection in biological products, incorporated, for the first time in an RTA, extensive trade secrets

obligations, and elevated criminal enforcement in IPR. More emphasis is placed on technical assistance in IPR and knowledge sharing.

For its part, the EU has large numbers of RTAs and Partnership Agreements with various partners. Over time, these agreements increasingly have focused on IPR issues, with pharmaceuticals, digital copyrights, and geographical indications taking center stage. Recent EU Partnership Agreements with developing economies have extensive language on the importance of knowledge sharing and facilitating the participation of partners in international technology networks (Maskus, 2013).

It is important to note that this growing emphasis on IPR chapters is a component of the growing complexity of RTAs, negotiations of which are more capable of incorporating wider ranges of "behind the border" regulations, such as services provision and access to procurement contracts. In fact, the most striking thing about the TPP as regards TT is not its IPR chapter (though the new and rigorous expectations on trade-secrets protection could be critical). Rather, it is the extensive roster of additional trade and regulatory issues it addresses that could heavily influence the productivity of intellectual property protection. Central among these are (1) additional access commitments in information technology (IT) markets and technology products; (2) liberalization pledges in producer and financial services; (3) investment protection and investor–state dispute resolution; (4) restraints on government policies seeking to force TT; (5) procurement rules; and (6) regulations on the market orientation of State-Owned Enterprises (SOEs).

In this context, one of the most important — and controversial — elements of megaregional trade agreements is the selection each agreement makes regarding such broader regulatory issues. They may be mutually inconsistent, especially where some countries may become partners to two or more agreements. For example, the proposed Regional Comprehensive Economic Partnership (RCEP), led by China, evidently says little about TT rules or SOEs and is markedly less rigorous about IPR in comparison with the TPP. How such differences may work out going forward is an interesting question. Perhaps more important is whether such differences will set up a dynamic pushing more regional countries to abandon one effort for another. Chinese authorities, for example, claim to be open to considering membership in TPP, which presumably would eviscerate RCEP. For present purposes, however, the interesting issue is how such regulatory provisions in megaregionals may interact with IPR protection.

3. Trade agreements and technology transfer

Existing literature strongly supports the view that international TT increases by growing trade, especially in capital-goods imports, FDI, and licensing of production rights that share access to intellectual property (Keller, 2004; Hoekman *et al.*, 2005). There is also evidence that access to global technologies is enhanced by incoming patent applications (Eaton and Kortum, 1996; Peri, 2005) and both temporary and permanent migration of technically skilled workers (Stuen *et al.*, 2012; Kerr, 2008; Hovhannisyan and Keller, 2015). Much of this transfer is direct and organized through market transactions, within or across firms. However, each of these flows offers considerable scope for technological spillovers, generating localized productivity gains that may in turn induce local innovation and export growth (He and Maskus, 2012; Keller and Yeaple, 2009).

Moreover, there is a solid evidentiary consensus that significant patent reforms, at least as undertaken by major middle-income and emerging economies, attracts increasing amounts of high-technology trade, FDI, and licensing (Ivus, 2010; Maskus, 2012). Studies using aggregate and industry-level data find mixed evidence about the potential for patent reforms to spur innovation in developing countries (Chen and Puttitanun, 2005; Schneider, 2008; Qian, 2007). However, firm-level studies suggest that such reforms can help stimulate local industrial development and export growth in such economies (Branstetter *et al.*, 2006, 2011).

These observations suggest that TT analysts should pose at least two fundamental questions about RTAs. First, it is essential to assess the impacts of proliferating RTAs (there are now over 400 trade agreements of various kinds notified to the WTO) and emerging megaregional agreements to study their potential for enhancing within-region technology flows and productivity spillovers. It is straightforward to claim that such effects are positive to the extent that RTAs create more trade, FDI, and technology transactions within the region than would have happened in their absence, an outcome that varies with agreement characteristics (size, extent of trade liberalization, scope for scale economies, factor prices, absorptive capacity, and other factors). Surprisingly, however, this simple conditional statement has been the subject of relatively little formal study, particularly as regards FDI and licensing. The limited available evidence indicates that vertical (factor-seeking) FDI between developed and middle-income countries is stimulated by RTAs, but the effects on horizontal (market-seeking) investments are ambiguous (Jang, 2011; Irarrazabal *et al.*, 2009; Bae and Jang, 2013). Bilateral investment treaties (BITs), however, do seem to induce increasing flows among partner countries (Egger and Pfaffermeyer, 2004). All of this suggests that RTAs can serve to increase technology spillovers through their impacts on increasing within-region economic activity. This finding recently has been supplemented by evidence that cross-border patent citations are significantly increased by trade agreements (Jinji *et al.*, 2013). To the author's knowledge, there are no available econometric studies of RTAs and licensing. Moreover, much of this analysis has been conducted with relatively aggregate data and basic specifications (e.g., gravity), leaving considerable room for improving and refining the available evidence.

The second question is: are there greater TT impacts in RTAs with strong rules on intellectual property than in other agreements? Put differently, does implementing a "TRIPS-Plus" regime among RTA partners seem to increase innovation, technology flows, and information spillovers more than would be the case in similar RTAs that do not require more than TRIPS standards? This possibility, if borne out in the data, would highlight an important complementarity between partial trade liberalization and patent rights (and perhaps digital copyrights) in enhancing technology diffusion, while informing thinking about potential megaregional effects. To date, there appear to be no studies of these basic questions.

Thus, the first suggestion arising from this chapter is to underscore the need for innovation and trade economists to undertake research on whether and how particularly protective IPR rules and other elements of RTAs interact to influence investments in knowledge and knowledge transfer. This research would require hard thinking about how to identify otherwise similar trade agreements and how to measure the stringency of IPR across RTAs, along with determining which industries (or, better, multinational enterprises and local firms) are more likely to be affected by this policy treatment. The kinds of questions that usefully could be posed include the following, among many others:

- Is there an exceptional increase in measures of innovation (R&D and patenting) within the higher-IPR RTA and across RTA partners? Does this vary by technology sector (e.g., pharmaceuticals, chemicals, and information technology) given their particularly elevated standards? Are there any indications of reduced relative use of IPR within the region from non-member countries? (One might refer to these outcomes as "intellectual property creation and diversion".)
- In tracking high-technology trade, affiliate activity, and patent citations, are there exceptional increases or decreases in apparent TT within the treatment RTAs? What other factors do such activities depend on?
- Do there appear to be exceptional changes in the productivity (TFP, market entry, product development, and the like) of local firms within treated RTAs? Do they invest in technology upgrading after RTA implementation, as noted, for example, in Bustos (2011) for Argentine firms with respect to MERCOSUR trade liberalization?
- Is there identifiable evidence of complementarity effects in RTAs that combine high-IPR standards with extensive liberalization commitments in IT goods, FDI, and producer services? By complementarity, the author means the possibility that RTAs with strong IPR chapters generate exceptional increases in within-agreement trade and FDI. Note that the opposite outcome is possible as well: the market power generated by strong patent rights could permit firms to cut trade, thereby diminishing within-agreement activity.
- To the extent that treatment RTAs generate higher inward flows of FDI to countries with larger policy changes, does this additional source of investment serve to help finance local R&D and innovation?

4. Broader initiatives

This volume is aimed largely at understanding the roles RTAs and especially megaregional agreements play in various important economic issues. The author argues that such agreements likely expand TT, though available analysis of exactly how that happens is frustratingly limited. Almost surely the great bulk of these information flows comes in the form of market transactions (trade, investment, and licensing) that are encouraged by relaxation of border restrictions within trade agreements. As noted, our knowledge of how IP chapters and other regulatory issues may matter is virtually absent and needs to be addressed by solid research.

This problem is all the more acute in light of the critical importance of knowledge transfers and information diffusion for addressing a range of important global public goods (GPGs) and competition issues in emerging and developing countries. Among the more prominent are affordable access to green technologies, the ability to foster and sustain a generic medicines industry, access to new genetically based and bioengineered plant varieties in agriculture, and encouraging the spread of scientific and cultural knowledge. Each of these areas could be the subject of numerous think-pieces and cannot be covered adequately in this chapter.[1] Rather, the balance of this chapter addresses three issues that illustrate the key tradeoffs between IPR protection and the achievement of public goods: patent policies in green technologies, greater skilled migration, and broader access to basic science and technology. Needless to say, we have little information about whether and how RTAs *per se* play a contributory or deterrent role in information flows in these areas, again a clear hole in our knowledge. Thus, the remainder of this piece departs from analyzing RTAs and megaregionals, except tangentially.

4.1. *Patents and transfer of green technologies*

No issue is more pressing than dealing with climate change by reducing emissions of greenhouse gases (GHGs). This is a complex challenge that will require extensive work beyond IPR policies. There must be considerable investments in clean energy resources and mitigation and adaptation technologies. It is also necessary to facilitate effective diffusion and adaptation of new technologies in developing countries, which account for a

[1] Among many other volumes, see the contributions in Cimoli *et al.* (2014). The author's views on these and other issues are spelled out in Maskus (2012).

rapidly rising share of GHGs emissions. Thus, sustainably transferring appropriate green technologies (GTs) to developing countries and adapting them to local needs form critical elements of addressing climate change. This need for enhancing TT to developing economies has long been a central issue in global negotiations. For example, a proposal was adopted at the December 2009 UNFCC meeting in Copenhagen to establish a $30 billion annual fund for TT and mitigation assistance, which never materialized. The Paris Agreement in December 2015 embodies similar exhortations for a $100 billion fund, albeit without specifying how it will come about.

The key question here is the role of IPR, in particular of patents, in this need for greater TT. There is a sharp divide between the developed economies (led by the US, Japan, and the EU), which remain the overwhelming providers of patented new GTs, and the developing and emerging economies (led by China and India), which need access to them. The former countries see little problem or need for limitations of scope or patentability in this area, while the latter nations prefer accelerated and affordable access through multiple means, including patent exceptions. Global patenting in GTs continues to rise rapidly, with growth rates averaging perhaps 20% per year between 1997 and 2008 in clean energy processes and other inventions aimed at environmental management (UNEP *et al.*, 2010). This trend likely has continued as patenting rates have recovered from a dip in the financial recession.

The limited evidence available to date suggests that patents have not been a significant barrier to international TT in key climate change technologies (Maskus and Okediji, 2010). One reason is that virtually no GT patents have been registered in the poorest countries. Another is that patent ownership within these technology groups tends to be spread widely across countries and firms, suggesting there is little risk of monopoly pricing or anti-competitive behavior (Copenhagen Economics, 2009). Put differently, patents have not significantly diminished within- or across-technology competition in global markets, leaving reasonable choices among substitutes available for potential users. In the coming years, this could change as investments are made in new forms of environmental technologies. For example, extensive patenting of the enzymes and micro-organisms that catalyze biofuels and synthetic fuels could expand, leading to potential patent thickets and transactions costs that increasingly impair access to biotechnological inventions. Perhaps more problematic is that solutions to many specific local ecological needs in poor countries could require particular patented technologies without available substitutes.

In any case, the fact that patents may not yet constitute an access problem does not mean that the current regime is the most effective means of facilitating sufficient innovation and diffusion of GTs. In fact, real private R&D spending by major developed countries on alternative energy sources and mitigation technologies has been flat for some time, though public investments through research grants and subsidies have increased (Barrett, 2009). Thus, the existing global IPR system apparently does not provide enough profit incentives to induce the extensive R&D needed to develop effective new technologies. Moreover, relatively little of either the public or private investment is aimed at the particular needs of developing countries, because their markets are too small. Finally, even where GTs could be deployed in those countries, the available resources to pay for the costs of adapting them to local soil, climate, production, and living conditions are insufficient. Indeed, the poorest countries have the least capacity to invest scarce development resources in costly solutions to domestic environmental concerns, implying a significant mismatch between funding and need.

It is evident from this description that the most effective solutions largely lie outside the IPR system: additional public R&D funding, a sustainable and internationally distributed tax on carbon use to incentivize new investments in GTs, and improved investment climates in poor countries to encourage specific R&D programs in local affiliates. Failing such initiatives, it is doubtful that working within the patent system can achieve adequate levels of innovation and technology diffusion to combat climate change.

Nonetheless, some systemic reforms could play a useful complementary role in raising the appropriate incentives. All of these ideas are consistent with TRIPS minimum standards and could potentially be implemented within RTAs, even on an experimental basis. Since space is limited, they are briefly described here.

First, there may be gains from modifying patenting conditions to expand access. For example, patent offices in major jurisdictions could be instructed to differentiate application and renewal fees if the applicants commit to effective TT and adaptation. Expedited examinations and mutual recognition of granting decisions would increase returns on R&D investments. Further, limited extensions to patent duration in developed and major emerging markets, granted midway through an invention's term and with a high renewal fee, could be offered in return for a commitment to low-cost licensing in poor countries. The obvious problem is that such differentiation could invite misclassification of technologies in patent applications. However, this issue could be managed through appropriate fee structures. For example, fees could be effectively differentiated through a partial rebate upon demonstration that the technology has been made available for licensing on reasonable and open terms (such as FRAND) in designated markets.[2]

Second, a complementary approach would be to support voluntary patent pools or networks. Users could acquire the licenses needed for specific mitigation and adaptation technologies from the network in return for royalties paid at *ex ante* determined rates, differentiated in favor of developing nations. Patent cooperatives can markedly reduce the costs of licensing to multiple markets (Lerner and Tirole, 2004). They are especially helpful when there are many patents on complementary inputs in cumulative inventions, such as solar photovoltaic energy and biofuels. Patent networks could be structured to sustain open licensing in return for an agreed payment, effectively a liability rule regime. To overcome the potential reluctance of patent holders to participate, public subsidization of license fees may be needed to build markets where there are external environmental benefits. Even if voluntary pools failed to attract sufficient private participation, universities and public research institutes could be encouraged to place their technologies into public repositories, accessible to all users in return for licensing fees differentiated by development levels and needs.

Third, as noted earlier, the need for significant funding to pay for TT and adaptation investments in poor countries is paramount. In this context,

---

[2]The acronym FRAND refers to "Fair, Reasonable, and Non-discriminatory," and this standard is commonly applied to the licensing of patents deemed essential for making information technologies interoperable across products.

there is scope for RTA partners to develop their own environmental funds in the event that global participation falls short. Such programs need to adopt principles for managing intellectual property in ways that ensure access on reasonable terms, encourage local adaptation, and incentivize participation by private partners, universities, and NGOs. For example, contractual arrangements could ensure that technology developers can profit from protected sales in non-targeted markets.

4.2. *Enhancing knowledge flows*

As noted elsewhere, there are several organizational forms that increasingly drive global technology flows (Ernst, 2006; Maskus and Saggi, 2013). Among these are vertical supply chains (VSCs), within which firms share technological processes and standards; global innovation networks (GINs), which feature cross-border collaborative efforts among private enterprises, universities, NGOs, and government institutions; and increasing amounts of temporary movement of skilled technical workers. Each of these is an important means by which firms and institutions in emerging and developing economies can potentially gain greater access to international technological information. Indeed, with VSCs and GINs, these institutions can participate in learning and innovation within networks, with potentially large payoffs. Governments in developing countries increasingly emphasize establishing such linkages through innovation and investment policies.

In this context, at least two additional policy ideas could be advanced to strengthen such linkages, and RTAs could play an important role in both. First, public agencies that support research should consider opening competition for research grants more widely to permit proposals from international research teams, especially those including participation by scientists from developing countries. To some degree, this idea is anticipated in recent EU Partnership Agreements, which sometimes contain provisions for sharing research tasks and building connections (Maskus, 2013).

Second, even relatively modest relaxation of migration restrictions facing skilled workers could generate large international income gains and transfer technologies more effectively. Using a simple framework, the author recently calculated the potential effects of a "Western Hemisphere Innovation Zone," in which each country would increase its work visas for managerial and technical workers by 20%, phased in over time (Maskus, 2014). With these ten-year visas permitting free circulation of such workers, the potential within-region discounted benefits could reach $158 billion over 25 years, against costs of about $8 billion. About 4% of these gains, or perhaps $6 billion, would come as technology spillovers into developing countries in the region.

4.3. *Access to basic science*

The importance of widespread access, including across borders, to the fruits of basic scientific research can scarcely be exaggerated. Basic science generally can be defined as efforts producing knowledge that may have little direct commercial utility but serves as a platform for multiple directions of applied research. Nearly without exception such work is financed by public grants to universities and research institutions in the US, Europe, Japan, Canada, and other developed economies. However, much of the fruits of basic research have been at least partially privatized through patents, generating concerns about the availability and costs of licenses, while keeping scientists in developing economies detached from its use (Maskus, 2012).

Because of such concerns, some analysts now call for the global community to consider negotiating a treaty on access to basic science and technology (ABST; see Barton and Maskus, 2006; Maskus, 2012). The basic idea, which attempts to complement the global IPRs system, is to preserve and expand the global basic knowledge commons without overly limiting private incentives to develop commercial technologies. Under an ABST, governments would agree to place the patented results of publicly funded research, defined as knowledge capable of supporting applied science and R&D, into access pools. This would be especially the case in areas of common global concern, such as climate change and genetic research tools. To implement this notion, the funding agencies in participating nations would certify that, as a condition for receiving a grant in specific areas of fundamental science, universities and scientists must agree to place the resulting patents into common resource pools. These patents would then be available for license to all competent agents from other member countries under terms worked out in advance, presumably something like FRAND provisions.

Note that an ABST could offer concessional terms to researchers from poor countries, so long as there is no discrimination among applicants within certain development tiers, which could be graduated. Further, the agreement could encourage researchers from member countries to participate in, or apply for, public research grants and related subsidies. It could also permit additional opportunities for temporary migration of scientific researchers in order to facilitate knowledge production. An ABST would also give researchers in other countries access to scientific knowledge and data produced from publicly funded research by any participant.

For reasons discussed elsewhere, such an agreement would fit best within the WTO.[3] However, RTAs could catalyze interest in it by experimenting with regional versions. Such experimentation could involve models of "input liberalization" (scientists sharing access to grants and having increased migration opportunities), "output liberalization" (sharing access to scientific results and data), or both. To reach agreement, however, countries may have to adopt a GATS approach, permitting governments to reserve sensitive areas of technology and to designate different levels of open access. Additional safeguards would have to be worked out, including equitable and efficient sharing of intellectual property.

5. Concluding remarks

This chapter has set out a series of ideas that might achieve the dual goals of increasing international TT and expanding the global commons for addressing public goods. At present, none of them seems feasible in political terms, except perhaps for the idea of greater funding pools for adapting technologies in poor countries to combat climate change. However, each of them bears promise for achieving such goals, both within RTAs and on a global scale. Some of them, including the idea of an ABST, have achieved initial momentum at forums such as the WIPO Development Agenda.

[3] See Maskus (2012), which also discusses related architectural problems, including the scope of subject matter and the need for protecting confidentiality and security.

Ultimately the question posed here is whether to engage in rebalancing the global patenting system more toward access models, at least as regards technological flows and, especially, basic knowledge. To a first approximation, this must be a global endeavor, given the public-goods aspects of knowledge and the need for international solutions. However, in certain respects, there may be scope for RTAs to weigh in. It is evident that the larger the economic weight of an RTA, the greater will be the impact on technology diffusion. Here, then, lies a potential role for emerging megaregional agreements. If they pay more attention to policy frameworks for TT, they could have significant impacts on global policy making. Unfortunately, the first such agreement, the TPP, does not move far in that direction.

References

Arora, A., A. Fosfuri, and A. Gambardella (2001). *Markets for Technology: The Economics of Innovation and Corporate Strategy* (MIT Press, Cambridge, MA).

Bae, C. and Y. J. Jang (2013). "The Impact of Free Trade Agreements on Foreign Direct Investment: The Case of Korea," *Journal of East Asian Economic Integration*, 17, pp. 417–445.

Barrett, S. (2009). "The Coming Global Climate-Technology Revolution," *Journal of Economic Perspectives*, 23, pp. 53–75.

Barton, J. H. and K. E. Maskus (2006). "Economic Perspectives on a Multilateral Agreement on Open Access to Basic Science and Technology." In *Economic Development and Multilateral Trade Cooperation*, S. J. Evenett and B. M. Hoekman (eds.) (Palgrave MacMillan, London).

Branstetter, L., R. Fisman, and C. Fritz Foley (2006). "Do Stronger Intellectual Property Rights Increase International Technology Transfer? Empirical Evidence from US Firm-Level Panel Data," *Quarterly Journal of Economics*, 121, pp. 321–349.

Branstetter, L., R. Fisman, C. Fritz Foley, and K. Saggi (2011). "Does Intellectual Property Rights Reform Spur Industrial Development?" *Journal of International Economics*, 83, pp. 27–36.

Bustos, P. (2011). "Trade Liberalization, Exports, and Technology Upgrading: Evidence of the Impact of MERCOSUR on Argentinian Firms," *American Economic Review*, 101, pp. 301–340.

Chen, Y. and T. Puttitanun (2005). "Intellectual Property Rights and Innovation in Developing Countries," *Journal of Development Economics*, 78, pp. 474–493.

Cimoli, M., G. Dosi, K. E. Maskus, R. L. Okediji, J. H. Reichman, and J. Stiglitz (eds.) (2014). *Intellectual Property Rights: Legal and Economic Challenges for Development* (Oxford University Press, Oxford).

Copenhagen Economics A/S (2009). "Are IPR a Barrier to Transfer of Climate Change Technology?" Report to Directorate General of Trade, European Commission, Brussels, January 19.

Eaton, J. and S. Kortum (1996). "Trade in Ideas: Patenting and Productivity in the OECD," *Journal of International Economics*, 40, pp. 251–278.

Egger, P. and M. Pfaffermeyer (2004). "The Impact of Bilateral Investment Treaties on Foreign Direct Investment," *Journal of Comparative Economics*, 32, pp. 788–804.

Ernst, D. (2006). "Innovation Offshoring: Asia's Emerging Role in Global Innovation Networks." Special Report, No. 10, East-West Center, Honolulu, HI, USA.

He, Y. and K. E. Maskus (2012). "Southern Innovation and Reverse Knowledge Spillovers: A Dynamic FDI Model," *International Economic Review*, 53, pp. 281–304.

Hoekman, B. M., K. E. Maskus, and K. Saggi (2005). "Transfer of Technology to Developing Countries: Unilateral and Multilateral Policy Options," *World Development*, 33, pp. 1587–1602.

Hovhannisyan, N. and W. Keller (2015). "International Business Travel: An Engine of Innovation?" *Journal of Economic Growth*, 20(1), pp. 75–104.

Irarrazabal, A., A. Moxnes, and L. Opromolla (2009). "The Margins of Multinational Production and the Role of Intra-firm Trade." CEPR Discussion Paper No. 7145.

Ivus, O. (2010). "Do Stronger Patent Rights Raise High-Tech Exports to the Developing World?" *Journal of International Economics*, 81, pp. 38–47.

Jang, Y. J. (2011). "The Impact of Bilateral Free Trade Agreements on Bilateral Foreign Direct Investment among Developed Countries," *The World Economy*, 34, pp. 1628–1651.

Jinji, N., X. Zhang, and S. Haruna (2013). "Do Regional Trade Agreements Enhance Technology Spillovers?" Kyoto University, Graduate School of Economics, Discussion Paper No. E-12-010.

Keller, W. (2004). "International Technology Diffusion," *Journal of Economic Literature*, 42, pp. 752–782.

Keller, W. and S. R. Yeaple (2009). "Multinational Enterprises, International Trade, and Productivity Growth: Firm-Level Evidence from the United States," *Review of Economics and Statistics*, 91, pp. 821–831.

Kerr, W. R. (2008). "Ethnic Scientific Communities and International Technology Diffusion," *Review of Economics and Statistics*, 90, pp. 518–537.

Lerner, J. and J. Tirole (2004). "Efficient Patent Pools," *American Economic Review*, 94, pp. 691–711.

Maskus, K. E. (2012). *Private Rights and Public Problems: The Global Economics of Intellectual Property in the 21st Century* (Peterson Institute for International Economics, Washington, D.C.).

Maskus, K. E. (2013). "Assessing the Development Promise of IPRs Provisions in European Union Partnership Agreements." In *EU Bilateral Trade Agreements and Intellectual Property: For Better or Worse?* J. Drexl, H. Grosse, R. Khan, and S. Nadde-Phlix (eds.) (Springer-Verlag, Heidelberg).

Maskus, K. E. (2014). "Benefits and Costs of the Science and Technology Targets for the Post-2015 Development Agenda." Working Paper Prepared for the Copenhagen Consensus Project.

Maskus, K. E. and R. L. Okediji (2010). Intellectual Property Rights and International Technology Transfer to Address Climate Change: Risks, Opportunities and Policy Options, International Center for Trade and Sustainable Development, Program on Intellectual Property Rights and Sustainable Development, Geneva, Issue Paper No. 32.

Maskus, K. E., and K. Saggi (2013). Global Innovation Networks and Their Implications for the Multilateral Trading System. International Center for Trade and Sustainable Development, The E-15 Initiative, Geneva, Working Paper.

Peri, G. (2005). "Determinants of Knowledge Flows and their Effects on Innovation," *Review of Economics and Statistics*, 87, pp. 308–322.

Qian, Y. (2007). "Do National Patent Laws Stimulate Domestic innovation in a Global Patenting Environment? A Cross-Country Analysis of Pharmaceutical Protection, 1978–2002," *Review of Economics and Statistics*, 89, pp. 436–453.

Schneider, P.H. (2005). "International Trade, Economic Growth, and Intellectual Property Rights: A Panel-Data Study of Developed and Developing Countries," *Journal of Development Economics*, 78, pp. 529–547.

Stuen, E., A. Mobarak, and K. E. Maskus (2012). "Skilled Immigration and Innovation: Evidence from Enrolment Fluctuations in U.S. Doctoral Programmes," *Economic Journal*, 122, pp. 1143–1176.

United Nations Environmental Program, European Patent Office, and International Center for Trade and Sustainable Development (2010). *Patents and Clean Energy: Bridging the Gap between Evidence and Policy* (UNEP, Geneva).

Yang, G. and K. E. Maskus (2001). "Intellectual Property Rights and Licensing: An Econometric Investigation," *Weltwirtschaftliches Arhiv*, 137, pp. 58–79.

Chapter 7

Learning from Past Mistakes — The US Patent System and International Trade Agreements

Robin Feldman

Institute for Innovation Law
College of the Law
University of California Hastings,
San Francisco, CA 94102, USA

Abstract

To ensure that the Trans-Pacific Partnership Agreement (TPP) works properly in practice with regard to patents and the governance of pharmaceuticals and semiconductors, strong mechanisms for enforcement and rooting out system abuse are needed. To make this happen, regulators should learn from past mistakes of the US patent system: the proliferation of non-practicing entities (NPEs) and strategic behavior that blocks or delays generic competition in the pharmaceutical industry. This chapter presents key findings of the author's research on the complexity of the US litigation and regulatory frameworks that have provided fertile ground for strategic behavior that can have negative effects on innovation and may distort the expected gains from trade for innovation.

1. Introduction

In the process of negotiating international trade agreements such as the Trans-Pacific Partnership (TPP), new regional standards will be set in areas ranging from textiles to intellectual property. These standards may serve as a baseline for greater global governance structures such as through the World Trade Organization. In the regional talks, intellectual property (IP) discussions have focused on debates over patent term lengths and data exclusivity rights. For example, based on US law, the US sought a 12-year period of data exclusivity for complex biological drugs derived from living organisms, while other countries argued that a shorter period is necessary to curb the climbing drug prices. The final length was a seven-year period. No matter what balances are chosen in international agreements, however, we need to ensure that these carefully calibrated choices function properly in practice and have both strong mechanisms for enforcement and for rooting out system abuse.

In particular, with patents and pharmaceuticals driving much of the IP debate in TPP negotiations, we should ensure that regulators learn from our past mistakes. Merely coming to a conclusion on issues such as patentability and term length is not enough — flexible yet firm enforcement initiatives are needed to ensure that the TPP becomes a strong instrument of trade, not another regulatory mechanism to be picked apart by those wishing to exploit it. And as pharmaceutical game-playing hits the press with the Turing and Valeant cases, where companies dramatically increased drug prices and attempted to obstruct generic competition,[1] it is important to consider the complexities in the US system that create opportunities for these activities.

With this concern in mind, this chapter will focus on two areas of US law, both of which are creating challenges within the US system. These are: (1) the rapid emergence of NPEs, entities that do not make products but whose focus centers on licensing and litigating patents, and (2) strategic behaviors to block the emergence of generic competition in the pharmaceutical industry. In both of these areas, complexity breeds

[1]For a summary of the Turing and Valeant controversies, see Feldman and Frondorf (2016).

opportunity. The complexity of various litigation and regulatory schemes has created opportunities for sub-optimal strategic behaviors. Many companies and patent lawyers in the US, however, have become accustomed to these intricacies and the accompanying strategic behaviors. And the solutions are not easy to identify. One has to choose carefully to avoid dampening the fires of innovation. The focus for all economies, however, should remain on product creation and production. Whatever type and stage of innovation industry may exist in an individual nation or in a region, policies that encourage the creation and production of innovative products will be more beneficial than policies that encourage intellectual property itself as a disembodied industry.

Although the lessons of this chapter focus on the patent system, they have broader implications for issues addressed in the TPP. A major section of the legislation relates to Trade Secrets,[2] which has long been an area of contention among the signatory nations and for which both national and provincial rules vary considerably. Thinking more broadly, the entire notion of national treatment and market access for goods[3] should be contemplated in the context of what goods US industries might export. To the extent those industries involve traditional goods, the classic benefits and tradeoffs apply. If those industries constitute monetization of patents or other types of intellectual property, however, additional sets of concerns may apply.

2. Problems and responses

The US patent system has helped to drive innovation in various fields, from computers to pharmaceuticals and other life science inventions to semiconductors. Although a system of patent rights could be based on an inventor's moral or natural right to the invention, the approach of the US system is explicitly utilitarian, that is, the US grants patent rights for the purpose of promoting the "useful arts",[4] in other words, products.

---

[2] See TPP Chapter 18, Article 18.78 (trade secrets).

[3] See TPP Chapter 2 (National Treatment and Market Access for Goods).

[4] US Constitution, Article 1, §8.

In addition to the Constitutional language and history, the importance of the creation of new products to patent policy is evident in the Bayh–Dole Act, passed by the US Congress in 1980. The Bayh–Dole Act gave universities the power to file for patents on federally funded inventions and also permitted them to license those inventions. The hope was that this licensing and commercialization power would spur the innovation of new products, furthering the societal benefit of spending federal dollars on research. Commentary at the time of the passage of the Act focused on the importance of the value proposition for taxpayers whose money is going to funding research at universities. In return, society should benefit in the form of the introduction of new products, and universities were judged to be better facilitators of commercialization than allowing patent rights to remain with the federal government itself.[5]

New patents, however, have not always meant new products. Until the last decade, most patent holders never succeeded in commercializing their inventions, and most patents garnered no returns. At the same time, although many patents contain claims of questionable quality, the fact that most patents sat on the shelf meant that the courts could focus on the limited number of patents that would have an impact in the market. That happy circumstance — limited quality of patent claims combined with limited numbers of patents in play — meant that the system would not be overly burdened.

The status has changed in the last five to seven years, however, with the advent of the modern NPE business model. NPEs, or "non-practicing entities", are those entities whose core activity involves licensing or litigating patents, rather than making products. Some NPEs have undertaken behavior such as indiscriminately suing or demanding payment from thousands of defendants, regardless of the merits of the claim, with the hopes that companies will choose to settle rather than face costly litigation. Such lawsuits originating from NPEs have risen significantly over the last decade. Patent lawsuits increased from about 2,500 in 2007 to over 5,000 in 2012, and NPEs have been responsible for the majority of this growth.[6] Further, the number of defendants sued has climbed along

[5]For more discussion of the Bayh–Dole Act, see Feldman and Lemley (2015).

[6]Feldman *et al.* (2013); see also Sag (2016) (finding that patent litigation volume doubled from 2010 to 2013).

with the number of lawsuits filed. Recent Supreme Court decisions led to a decline in the amount of patent litigation in 2014,[7] but activity has remained extremely high compared to pre-2007 levels. And levels appear to be rising once again — November 2015 was an all-time record month for lawsuit filings, with 790 filed.[8] New regulations have not stopped new NPEs and novel models of NPE activity from appearing, with one private study concluding that 143 new NPEs filed patent lawsuits in 2014.[9] While there are numerous definitional and methodological choices made in the studies of NPE activity, conclusions across studies been remarkably consistent — patent litigation activity in the US has exploded since 2007, and the large majority of this proliferation can be traced to NPEs.

Many of these cases are filed to exploit the cost and uncertainty of patent litigation in an effort to obtain a settlement, regardless of the merits of the patent claims. Patent litigation can easily cost $1 million or more, while the price of a settlement is often a mere fraction of litigation costs.[10] For many rational companies, settling is the logical choice, even in the face of serious concerns about whether the patent being wielded is valid or whether the company is actually infringing. NPEs are aware of this incentive to settle, particularly for small companies. For example, one study suggested that more than half of unique defendants in NPE suits have less than $10 million in annual revenue, and more than three-quarters have annual revenues less than $100 million.[11]

The question remains, however, whether NPE activity is actually hampering innovation. Few patent demands actually reach the litigation stage, with most early-stage threats and licensing negotiations taking place outside of the courtroom. Information about this widespread activity, perhaps 90% of all patent demand activity, is not captured by large-scale, generalizable data sets similar to those available for litigation. Instead, the interaction might be as simple as: "You are infringing my patents; if we take this to court, it will cost you millions in legal fees; many other companies have been agreeable and accepted a licensing deal,

---

[7] For a description of key Supreme Court decisions in 2014, see Feldman (2014).

[8] Mullin (2015). (Citing data from Lex Machina.)

[9] RPX (2014).

[10] Chien (2014)

[11] *Ibid.*, p. 471.

so there must be a way for us to work something out. To get more details, we also need you to sign a non-disclosure agreement." The complexity and incentives of the US legal system have helped create this mostly unregulated market of demands and settlements, with little data collected to track the phenomenon.

The data that are available, often from small-sample surveys, are not encouraging. These studies observe a disproportionate effect of NPEs on start-ups, little evidence of innovation transfer through NPE licensing, substantial legal costs, and additional harm to business and product development, including minimal "trickle down" of NPE revenue to original patentees. Moreover, other studies have reported the use of pressure sales tactics and opportunistic behavior by NPEs.[12]

Despite evidence that NPEs are hampering, and at times downright abusing, a system meant to spur innovation, patent rights in general are essential to innovation. These rights would be useless if they cannot be enforced. Efforts to curb strategic behavior have been met with resistance, as universities, NPEs, and pharmaceutical companies have argued that such reform efforts would diminish the ability of patent holders to legitimately assert their rights. Of particular note, the US semiconductor industry has been a vocal and active opponent of patent reform. Key semiconductor companies do not actually manufacture products but instead invent and create prototypes that the companies then license to others to manufacture. Thus, semiconductor companies occupy an unusual space, one not purely like large patent-holding companies or like traditional product-producing companies — although some have complained that the semiconductor company behavior at times strays across the line toward classic NPE tactics.

Given the state of the global semiconductor industry, entities across the world have a significant stake in the result of US patent reform. The US and China, for example, are particularly interdependent in the electronics space: semiconductor firms such as Intel and Qualcomm are dependent on China for one-fifth to a full half of their revenue, while China currently imports 80% of the semiconductors it uses in electronics

---

[12] For more detail about these studies, see Feldman (2016).

manufacturing.[13] As the world's largest electronics exporter, China then exports a large proportion of the finished goods back to the countries where the semiconductor components were first created. China is trying to grow its own domestic semiconductor industry; US and other international companies are clamoring to maintain their Chinese market share through joint ventures and technology transfer with Chinese firms that may well become competitors. To the extent that US semiconductor companies, Chinese firms, and government entities establish partnerships, these nations are likely to face challenges and pressure to skirt legal rules that might hamper the semiconductor approach to patent licensing and enforcement.

In addition to the proliferation of the NPE business model, and the resulting fallout, the US patent system has struggled with another problem in the pharmaceutical industry: manipulations that block or delay the introduction of generic drugs.[14] Competition from generic drugs is an immensely powerful factor in lowering the price of prescription medications in the US. When a generic enters a monopoly market, the drug is normally priced at 80% of the cost of the brand-name drug within six months of launch.[15] Over time, the price of generics generally falls to just 15% to the 20% of the name-brand cost,[16] and as low as 10% of the original price when many generic entrants are present. In 2012 alone, the FDA estimated that generics saved consumers over $217 billion.[17]

On the other side of the coin, generic introduction is devastating to the revenues of the brand-name drug maker. In the year after generic entry, the brand-name drug can lose between 80% and 90% of its market share.[18] With many popular brand-name drugs bringing in over $1 billion in US revenue a year for drug manufacturers, the loss of a monopoly market — often

[13] Ernst (2015).

[14] For a full discussion of original and new tactics for blocking or delaying generic competition, see Feldman and Frondorf, *supra* note 1.

[15] Berndt and Aitken (2010, pp. 4, 9–10, Figure 2).

[16] US Food and Drug Administration, "Facts About Generic Drugs". Available at: http://www.fda.gov/Drugs/ResourcesForYou/Consumers/BuyingUsingMedicineSafely/UnderstandingGenericDrugs/ucm167991.htm. Accessed December 9, 2015.

[17] Generic Pharmaceutical Association (2013) (data supplied by IMS Health).

[18] See Berndt and Aitken (2010, p. 9–10, Figure 2) and Grabowski *et al.* (2011).

known as the "patent cliff" when it coincides with patent expiration — can lead to the near-instantaneous loss of a company's largest revenue streams. This balance, however, is written into the US patent system itself: in return for publicly disclosing its invention, a pharmaceutical company receives a *limited-time* grant of exclusivity to help recoup the costs of research and development and (hopefully) incentivize further innovation. For pharmaceutical companies, all good things must come to an end.

However, complexities in the US regulatory pathway for generic entry, known as Hatch–Waxman, have changed the calculus for many pharmaceutical companies. It appears that some companies are no longer solely competing on the basis of innovation but also on their ability to manipulate existing regulation. Rather than placing bets on risky, expensive drug development pipelines, companies are driven to block generic entry by any means possible, reclaiming enormous profits and expending far less money and effort than is required to develop new drugs. Even a few months of delay — and thus a few additional months of monopoly profits — can be worth hundreds of millions of dollars, so the temptation is clear.

To block generic entry, companies are using ever more clever and complicated strategies to extend periods of monopoly and duopoly sales. Not all generic entry disputes are concerning or even malicious in nature. Hatch–Waxman provides a system by which prospective generic manufacturers can challenge brand-name patents even before patent expiration, and when companies preserve their existing exclusivity by reasserting the validity of their patents, that behavior is appropriate and fully in line with the goals of intellectual property. After all, if patent rights were found to have no actual enforcement power in practice — and generics could enter at any point — no one would see patents as providing a legitimate incentive for innovation. However, when pharmaceutical firms use tactics to unlawfully prevent generic entry — or block generics from even challenging brand-name companies — the behavior undermines the patent system by imposing costs on society. The public loses billions of dollars in savings, patients are unable to afford prescriptions that would otherwise have generic equivalents, and resources are diverted away from actual innovation activities such as drug development.

At first, strategies to prevent generic entry took the form of "pay-for-delay" settlements: brand-name pharmaceutical companies would pay generic competitors to *stay out* of the market, essentially providing the generic a share of the monopoly profits greater than the generic would earn in a duopoly or competitive market. While pure cash pay-for-delay deals have begun to lose prominence, especially as they face suspicion from government agencies, pharmaceutical companies have made settlements involving generic delay by using more opaque forms of payment. These deals can involve settling multiple cases at once to mask the value being transferred or deals where a generic company is overpaid to promote, manufacture, or supply materials for drugs sold by the brand-name company in exchange for generic delay.

New tactics no longer focus on making deals with would-be generics; instead, the goal is to obstruct generics by using trivial or minor drug modifications, US Food and Drug Administration (FDA) processes, and regulatory abuse to extend monopoly franchises and prevent generic access. In many instances, the timing and deployment of these tactics seem clearly geared toward generic delay, hidden behind ulterior motives of safety concerns or improved drug efficiency. Strategies include filing "citizen petitions" with the FDA, cautioning against generic approval with the knowledge that even a baseless petition will take months to review and deny. Other strategies include "REMS"-based delay where a pharmaceutical company cites FDA-imposed safety restrictions as a reason for not providing samples to generic manufacturers. Others involve "product hopping" schemes, where a company switches consumers to a slightly different form of an existing drug (e.g., extended-release, a new tablet formulation) just before generic entry to severely constrain the market for the generic. And finally, companies engage in "multiplicity tactics", whereby a number of these strategies are deployed at once. These mechanisms, and the complex legislative and regulatory framework that allow for their development, help to maintain pharmaceutical pricing and avoid competitive entry. Further, Hatch–Waxman does not have the proper enforcement safeguards against abuse. Rather, most changes to the pathway have been "hole-plugging" measures that patch over long-standing problems, rather than forward-looking measures that prevent further exploitation.

3. Conclusion

As described earlier, the complexity of US litigation and regulatory frameworks have provided fertile ground for strategic behavior that can have negative innovative effects. These types of complexity tend to reward large players and repeat players, at the expense of smaller — and sometimes more innovative — entities. Regardless of the level of innovative industry for any nation, creation of new products is the optimal developmental approach. As we develop regional and global intellectual property frameworks, the focus should be on streamlining complexity and the creation of new products, rather than the creation of new intellectual property games and stripped markets. In this era of megaregionalism, we want to export the strengths of the US intellectual property system and not its flaws to the world.

References

Berndt, E.R. and A. M. Aitken (2010). "Brand Loyalty, Generic Entry and Price Competition in Pharmaceuticals in the Quarter Century after the 1984 Waxman-Hatch Legislation," NBER Working Paper No. 16431, pp. 4, 9–10 & Figure 2, October. Available at: http://www.nber.org/papers/w16431.pdf.

Chien, C. (2014). "Startups and Patent Trolls," *Stanford Technology Law Review*, 17, pp. 461, 472–473, Table 1. Available at: http://journals.law.stanford.edu/sites/default/files/stanford-technology-law-review/online/startupsandpatent-trolls.pdf.

Ernst, D. (2015). "Xi's Visit Highlights U.S. and Chinese Expectations in the Semiconductor Industry," East-West Center, September. Available at: http://www.eastwestcenter.org/news-center/east-west-wire/xi's-visit-highlights-us-and-chinese-expectations-in-the-semiconductor

Feldman, R. (2014). "Theme of Restraint in Term's IP Cases," *Daily Journal*, July 8.

Feldman, R. (2016). "Federalism, First Amendment, & Patents," *Columbia Science & Technology Law Review*, pp. 11–13. Available at: http://ssrn.com/abstract=2616519.

Feldman, R. and E. Frondorf (2016). "Drug Wars: A New Generation of Generic Pharmaceutical Delay," 53, *Harvard Journal on Legislation*, pp. 52–57. Available at: http://ssrn.com/abstract=2659308.

Feldman, R. and M. A. Lemley (2015). "Do Patent Licensing Demands Mean Innovation?" *Iowa Law Review*, 137, pp. 174–175. Available at: http://ilr.law.uiowa.edu/files/ilr.law.uiowa.edu/files/ILR_101-1_Lemley%26Feldman.pdf

Feldman, R., T. Ewing, and S. Jeruss (2013). """The AIA 500 Expanded: The Effects of Patent Monetization Entities," *UCLA Journal of Law & Technology*, 1, p. 42. Available at: http://www.lawtechjournal.com/articles/2013/041024-Feldman.pdf.

Generic Pharmaceutical Association (2013). Generic Drug Savings in the U.S. Available at: http://www.gphaonline.org/media/cms/2013_Savings_Study_12.19.2013_FINAL.pdf.

Grabowski, H. G. *et al.* (2011). "Evolving Brand-Name and Generic Drug Competition May Warrant a Revision of the Hatch-Waxman Act," *Health Affairs*, 30, p. 2157, 2163.

Mullin, J. (2015). "Patent Trolls Filed Hundreds of Lawsuits to Beat Dec. 1 Deadline," *Ars Technica*, December 2. Available at: http://arstechnica.com/tech-policy/2015/12/patent-trolls-filed-hundreds-of-lawsuits-to-beat-dec-1-deadline/

RPX (2014). NPE Litigation Report 6. Available at: http://www.rpxcorp.com/2015/01/09/2014-npe-litigation-new-and-smaller-targets-2/.

Sag, M. (2016). "IP Litigation in United States District Courts: 1994 to 2014," *Iowa Law Review*, 101(03). Available at: http://papers.ssrn.com/sol3/papers.cfm?abstract_id=2570803.

Chapter 8

Harmony and Disharmony in International Intellectual Property Law

Colleen Chien

Professor, Santa Clara University School of Law
Visiting Scholar, Stanford Computational Policy Lab, 2013-2015
Obama White House Senior Advisor, Intellectual Property and Innovation

Abstract

Global trade agreements aim to harmonize standards and create a uniform climate for trade and investment. As potential provisions are negotiated, governments and stakeholders must consider the long-term impacts of particular terms. As they do so, they should keep in mind that harmonized treaty standards often translate into significantly disharmonious country standards, due to the different ways in which countries comply with treaties, translate these treaties into laws, and administer the laws. This chapter provides a case study of the harmony and disharmony in substantive patent law that followed ratification of the TRIPS agreement. Using the (perhaps) moribund Trans-Pacific Partnership (TPP) agreement as a point of comparison to TRIPS, the author discusses and provides examples of three types of differences: (1) differences between the minimum standards that are negotiated and compliance with them;

(2) differences between the substantive principles agreed to and the actual laws that implement them (and the procedural contexts in which they operate); and (3) differences between the laws on the books and the laws as administered.

1. Introduction

The purpose of trade agreements is to reduce barriers to trade in order to increase overall economic efficiency and welfare. The rationale for including intellectual property (IP) provisions within such agreements is that, although patent and copyright laws are unlike tariffs, any differences between IP laws of various countries will discourage trade as it increases the cost of complying with varied, non-harmonized regimes. If countries agree, for example, that changing the formulation of a pharmaceutical drug is a patentable advance, a pharmaceutical company can rely on that knowledge to make long-term investments in such improvements and not worry about the availability of exclusive rights. A single, international level of protection can reduce uncertainty and support differential pricing strategies that enable lower prices to be offered to lower-income consumers.

Efforts to harmonize IP law internationally date back to at least 1883, with the adoption of the Paris Convention for the Protection of Industrial Property ("Paris Convention") and application of the principle of national treatment to intellectual property.[1] The Trade-Related Aspects on Intellectual Property Rights (TRIPS) agreement is the most important and comprehensive modern international IP treaty, and establishes minimum, substantive intellectual property international standards. Its adoption in 1995 in many ways was a landmark achievement concordance, encompassing over 150 countries and establishing mechanisms for dealing with non-compliance.[2] Over two decades later, the IP Chapter of the Trans-Pacific Partnership (TPP) proposed to harmonize policies relating to patents, regulatory test data, trademarks, geographical indications, copyrights, trade secrets, IP enforcement, criminal penalties for trade secret theft, IP liability for state-owned enterprises, anti-counterfeiting provisions,

[1] Dreyfuss

[2] Davey (2005).

and additional subject matter,[3] among over a dozen countries.[4] Though the Presidency of Donald Trump significantly reduces resulted in US withdrawal from the chances of TPP ever becoming law, diminishing its significance, its formation included a number of parallels with the formation TRIPS. Both agreements were devised in secret, through rounds of negotiations largely only accessible to trade negotiators and partly accessible to the industry and stakeholder groups whose interest they represent but would required significant public action and changes in law in order to comply with their mandatory obligations. created under the agreement There is no reason to think future agreements will be any different.

However high, the costs might be worth it, at least to proponents of global trade, if the goal of uniform standards is achieved, not only through treaty commitment but also in national laws, and in practice of the law. Now is an opportune time to look back and assess the level of harmonization actually achieve in the 20 years since the passage of the TRIPS Agreement. Public sentiment has been mixed — on one hand, TRIPS, notwithstanding its flexibilities, has been criticized for discouraging experimentation among public policies in a way consistent with flexibilities provided by the Act.[5] On the other hand, rights-holders, disappointed with how TRIPS has been implemented, particularly with respect to compulsory licensing, have shifted their focus to other policy regimes and venues.[6] The failures of many of these attempts (e.g., the Substantive Patent Law Treaty (SPLT) and Anti-Counterfeiting Trade Agreement (ACTA)) to enact new global norms suggest that the consensus that TRIPS achieved on paper was remarkable. The question that remains is how has (or has not) this consensus translated into harmony in national laws, national contexts, and day-to-day practices. Section 2 discusses After discussing the TRIPS and argues that its terms support disharmony by design through the framework accommodation it

---

[3]United States Trade Representative, *TPP: Made in America. Executive Summary of Chapter 18* (official overview of the TPP IP Chapter encompassing these and other topics, like cybersquatting). Available at: https://ustr.gov/sites/default/files/TPP-Chapter-Summary-Intellectual-Property.pdf

[4]Including the US, Australia, Brunei Darussalam, Canada, Chile, Japan, Malaysia, Mexico, New Zealand, Peru, Singapore, and Vietnam.

[5]See, e.g., Land (2012).

[6]See generally, Yu (2011).

makes for country-level differences in implementation for agreement and disagreement, this chapter considers the record of the last 20 years. Section 3 explores disharmonies that stem from country level non-compliance with TRIPS' terms. Although disputes have been few and far between, they illustrate the variety of domestic laws that may implement the TRIPS terms. Section 4 explores disharmonies that stem from the different ways in which nationals laws are constructed and applies, focusing in particular on the inventive step requirement and the design of post-grant patent challenges. It reviews trilateral patent application studies, including the author's own, that trace what happens when the same patent application is filed in multiple jurisdictions, demonstrating that even standards that are harmonized can lead to dramatically different outcomes as administered. Section 5 provides the conclusion

2. Disharmony by design

Some of the disharmony among country implementations of TRIPS has been by design — as recounted below, the TRIPS Agreement (and subsequent amendments) incorporates safeguards that affirm participating countries' sovereignty, the relevance of competitive interests and frameworks for deviating from the agreement, and commitments to access to medicines. Any difference in the implementation of TRIPS standards, therefore, might have been predicted, for example, by its opening Article, which states that member-countries

> "...shall give effect to the provisions of this Agreement. Members may, but shall not be obliged to, implement in their law more extensive protection than is required by this Agreement, provided that such protection does not contravene the provisions of this Agreement. Members shall be free to determine the appropriate method of implementing the provisions of this Agreement within their own legal system and practice".[7]

TPP's IP chapter uses exactly the same language.[8] In addition, TRIPS does not purport to replace substantive IP laws; rather it only imposes

[7]TRIPS Article 1.1.

[8]TPP Art. 18.5.

minimum standards upon all member countries of the WTO. The language of the agreement specifically allows TRIPS members, when shaping their domestic laws and regulations, to "adopt measures necessary to protect public health and nutrition, and to promote the public interest in sectors of vital importance to their socio-economic and technological development, provided that such measures are consistent with the provisions of this Agreement",[9] as well as to take "appropriate measures, provided that they are consistent with the provisions of this Agreement, to prevent the abuse of IP rights by right holders or the resort to practices which unreasonably restrain trade or adversely affect the international transfer of technology".[10] Article 18.3 of the TPP incorporates this language verbatim and extends it, as described below.

The principle that exceptions and limitations to the agreement are permitted under certain limited circumstances is enshrined in both agreements.[11] So, too, is the Doha Declaration, which in 2001 confirmed that "patents would not prohibit countries' ability to use compulsory licensing to address public health needs",[12] and the 2005 amendment to Article 31 of the TRIPS Agreement, which enabled countries lacking the capacity to manufacture generic pharmaceuticals under domestic compulsory licenses to import them.[13]

In 2014, prominent IP scholars from across the world drafted the *Declaration on Patent Protection: Regulatory Sovereignty under TRIPS*, in order to "clarify some of the regulatory options states still retain under international law, in particular the TRIPS Agreement."[14] Drawing upon some of the flexibilities cited earlier, as well as other provisions of the law, the Declaration interprets, potentially too expansively,[15] the regulatory discretion

---

[9]TRIPS Art. 8.

[10]*Ibid.*

[11]Otherwise known as the "three-step test", *cf.* TRIPS Arts. 17, 26, and 30 and TPP Arts. 18.21, 18.40, and 18.65.

[12]Harris (2011).

[13]*Ibid.*

[14]Lamping *et al.* (2014).

[15]Dreyfuss and Donoso (2016) (calling the Declaration's interpretation of the flexibilities afforded by TRIPS "aggressive" and failing to give guidance to countries that seek to maximize innovation, as opposed to consumer, interests).

that national legislators have in implementing their own patent systems.[16] However, some provisions leave less room for discretion. Article 33 specifies that "the term of protection available shall not end before the expiration of a period of twenty years counted from the filing date", seemingly foreclosing proposals for differential terms, e.g., shorter for certain technology patents, and longer for certain drug patents,[17] for different types of inventions.

3. Disharmony due to TRIPS (non-) compliance

Countries that do not comply with TRIPS provisions are subject to dispute resolution procedures to ensure their compliance. One way to assess the extent to which TRIPS has achieved its harmonization aims is through the dispute settlement system within the WTO. Reviewing 15 years of dispute settlement proceedings, Pauwelyn has concluded that TRIPS was "the dog that barked but did not bite",[18] by which he means that concerns that TRIPS would significantly constrain national policy formation have largely not materialized. Pauwelyn reports that 23 of the 27 TRIPS disputes filed in the first 15 years following TRIPS enactment were lodged during the first five years of TRIPS (at an annual rate of between three and six cases), but that since 2000, less than one case has been brought per year, on average. There has been no "flood of IP disputes", no aggressive enforcement by developed against developing countries, and TRIPS disputes settled more frequently than other WTO disputes (67% vs. 58%).[19]

Edward Lee's parallel study does not dispute these findings, but does point out that fears that developed countries would use the Agreement to force changes in others' laws have not been entirely unfounded.[20] While there have been few IP disputes, the majority of them have been initiated by the US and Europe.[21] However, the targets of the disputes have also been developed countries, again, with the US, Europe, and Canada all

[16] Lamping *et al.*, *supra* note 14.
[17] See, e.g., Tabarrock (2002), Roin (2014), and other references described in Love (2016).
[18] Pauwelyn (2010).
[19] *Ibid.*
[20] Lee (2011).
[21] *Ibid.*, p. 410.

Table 1. Time taken by member countries to comply with decisions finding TRIPS violations.

Member and dispute	**Time to comply**
India — Pharmaceutical Patents ("mailbox rule")	15 months fixed
Canada — Pharmaceutical Patents	4 months fixed
Canada — Patent Term	9 months fixed
EU — Trademark and GIs	1 year fixed
China — IP Rights	1 year fixed
U.S. — Section 110(5)	10 year + counting*
U.S. — Havana Club Run	9 year + counting*

Note: *Compliance times for the US calculated as of 2011.

Source: Lee (2011).

found to have been in violation.[22] Strikingly, the US is the only country to have been found in violation of TRIPS provisions that has, of yet, failed to change its laws in order to comply (see Table 1).

However, to initiate a dispute requires significant governmental resources, and much of the shaping of national laws happens in the shadow of dispute resolution, including thorough technical assistance. In addition, commentators have noted that US enforcement efforts have largely moved away from the multilateral WTO dispute settlement system and reverted to unilateral efforts such as Special 301, a mechanism that allows for a possible retaliatory trade action against a country which fails to provide adequate protection for the US IP rights, as well as through the pursuit of commitments that go beyond TRIPS ("TRIPS-plus"), in TPP and other recent free-trade agreements.[23] As a result, dispute resolution outcomes provide, at best, an incomplete indicia of the extent to which TRIPS has succeeded or failed in its aims.

Another way of measuring the harmonizing power of international IP agreements is the degree to which agreements have been the "but-for" cause of domestic activities — compelling countries to take certain steps

[22] *Ibid.*

[23] Harris, *supra* note 12, p. 374.

because of an agreement, and conversely, discouraging the adoption of measures that, at least in theory, contravene its provisions. A review of the experience of the US provides examples of all three dynamics. For example, over the objections of a number of stakeholders, including independent inventors and Nobel Laureates,[24] Congress passed provisions of the America Invents Act "with an eye toward harmonization,"[25] by moving the US in the direction of the rest of the world through its substantial adoption of a first-to-file system, expansion of source of prior art to all geographies, and limiting of the best mode requirement.[26] The TRIPS Agreement's requirement that countries make patents eligible for 20-year terms[27] has made it impossible to enact shorter durations for software or internet patents.[28] The TRIPS Agreement's commitment to technology neutrality, which requires patents to be "available for any inventions, whether products or processes, in all fields of technology,"[29] has also discouraged differential treatment.[30] However, it has not prevented the US Supreme Court from taking significant steps to curtail the patenting of abstract inventions[31] and diagnostic methods,[32] or prevented Congress from enacting special provisions to be developed to challenge patents over financial inventions, specifically excluding "technological inventions"[33] to avoid running afoul of TRIPS. These "exceptions" to technology neutrality in patents join numerous other technology-specific domestic provisions. For example, in the US, 35 U.S.C. § 287(c) prevents surgeons from being prosecuted for infringement of diagnostic patents.[34] Likewise, surgical methods cannot be patented in Japan and the UK.[35] EPC Article

[24] See Chien (2016).
[25] Dreyfuss, *supra* note 1, p. 679.
[26] *Ibid.*, pp. 680–681.
[27] TRIPS Art. 33.
[28] As suggested, e.g., by Amazon CEO Jeffrey Bezos, as described in Thurm (2000).
[29] TRIPS Art. 27.
[30] See Chien (2012).
[31] See, e.g., Bilski v. Kappos, 561 U.S. 593 (2010); Alice v. CLS Bank, 134 S. Ct. 2347 (2014).
[32] Ass'n for Molecular Pathology v. Myriad, 133 S. Ct. 2107 (2013).
[33] Pub. L. 112 29, § 18, 125 Stat. 329–331 (2011).
[34] 35 U.S.C. § 287(c) (2012).
[35] Described, e.g., in Takenaka *et al.* (2010).

52(c) and (3) prevents the patenting of computer programs "as such" in Europe,[36] though novel and non-obvious solutions to technical problems can still qualify.

4. Disharmonies in the implementation of treaties through national laws and procedures

TRIPS and TPP explicitly state that "[m]embers shall be free to determine the appropriate method of implementing the provisions of this Agreement within their own legal system and practice."[37] This clearly contemplates that specific national laws implementing the same legal principle may look quite different from country to country. In addition, the agreements, as described above, sanction the adoption of measures to address other, non-innovation priorities, including public health and technology transfer, although only to the extent that they are consistent with the other terms of the agreements.[38] Article 18.4 of the TPP includes additional language that acknowledges that the agreement is formed in a broader context in which Parties need to "(a) promote innovation and creativity; (b) facilitate the diffusion of information, knowledge, technology, culture and the arts; and (c) foster competition and open and efficient markets."[39] All of these factors contribute to another level of disagreement among countries that comply with TRIPS — disharmony among the legal rules that implement the TRIPS standards, and the context in which these rules are implemented.

While we are aware of no comprehensive public survey that compares laws of various countries implementing TRIPS, some level of variation is natural and to be expected, with countries clustering around a certain set of standards. Because TRIPS sets a floor, not a ceiling, countries are free to ratchet up from the minimum the amount of protection provided. Thus, for example, TRIPS requires countries to grant copyright for at least the life of the author plus 50 years, but many countries provide

[36] *See* EPC Art. 52.
[37] TRIPS Article 1.1.
[38] See discussion *supra* Disharmony by Design.
[39] TPP Art. 18.4.

for copyright that is life plus 70 years, with a number of other countries, including Mexico which gives life plus 100 years,[40] granting even more. Though rights-holders are unlikely to complain about the variation *above the minimum* that countries have given, it is notable that it hardly solves the problems of differential standards that trade agreements purport to solve.

In the below sub-sections, I provide an example of how countries have used their discretion to implement one particular requirement — the "inventive step" requirement — in four different jurisdictions. I then consider how differences in the procedural contexts in which treaty standards are applied, focusing specifically on post-grant challenges to patent validity, have created another source of disharmony.

4.1. *Implementation of the "Inventive Step" requirement in the US, India, Japan, and Europe*

The TRIPS Agreement requires countries to grant patents to inventions that are "new, involve an inventive step and are capable of industrial application".[41] However, countries have tailored their implementation of this international requirement in response to domestic pressures and needs.[42] In the US, according to Section 103 of the Patent Act, a patent cannot be awarded, "if the differences between the claimed invention and the prior art are such that the claimed invention as a whole would have been obvious before the effective filing date of the claimed invention to a person having ordinary skill in the art to which the claimed invention pertains".[43] In Europe, according to Article 56 of the European Patent Convention, the focus is on whether or not "an invention shall be considered to involve an inventive step". The Japanese standard resembles the US approach, asking: (1) what is the prior art? (2) what are the differences

[40] Smith (2011).

[41] TRIPS Art. 27(1).

[42] See Kapczynski (2009) (commenting on India's decision to adopt "an exceptionally high threshold for inventive step"); Minssen (2010) (comparing application of the inventive step/obviousness analysis in Europe and the US).

[43] 35 U.S.C. § 103(a) (2011).

between the claimed invention and the prior art? And (3) would the invention be obvious to a person with ordinary skill in the art?[44]

India's inventive step standard has been described as "exceptionally high".[45] It requires inventions to contain "a feature that involves technical advance as compared to the existing knowledge or having economic significance or both and that makes the invention not obvious to a person skilled in the art",[46] in order to be patentable. In addition to specifying that an invention be "not obvious to a person skilled in the art", the US standard, India erects an additional hurdle — that the invention include a technical advance or have economic significance — in order to be patentable. Depending on how it is applied, this standard could likely be used to exclude a host of inventions that might otherwise be patentable based on standards used by other countries, Kapczynski has observed.[47] She characterizes this standard, as well as a host of other substantive provisions of the Indian Patent System, including novel limitations on subject matter, compulsory licensing provisions, patent misuse standards to constrain voluntary licensing activity, and limits on injunctive remedies, as evidence of the extensive and underappreciated flexibilities that TRIPS affords.[48]

4.2. *The procedural context — Patent challenges*

The impact of harmonized patent standards depends on the procedural context in which they are administered, and in particular, processes available to challenge patents. In Europe, post-grant proceedings have long been available way to correct Patent Office errors. While individual European countries have their own proceedings, the primary means for challenging European patents is through opposition proceedings at the European Patent Office (EPO) that take place within nine months of a patent's grant. Opposition proceedings are used to review 5–8% of issued

---

[44] Nelson (2004).

[45] Kapczynski, *supra* note 42, p. 1579.

[46] The Patent Amendment Act, No. 15 of 2005, § 2(ja), India Code (2005).

[47] Kapczynski, *supra* note 42, p. 1593.

[48] *Ibid.*, pp. 1589–1616.

patents and have resulted in the partial or total invalidation of patents close to 70% of the time.[49] The outcomes in German revocation proceedings, which take place later in a patent's life, are similar — 73% of patents are partially or fully revoked.[50] There appears to be variance in revocation rates among other European courts adjudicating patents in European countries,[51] but selection effects make it difficult to compare outcomes directly.

The US experience with post-grant challenges in recent years provides a good example of how much procedure matters. Since 1981, patentees and third parties have been able to seek reexamination of issued US patents upon presentation of a "substantial new question of patentability,"[52] with respect to novelty or non-obviousness, for example, due to a new piece of prior art or a change in the law. However, reexamination procedures have never been used much, due in part to their uncertain review periods and the inability to consider outside evidence.[53] In the last decade, concerns about low-quality patents[54] led to the introduction of new procedures. The America Invents Act of 2011 created three new ways for members of the public to challenge issued patents: *inter partes* review (IPR), the covered business method transitional program (CBM), and post-grant review (PGR), which were more streamlined and predictable than reexamination. These redesigned procedures for challenging patents, in particular IPR, have been much more popular than their predecessors. While the USPTO received an average of 10 requests per year in the first five years of *inter partes* reexamination, in the second full year of IPR, the Patent Trial and Appeal Board (PTAB) of the USPTO received 1,310 requests, a massive increase.[55] An estimated 12% of litigated patents have been subject to petition,[56] and about 25% of all patent claims put before

[49] Hall and Harhoff (2004).
[50] For statistics, see Chien and Helmers (2018, forthcoming).
[51] Cremers *et al.* (2013).
[52] 35 U.S.C. 303(a) (2012).
[53] Matal (2012).
[54] Matal, *supra* note 53, p. 630.
[55] Chien and Helmers, *supra* note 50, pp. 3–4.
[56] Unified Patents (2015).

the USPTO have been invalidated.[57] Small changes in procedure have made a big difference, proving the point that the procedural context in which laws are evaluated can matter as much if not more than the substantive rules that apply.

Glimpsing at the post-grant opposition systems of two other countries, India and China, illustrates the diversity of procedures in effect. India features two ways to challenge the validity of issued patents — post-grant oppositions before the patent offices and patent revocation proceedings that may be initiated before the Intellectual Property Appeal Board or via a counterclaim in an infringement suit. It also offers pre-grant opposition procedures that can be initiated by any third party on a broad set of grounds.[58] In China, PGR is the only option for challenging a patent, and the courts strictly weigh in on whether a patent is infringed. Thus, the only way to invalidate a patent is through the Chinese Patent Office (SIPO)'s Patent Review and Adjudication Board (PRAB) which corresponds to the PTAB in the US. Uniquely, SIPO can also decide on infringement.[59]

4.3. *Disharmonies in the administration of the law (the example of patent quality)*

The final and perhaps least appreciated form of disharmony is also the most important to individual rights-holders — what happens when the same claim to rights is evaluated by different jurisdictions? Ultimately, what matters to a rights-holder is whether, when countries apply harmonized standards, they produce consistent or inconsistent outcomes, the most important of which are the rights and remedies available. In this section, we consider patent application outcomes, focusing in particular on the submission of a single patent application to multiple patent offices, as a way to measure the extent of harmony or disharmony between the laws on the books and the laws applied. The results underscore that although government negotiators can agree with each other on

[57] USPTO (2015); see also Chien (2015).
[58] Kapczynski, *supra* note 42, pp. 1598–1600.
[59] Chien and Helmers, *supra* note 50, p. 13.

the provisions of international agreements, many other factors shape the extent to which countries can in fact rely on the consistent and uniform availability of rights.

Take, for example, US Patent 6,032,137, which covers a way of depositing a check by imaging and sending it, rather than physically transferring it to the bank.[60] The inventor, Claudio Ballard, tried for several years to develop the invention. Though he failed, the technology thrived, and his company, DataTreasury, sued dozens of banks for their use of check imaging.[61] In 2013, Fidelity National Information Services, after being sued by Ballard, asked the USPTO to take a second look at the Ballard patents, to which it agreed. In 2015, a panel of patent judges revoked the 137 patent as overly broad and vague, as well as likely already practiced, and therefore invalid.[62] By that time DataTreasury and its two employees had collected $350 million in licensing fees.[63] There is no counterpart to this now-invalidated patent in Europe, though it was not for a lack of trying. The application that matured into the US patent was also the basis of seven patent applications at the EPO,[64] but none were granted.[65] While some applications were withdrawn from EPO prosecution, others were denied.[66] Though in both cases, both European and US Patent Offices ultimately decided that the applications were not worthy of patents, US regulators reached this conclusion decades and hundreds of millions of dollars after the Europeans did. This extreme example shows how even countries whose negotiators agree, and whose standards are arguably close, can still in individual cases reach vastly different outcomes.

[60] U.S. Patent No. 6,032,137 (issued February 29, 2000).

[61] For a longer description and cites, see Chien, Cotter, and Posner, *supra* note at Chapter 3.

[62] Bultman (2015).

[63] Lawson (2013).

[64] See "also published as" section at http://www.google.com/patents/US6032137 (listing related applications EP 1008086 A2, EP 1008086 A4, EP 1688876 A2, EP 1688876 A3, EP 1986148 A1, EP 2267652 A1, EP 2267653 A1).

[65] *Ibid.* (showing no "B" or granted EP publications).

[66] Author's analysis is based on the EP Register.

Previous efforts to measure what happens when the same patent application is submitted to different patent offices report wide variability.[67] Analyzing a cohort of patent applications submitted to the Australian, European, and Japanese Patent Offices from 1990 to 1995 that matched as equivalents to 9,618 US patents,[68] Jensen and his colleagues found that the Australian Patent Office granted almost all (86%) of these applications, while the JPO granted less than half of them (42.6%), and the EPO grant rate was between (74%) these two figures.[69] In a related analysis, Harhoff and Graham analyzed the EPO counterparts of a sample of 2,474 US patents litigated from 1963 to 2003 and their non-litigated counterparts, and reported comparable grant rates, between EPO grant rates of 68% (among counterparts to non-litigated US patents) and 80% (among counterparts litigated US patents), as compared to US patents (100% grant rate).[70]

For a related analysis,[71] I created a set of matched EPO–USPTO patent applications from 2002. Patent rights are territorial, so an inventor seeking protection over the same invention in multiple jurisdictions must file multiple applications. If an applicant for a US patent also seeks protection in Europe, she will typically file the same application, with slight modifications,[72] within a year to the EPO or World Intellectual Property Office (WIPO). While there are numerous ways to associate US and EPO patent applications, in our analysis, I applied the most conservative approach and matched US patent documents to EPO patent documents

---

[67] Jensen *et al.* (2005), Webster *et al.* (2007), Palangkaraya *et al.* (2011) and Webster *et al.* (2014) (finding that local offices tend to favor local applicants).

[68] Jensen *et al.* (2008).

[69] *Ibid.*, Figure 2. These estimates are slightly deflated, as 1.7–13% of applications were still pending at the time of the analysis.

[70] Graham and Harhoff (2009). In other work by the same authors, they found that EPO equivalents of US-litigated patent applications were more likely to be awarded EPO patent protection than were equivalents of unlitigated patents (Graham and Harhoff, 2006).

[71] Chien (forthcoming 2018).

[72] Graham and Harhoff, *Separating Patent Wheat from Chaff*, *supra* note 70.

with identical priority claims.[73] I included in our set all available EPO and US application pairs from 2002[74] ($N = 99{,}221$), and traced the fate of each application through the two jurisdictions. I focused first on whether or not the application had been "Granted," a status designated in PATSTAT.[75]

Of applications filed in both jurisdictions, 77% were granted in the US, while only 52% were granted in the EPO. The difference in grant rates was robust across all five invention sectors tracked under the WIPO — in each case, there were more US than EPO patents granted. Among technology sectors, the differences were most pronounced for electrical engineering (or "technology") patents (Fig. 1). While electrical engineering applications had a less than one in two chance of becoming a patent granted by the EPO, it had a greater than three out of four chance of maturing into a US patent,[76] a 34% difference (Fig. 1). The difference was less pronounced, but still significant for the other classes of patents. Chemistry and mechanical engineering applications and instruments

---

[73]Under the Patent Cooperation Treaty, a claim of "priority" to the first application by the second application confers the important benefit that the second application is treated as if it were filed on the same day as the first application. Subsequent applications can claim priority to one or more applications, within the same or different jurisdictions, creating the possibility of multiple applications within a single patent "family." A patent family, in turn, can be either "simple" or "extended" — the members of a simple family share at least one common priority or "parent," while members of an extended family include documents linked directly or indirectly through common priority claims. Patent families may also be formed through "expert-review," for example, through priority information as well as the review of content by experts as performed by the Derwent World Patent Index. See Graham and Harhoff, *supra* note 70; Martinez (2011). We obtained these matches, called "equivalents," from PATSTAT, which publishes such matches through the DOCDB database.

[74]That is, with a US filing date of September 2002.

[75]While the "granted" variable is viewed as very reliable by PATSTAT researchers, in two cases, it may not reflect the current status — first, because PATSTAT is not updated in real time, it does not always reflect the most recent months of grants. Second, where there is a subsequent revocation of a patent, the granted status does not reflect the change in status. These divergences should have limited, if any impact, on our reported results.

[76]Of the 2,784 applications classified as "electrical engineering," 76%, or 2,105 became US patents, while 43%, or 1,273, became EPO patents.

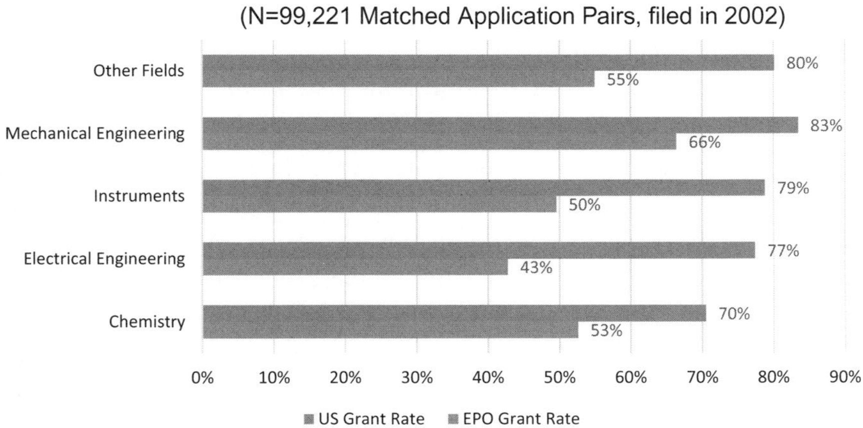

Figure 1. Comparative grant rates.
Source: PATSTAT (2015), Smoch/WIPO, author's analysis.[77]

applications were 17% and 29%, respectively, more likely to be awarded in the US as compared to the EPO (Fig. 1).

In every one of the 35 sector subcategories defined by WIPO, I found that the US was more likely to grant a patent than the EPO.[78] However, it is possible that the results were an artifact of the time period tested, as the fluctuation in US grant rates over time has been well-documented. To rule out this possibility, I expanded my analysis to a sample that included EPO applications matched to US grants in the period from 1975 to 2014. I found the relatively lower EPO grant rate to persist over time, consistent with the findings of Quillen and Cotropia,[79] and observed that the relative EPO grant rate, in fact, declined over the tested period, though I believe

[77]Chien, *Comparative Patent Quality*, *supra* note 71. Webster also performed parallel calculations. *Cf.* Webster *et al.*, *Characteristics of International Patent Application Outcomes*, *supra* note 67, Figure 3.

[78]We performed T-tests on each difference, finding *p*-values between 0 and 2.49341704573321E-32, allowing us to reject the null hypothesis that the difference in grant rates was the result of chance.

[79]Cotropia *et al.* (2015, Figure 9) (documenting a persistently lower EPO grant rate over the period of time 1996–2013).

some of the decline is due to time effects. These results confirm and expand upon earlier results.[80]

The pervasiveness of the gap in grant rates across technology areas between the two jurisdictions is striking. They contribute to a significant level of disharmony in the protections available in specific countries, despite similar, harmonized standards. One major difference that contributes to this difference is the bifurcation of search and examination in Europe, as compared to the unification of these steps in the US.

Though my analysis was limited to US and European applications, Sampat and Shadlen, in a forthcoming study, follow the filing and grant of about 277 "twin" drug patent applications filed in six jurisdictions: India, Brazil, the US, Japan, EPO, South Africa, Mexico, and Argentina.[81] They coded the claims of each application as primary (covering new molecules) or secondary (covering alternative forms of existing molecules) and examine how national grant rates for these types of patents differ. Since overall grant rates can be influenced by the quality of filings, their analyses focus on "twin" applications filed in all six jurisdictions.

They find that, consistent with the results reported earlier, the US has the highest grant rate among patent applications, being the most likely, after South Africa, to grant patents on both primary and secondary drug patents. They also find that although India and Brazil have enacted specific measures to restrict patenting of secondary drug inventions,[82] these

[80] See Jensen *et al.*, *supra* note 68; Harhoff and Graham, *Can Post-Grant Reviews Improve Patent System Design? supra* note 70 (finding the EPO to grant a fraction of US patents in their samples); and Quillen and Cotropia, *Ibid.* (documenting a declining EPO grant rate over time). Jensen also reported comparative grant rates by select international patent classifications. While they do not correspond with the industry sectors from WIPO that we considered, they also found relatively smaller differences in the treatment of mechanical and instruments patent applications and relatively larger differences in grant rates of hardware and communications applications. See Jensen *et al.*, *supra* note 68, p. 13, Figure 3.

[81] Sampat and Shadlen (2017).

[82] *Ibid.*, pp. 3–4 ("Section 3(d) of India's post-TRIPS patent law states that new forms of old substances are not patentable (unless they show improved efficacy). In Brazil, pharmaceutical patents cannot be granted by the patent office unless the Brazilian health ministry also approves, and the agency responsible for making these decisions (ANVISA) gives prior consent.").

measures have had little direct effect on patent examination outcomes. Though grant rates in these countries are lower than in the US, grant rates on secondary patents are not measurably lower than grant rates on primary drug patents.

These results underscore that although government negotiators can agree with each other on the provisions of international agreements, many other factors shape the extent to which countries can in fact rely on the consistent and uniform availability of rights.

5. Conclusion

Though trade agreements are negotiated at the international level, they must be translated into national laws, administered in national courts and halls of justice, and implemented through local and national procedures. The record of the TRIPS Agreement shows that along the way, much can be lost in translation, but in many ways, that is a feature, not a bug. National and local laws, context, and procedures can be tailored and evolve in response to domestic priorities, which preserves the country's sovereignty and ensures that the law continues to be relevant and responsive, rather than unduly rigorous. At the end of the day, inclusion of TRIPS flexibilities did not save TPP, which suffered from numerous defects. But including them in the future, and doing so in the open, may give them the legitimacy[83] they need to at least restart the conversation.

References

Bultman, M. (2015). "Data Treasury Patents Nixed by PTAB in AIA Review," *Law360*, April 29. Available at: http://www.law360.com/articles/649511/datatreasury-patents-nixed-by-ptab-in-aia-review.

Chien, C. (2013). "Tailoring the Patent System to Work for Technology Patents" (unpublished short essay). Available at: https://papers.ssrn.com/sol3/papers.cfm?abstract_id=2176520.

Chien, C. (2015). "The Scariest Number in the Patent Reform Debate is also Wrong," *Politico*, June 3. Available at: http://www.politico.com/agenda/

[83] Dreyfuss, *supra* note 1, p. 677.

story/2015/06/the-scariest-number-in-the-patent-debate-is-also-wrong-000068.

Chien, C. (2016). "Exclusionary and Diffusionary Levers in Patent Law," *Southern California Law Review*, 89, p. 793.

Chien, C. (forthcoming 2018). "Comparative Patent Quality," *Arizona State Law Journal*. Available at: https://papers.ssrn.com/sol3/papers.cfm?abstract_id=2833980.

Chien, C., and C. Helmers (2018). "Inter Partes Review and the Design of Post-Grant Patent Reviews," working paper.

Cotropia, C. A., C. D. Quillen, Jr., and O. H. Webster (2015). Patent Applications and the Performance of the U.S. Patent and Trademark Office as of FY 2014 (unpublished manuscript), Figure 9. Available at http://law.richmond.edu/docs/FY-2014-Update-Figures-and-Tables-1996-2014-29.pdf.

Cremers, K. *et al.* (2013). Patent Litigation in Europe, Discussion Paper No. 13-072, ZEW Centre for European Economic Research. Available at: http://orca.cf.ac.uk/65389/1/Patent%20Litigation%20in%20Europe.pdf.

Davey, W. J. (2005). "The WTO Dispute Settlement System: The First Ten Years," *Journal of International Economic Law*, 8, p. 17, 32.

Dongwook, C. (2012). "Justifying Patent Harmonization," *Asper Review of International Business & Trade Law*, 12, pp. 99–100.

Dreyfuss, R. C. (2012). "The Leahy-Smith America Invents Act, A New Paradigm for International Harmonisation?" *Singapore Academy of Law Journal*, 24, p. 669.

Dreyfuss, R., and E. Donoso (2016). "On Aiding Technical Development: The Max Planck Declaration on Patent Protection, *University of California Irvine Law Review*, 6, p. 321.

Graham, S. J. H., and D. Harhoff (2006). "Can Post-Grant Reviews Improve Patent System Design? A Twin Study of US and European Patents," Discussion Paper No. 5680, Centre for Economic and Policy Research.

Graham, S. J. H., and D. Harhoff (2009). "Separating Patent Wheat from Chaff: Would the U.S. Benefit from Adopting a Patent Post-Grant Review?" *Research Policy*, 43, p. 1649, 1659.

Hall, B. H. and D. Harhoff (2004). "Post-Grant Reviews in the U.S. Patent System — Design Choices and Expected Impact," *Berkeley Technology and Law Journal*, 19, p. 989.

Harris, D. (2011). "TRIPS After Fifteen Years: Success or Failure, as Measured by Compulsory Licensing," *Journal of Intellectual Property Law*, 19, p. 367, 386.

Jensen, P. H. *et al.* (2005). "Disharmony in International Patent Office Decisions," *Federal Circuit Bar Journal*, 15, p. 679.

Jensen, P. H. *et al.* (2008). "Application Pendency Times and Outcomes Across Four Patent Offices," Working Paper No. 06/08, 2008, Melbourne Institute of Applied Economics & Social Research. Available at: http://melbourneinstitute.unimelb.edu.au/downloads/working_paper_series/wp2008n06.pdf.

Kapczynski, A. (2009). "Harmonization and Its Discontents: a Case Study of TRIPS Implementation in India's Pharmaceutical Sector," *California Law Review*, 97, p. 1571, 1589.

Lee, E. (2011). "Measuring TRIPS Compliance and Defense: The WTO Compliance Scoreboard," *Journal of Intellectual Property Law*, 97, p. 401.

Lamping, M. *et al.* (2014). "Declaration on Patent Protection — Regulatory Sovereignty under TRIPS," *International Review of Intellectual Property & Competition Law*, 45, 679.

Land, M. (2012). "*Rebalancing TRIPS*," *Michigan Journal of International Law*, 33, pp. 433–434.

Lawson, A. (2013). "Controversial DataTreasury Patents Face Biz Method Review," *Law360*, November 8. Available at: http://www.law360.com/articles/487185/controversial-datatreasury-patents-face-biz-method-review.

Love, B. J. (2016). "Patent Duration." In *Research Handbook on the Economics of Intellectual Property Law, Analytical Methods*, Chapter 2, P. S. Menell and D. L. Schwartz (eds.) (Edward Elgar Publishing).

Martinez, C. (2011). "Patent Families: When Do Definitions Really Matter," *Journal of Scientometrics*, 86, p. 39.

Matal, J. (2012). "A Guide to the Legislative History of the America Invents Act: Part II of II," *Federal Circuit Bar Journal*, 86, p. 539, 602.

Minssen, T. (2010). "Meanwhile on the Other Side of the Pond: Why Biopharmaceutical Inventions that were "Obvious to Try" Still Might be Non-Obvious Part I," *Chicago-Kent Journal of Intellectual Property*, 9, pp. 60–61.

Nelson, A. (2004). "Obviousness or Inventive Step as Applied to Nucleic Acid Molecules: a Global Perspective," *NC Journal of Law & Technology*, 6, p. 1, 30.

Palangkaraya, A. *et al.* (2011). "Misclassification between Patent Offices: Evidence from a Matched Sample of Patent Applications," *Review of Economics and Statistics*, 93, p. 1063, 1075.

Pauwelyn, J. (2010). "The Dog that Barked that Didn't Bite: 15 Years of Intellectual Property Disputes at the WTO," *Journal of International Dispute Settlement*, 10, p. 389.

Roin, B. N. (2014). "The Case for Tailoring Patent Awards Based on Time-to-Market," *UCLA Law Review*, 61, p. 672.

Sampat, B., and K. Shadlen (2017). "The Effects of Restrictions on Secondary Pharmaceutical Patents in Brazil and India," National Bureau of Economic Research Working Paper No. W23114, 2017.

Smith, B. (2011). "List of Different Countries and the Length of Their Standard Copyright in Years." Available at: https://commons.wikimedia.org/wiki/File:World_copyright-terms.jpg.

Tabarrock, A. (2002). "Patent Theory and Patent Law," *Contributions to Economic Analysis and Policy*, 1(1), p. 1.

Takenaka, T. *et al.* (2010). "Global Issues in Patent Law," pp. 73–74.

Thurm, S. (2000). "Amazon.com Chief Executive Urges Shorter Duration for Internet Patents," *Wall Street Journal*, March 10. Available at: https://www.wsj.com/articles/SB952650678789578387

Unified Patents (2015). "Outcomes at the PTAB (as of May 2015)," *Unified Patents Blog*, May 5. Available at: http://unifiedpatents.com/outcomes-at-the-ptab-as-of-may-2015/.

USPTO (2015). "Inter Partes Review Petitions Terminated to Date (as of 4/30/2015)." Available at: http://www.uspto.gov/sites/default/files/documents/inter_partes_review_petitions_%2004%2030%202015_0.pdf(2015)

Webster, E. *et al.* (2007). "Characteristics of International Patent Application Outcomes," *Economics Letters*, 95, p. 362, 368.

Webster, E. *et al.* (2014). "Patent Examination Outcomes and the National Treatment Principle," *The RAND Journal of Economics*, 45, p. 449, 469.

Yu, P. (2011). "TRIPS and Its Achilles' Heel," *Journal of Intellectual Property Law*, 45, p. 479.

Chapter 9

Shadow Patent Systems: Technology, Economics, and Geopolitics*

Brian Kahin

bkahin@gmail.com

1. Introduction

> "IP rights are the global currency for creating value for products and services, for all innovators, in all markets."
>
> Under Secretary of Commerce for Intellectual Property David Kappos[1]

The vision expressed by the former Under Secretary Kappos reflects an aspiration of commoditization and liquidity that would bind patents to the vast resources of global finance. But whereas currency is the ultimate commodity, patents are unique expressions of new knowledge. Each dollar is the same as every dollar, and all can be aggregated in a purely linear manner. By definition, a patent cannot be the same as another patent.

---

*This chapter is a substantially abridged version of a paper available directly from the author and on the Web (SSRN).

[1] Speech at the Center for American Progress, November 20, 2012.

Furthermore, it takes time and resources to find out whether the patents are valid — and then whether and how they are infringed by particular technology. Moreover, in a world economy characterized by specialization, global value chains, and high volumes of trade across borders, intellectual property rights remain territorial. In the case of patents in particular, states play many different roles. They design patent rights, evaluate applications, issue patents, enforce patent rights, and award damages, but they have been reluctant to require reporting on business uses and behavior once the patent is granted.

It is important to emphasize that patents are *negative* rights.[2] They do not authorize the inventor to exploit an invention. Instead, patents allow their owners to exclude others from using the invention, assuming that the patent is valid. In effect, a patent is an exotic instrument in the nature of an option to sue. Its value ultimately lies in its ability to construct a protective barrier — or to find and target an infringer who can be made to pay. In digital technology, an individual patent can often be designed around, and the barrier function is more effective at a portfolio level where patents can serve as thickets that limit "freedom to operate" by preventing or inhibiting use of large swaths of complementary or adjacent technologies and resources.

Despite their territorial limits, patents are routinely assigned and licensed across borders — as are digital products, services, and data, so from a great distance, they may look as plentiful and fluid as money, as Under Secretary Kappos suggests. A contrasting perspective is expressed by Jim Balsillie, who speaks as an entrepreneur and former CEO of RIM (Blackberry), a company from a mid-level economy (Canada) that fought hard to succeed on a global scale: "The land of intellectual property rights is manipulative, predatory and vicious. And here is the best part: It's managed at the national level."[3]

Both individual patents and patent portfolios function like trade barriers that protect against importation. In doctrinal terms, the ability to exclude is justified on the basis of dynamic efficiency: blocking direct

---

[2] "Patents confer negative rights, i.e. the right to exclude the making, selling, or using of the claimed subject matter." See: http://www.ncbi.nlm.nih.gov/pmc/articles/PMC526265/#fn1

[3] Balsillie (2014).

competition for a period provides an incentive to make investments in invention or complementary assets that will pay off over the long term. This rationale is analogous to the argument for protecting infant industries from competition through tariffs — a time-honored practice extending back to the principles propounded by Alexander Hamilton in his *Report on Manufactures* and promoted as the "American System" by Henry Clay.[4]

While strategic barriers can be monitored reasonably well at the level of tariffs and trade, the strategic uses and effects of both patents and patent portfolios remain confidential business information despite the public nature of the underlying patent instruments. This makes it difficult to separate out and piece together meaningful information on licensing that represents real technology flows, especially since many licenses are between affiliates.[5] In these shadow markets, patent owners are free to use patents secretly in ways that may or may not serve competition policies and the long-term economic goals of the system. There is no forum for probing deeper into the factors, imbalances, and differences in practice that shape international flows in patent rights — or the issues around state participation in these rights and flows. Nor is there a policy-level effort to improve transparency in a way that reveals territorial and sectoral differences and their economic consequences. No other domain of economic policy operates in a darker twilight zone.

In this uninformed, unfocused, but expanding policy space, there are disparate results. The system appears to work well enough for frontier firms in pharmaceuticals, where patents equate roughly to products. It appears to work poorly in software and Internet services,[6] where it is fraught with controversy. Recently, the push for patent reform in the US and the closer look at injunctions and damages for standard-essential patents under "Fair, Reasonable and Non-discriminatory" (FRAND) commitments has exposed growing differences across Information and Communication Technology (ICT) segments and value chain positions.

[4] See: https://en.wikipedia.org/wiki/Report_on_Manufactures;

[5] For a creative assault on this problem, see Robbins (2009).

[6] Federal Trade Commission (2003) (distilling the views of four innovative sectors); for a remarkable effort to aggregate rents and costs by sector, see Bessen and Meurer (2008) described below.

2. Diversity of patent sectors

In the popular mind, patents are associated with major breakthroughs, such as the telephone and the incandescent lightbulb. However, in the US, patents are awarded as a legal entitlement that is not related to economic or technological value. An application must be granted, unless the examiner can affirmatively show why it should not be granted (operating under time constraints — an average of 18 hours for processing and examination[7]). The operating philosophy was described by Judge Giles Rich, the former dean of Federal Circuit jurisprudence and principal draftsman of the 1952 Patent Act:

> [Patents] are not for exceptional inventors but for average inventors and should not be made hard to get.... Why must an invention be a commercially hot number to be patentable? If it is a total dud, how is the public injured by a patent on it? A monopoly on something nobody wants is pretty much of a nullity. That is one of the beauties of the patent system. The reward is measured automatically by the popularity of the contribution.[8]

This surprisingly permissive view may not be problematic where there is a close correspondence between patents and products and independent invention is unlikely, but if the patented function is one among tens of thousands, there is great potential for leveraging trivial patents against investments in whole products.

The FTC/DOJ hearings[9] in 2002 revealed stark differences in how four innovative sectors perceived the US patent system, ranging from deep discontent in the case of software and Internet services to general satisfaction in the case of pharmaceuticals.[10] Using a pioneering cost-benefit

---

[7] Actual search time may be less than two hours, with around 13 hours for the examination process. Van Pottelsberghe (2014)

[8] Rich (1960).

[9] See: https://www.ftc.gov/news-events/events-calendar/2002/02/competition-ip-law-policy-knowledge-based-economy-hearings

[10] Federal Trade Commission (2003).

analysis based in part on event studies (market response to patent litigation), economists James Bessen and Michael Meurer found that pharmaceuticals and chemicals benefited from the patent system while software and financial services experience it as a net tax on innovation (*Patent Failure*, 2008). Patent scholars Dan Burk and Mark Lemley published an entire book, *The Patent Crisis and What the Courts Can Solve It* (2010), on the problem of technological diversity in patent law. Despite the inviting sub-title and the credentials of the authors, the Court of Appeals for the Federal Circuit has completely ignored their analysis — indicative of the distance between judicial and academic thinking on patent policy.

In the meantime, differences in perspective and practice have been abundantly reflected in the US debate over patent reform. Given the differences among technologies, it should not be surprising that a unitary legal system leads to economic distortions. In adhering to a system optimized for marketed molecules, digital technology in effect subsidizes the pharmaceutical industries.

However, there is a practical market need to develop global standards for digital technology — especially in products, systems, and networks, where different components/products are expected to interoperate. Combined with a recurring need to develop markets for new products, this has led to a prominent role for standards institutions that can operate efficiently on a global basis. There has been a profound shift from officially sanctioned national bodies feeding into international organizations to industry-led consortia that are able to create globally accepted standards in relatively short order.

These relatively open consortia have avoided practices that have drawn anti-trust scrutiny in the past. By requiring that participants commit to licensing standard-essential patents (SEPs) on FRAND terms, they have created a contract-based environment in which competition agencies feel more comfortable and in which judges can rethink what reasonable royalties should be. However, the stakes are very high, since the global nature of the standard means that the entire industry — producers, distributors, and users — can be expected to pay royalties.

3. National, regional, and multilateral reform and harmonization

The US Constitution does not require that the US government grant patents; it merely authorizes the granting of patents.[11] By contrast, the TRIPS agreement (now part of the WTO charter) essentially imposes an overriding obligation to do so, and it specifically mandates that the patent term runs at least 20 years from the date of filing — for all "fields of technology," albeit without defining the term "technology." This one-size-fits-all mandate is immutable despite considerable agreement among scholars that, under an economic calculus, the appropriate term should vary depending on factors such as scale of investment, risk, and time to market.[12]

There has been virtually no progress on multilateral harmonization in the past two decades, in either WTO or WIPO. From the developed economy perspective, there was to be a one-way path to greater protection, an alignment that all would acknowledge and achieve in due course. This perspective is embodied in the Ginarte–Park index, in which 5.0 is a perfect score, approached only by the US and a few other developed economies — in effect, a Washington Consensus for patents.[13]

Ginarte–Park instantiates a simplistic view of patent systems along a unidimensional "weak" to "strong" scale that continues to overhang the geo-economics of patents. The index includes factors — availability of patents for software and availability of preliminary injunctive relief — that have proved especially problematic in the digital economy, however much they may please certain private interests. The US and Europe have sought to increase standards that benefit IP incumbents through bilateral and regional trade agreements ("TRIPS plus"). Yet discussions on substantive harmonization have stalled in WIPO, and major economies have

[11] In Article I, Section 8, the Constitution specifies only that Congress has the power "To promote the Progress of Science and useful Arts, by securing for limited Times to Authors and Inventors the exclusive Right to their respective Writings and Discoveries."

[12] See Roin (2014)

[13] The original 1997 index was updated by Park in 2008. See: http://fs2.american.edu/wgp/www/res_policy08.pdf.

taken independent paths in expanding or reforming the scope and direction of patent systems under their jurisdiction.

In 1997, the Business and Industry Advisory Council to the OECD expressed its aspirations[14]:

> A common message from industry is the desire for legal certainty accompanied by a substantial reduction in costs, prompt patent examination and continued efforts to ensure the highest levels of intellectual property protection and enforcement world-wide.

Demand for cheaper, faster, and stronger patents persists after two decades of wrestling with over-patenting, poor quality, backlogs, and uncertainty — problems largely occasioned by and centered in the ascendant digital economy.

Nevertheless, there have been significant but uncoordinated changes in the three largest markets — US, China, and Europe. On the whole, patents can be expected to be of most value in these markets because of the sheer size of the market they can or will protect, but the state of play in each of the three is unique. China's subsidized, hyperactive patent system with its prodigious volume of utility models and design patents remains on a meteoric trajectory. In 2015, SIPO granted 359,316 *invention patents* (equivalent to US *utility patents* — and surpassing the US number for the first time[15]), 876,217 utility models (no US equivalent), and 482,659 design patents (compared to a record 25,986 US design patents granted).

Europe's long-term effort to integrate its territorially bound patent systems via a new *unitary patent* with a *unified patent court* is a characteristically European market integration project. This highly specialized system will take time to work out its procedures and institutional anomalies, including its need to be self-supporting, i.e., free of customary public

[14] See: http://biac.org/wp-content/uploads/2014/05/IPR97FIN1.pdf

[15] USPTO utility patent grants totaled 298,407 in 2015, a slight decline from 2014 — and the first decline since 2007, at the time of Under Secretary Dudas's quality initiative. See: https://www.uspto.gov/web/offices/ac/ido/oeip/taf/us_stat.htm

support for courts and the judicial system. Thanks to Brexit, it is currently in limbo.

The politics of patent reform in the US reveals an almost unmanageable diversity of economic interests — from "Main Street" users of information technology to multinationals with different business models. Reform has become a perpetual process while controversies, old and new, have grown in scope and intensity. These controversies divide industry along many dimensions: digital versus molecular industries; large versus small; upstream versus downstream; incumbents versus startups; and operating companies versus patent assertion entities.

4. Subsidies and their costs

To understand how the patent system works, it is useful to examine the subsidization of patenting and the costs that result from it. For the entire process from filing through examination, the US Patent and Trade Office (USPTO) expends an average of only $3,713 per patent application.[16] Yet even this small expenditure is subsidized. Instead of paying for the cost of the service, which is the general rule for fees for government services in the US, applicants pay at most $1,600, no matter how big they are. Small firms and micro-enterprises get the same service for $800 and $400, respectively (50% and 75% discounts) under statutory subsidies prescribed by Congress. At this low price, the successful applicant gets a patent with an enhanced presumption of validity that can only be overcome in court by "clear and convincing evidence," a high burden of proof.

China also provides substantial subsidies, but rather than providing discounts for applications, direct subsidies for domestic applicants are provided from all levels of government — local, provincial, and national. The national government also provides direct subsidies for patent applications abroad (by Chinese inventors).[17] Such subsidies for domestic

[16] See: Federal Register (2013).

[17] This practice was attacked by the National Association of Manufacturers (US) in a 2011 submission (p.11) to USTR's Special 301 consultation. See: http://www.nam.org/Issues/Trade/NAM-2010-Submission-to-USTRs-Special-301-Process/.

Table 1. Average costs of full litigation (AIPLA 2013).

Amount in controversy	**Legal costs per side**	**× 2 (total for both)**	**Minimum legal costs *relative* to amount in controversy**
> $25 million	$5.9 million	$11.8 million	n/a
$10–25 million	$3.2 million	$6.3 million	> 25%
$1–10 million	$2.1 million	$4.2 million	> 42%
< $1 million	$0.968 million	$1.936 million	> 194%

applicants are suspect under the WTO, but payments for applications granted can be characterized as awards for technological achievement.

Subsidies of one form or another are pervasive throughout the patent system. For example, the court system is heavily subsidized almost everywhere, which means that the cost of hearing patent disputes once a suit is filed is born by the public.[18] However, as shown above, the costs of litigation are quite high and can be greater than the contested amount. This is particularly significant when the amount at stake is low, which is also the range within which small businesses are likely to litigate. While it is difficult to determine just how much will be expended in the course of a fully litigated dispute to search for and evaluate prior art, it will certainly far exceed the average $13,000 for a routine validity opinion with no guarantees.

5. Quality

The process by which patents are issued can create a "quality problem," in which patents that may eventually be invalidated in court are approved, placing an additional burden on the private sector both by adding an increased number of patents to search through, but also the high costs, shown earlier, of future litigation to invalidate the patent.

[18]The Congressional Budget Office (CBO), which is charged with scoring the economic burden of proposed legislation, does not take the cost of judicial resources into account. This is remarkable since the parties' costs of litigation are a major factor determining private behavior.

A major problem with the front-end subsidy of application and examination fees is that it provides the patent office with a strong incentive to grant patents. Only if applications are granted will issuance and maintenance fees be paid, and these fees are needed to make up for the initial subsidy. Frakes and Wasserman find that budgetary concerns affect the allowance rate.[19] The political sensitivity of the allowance rate is shown by a dramatic reversal between the later Bush years under Jon Dudas and the initial Obama years under David Kappos. After Dudas attempted to reverse permissive standards, to the consternation of the patent bar, the effective allowance rate shot up dramatically under Kappos[20]:

As Fig. 1 shows,[21] the effective allowance rate is considerably higher in the US than in Europe and Japan. This is consistent with the annual IAM survey, which show the USPTO significantly below the European Patent Office (EPO) and Japan Patent Office (JPO) in terms of perceived quality.[22] Concerns have been raised about the volume and quality of Chinese patents, and China has also institutionalized the granting of low-quality patents in the form of legions of utility model patents. While utility models are not unique to China (47 countries grant them in some form), China granted over 700,000 utility model patents in 2014, which was over 46 times the number awarded in third-ranked Germany. China also has roughly twice the volume of patent litigation as the US, although this appears to be mostly small-stakes litigation among Chinese individuals and small firms. China granted 15 times as many design patents as the US did in 2014.

[19] Frakes and Wasserman (2013).

[20] As reported in the *ABA Journal*, when Kappos arrived as Under Secretary, he sent an e-mail to all examiners addressing the culture of rejection in which a lower allowance rate became the measure of higher quality. "Let's be clear: Patent quality does not equal rejection" (see: http://www.abajournal.com/magazine/article/a_patent_on_problems/). While true in a limited sense, it is hard to conceive of a more objective measure. For a critique of the quality measures developed by the Office of Chief Economist under Kappos, see Kahin (2013).

[21] Cotropia *et al.* (2013) and 2017 update by Cotropia and Quillen, https://papers.ssrn.com/sol3/papers.cfm?abstract_id=3147056. The data also suggest corresponding trends at the EPO and JPO (perhaps taking policy cues from the US), although the shift is most dramatic at the USPTO.

[22] Clover (2016).

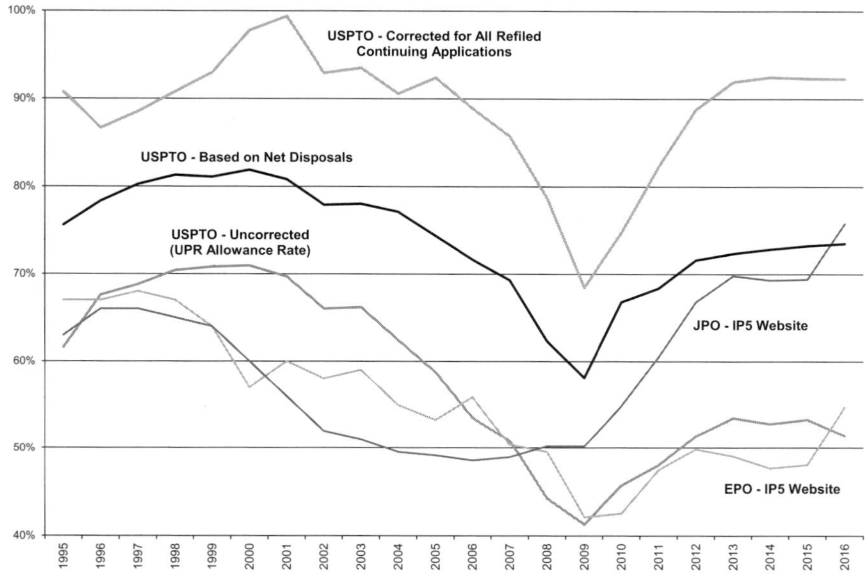

Figure 1. Comparative allowance rates (grant rates) (1996–2003).

Source: Cotropia *et al.* (2013); see Footnote 21.

6. Uncertainty

Like the financial system, patenting has expanded through extensions that are speculative, transaction-driven, and more remote from the real economy. In the extended digital environment, many factors contribute to the uncertainty of patent value: the vast amount of prior art; the ambiguous language of patent claims ("fuzzy boundaries"); the difficulty of applying the inventive step (non-obviousness) standard; and the many paths to performing the same function. Unlike innovative financial instruments, patents are a creation of the state, and because patent administration is transaction-based, there are incentives for patent agencies to expand the volume and scope of patenting. When patents are invalidated in court, patent offices do not ordinarily investigate what went wrong. There is no monitoring of how issued patents fare in practice. The cost of mistakes will be borne by unknown firms in the private sector, not by the patent office or its employees. With a rapid turnover of examiners in the US (average tenure is three years), it may be easy to think in the same terms that characterized volume-driven intermediaries in the financial sector: "I'll be gone; you'll be gone." The adverse consequences of low-quality and wrongly issued patents take years to materialize.[23] At a trade policy level, this means that the handicap for emerging economies may effectively be measured in decades rather than years.

At the same time, the explosion of patenting in digital technologies has brought fundamental changes to patent practice. The legal model of absolute exclusivity has given way to a practical framework of interdependence.[24] Accordingly, large patent portfolios are routinely cross-licensed on a global basis, but the terms are private. This interdependence parallels the stylized structure of international trade in which specialized inputs, products, and services are exchanged to mutual advantage. However, the structural interdependency of patent licensing is more complex because the blocking function of patents does not fit the cumulative assembly-line framework of value chains. This interdependency is far removed from public view — despite the strong state interest in the proper functioning of the patent system. Yet, the global patent community is reluctant to relinquish the paradigm of "strong" exclusivity, since doing so would lower the value of patents and portfolios.

Excessive patenting in ICT inevitably leads to a cognitive overload that burdens entrepreneurs, engineers, designers, and lawyers. Defensive aggregator RPX has estimated that as many as 250,000 patents may be implicated in a smartphone.[25] How do firms, especially upstarts in developing economies, locate/identify patents on this scale and then evaluate them properly? Digital start-ups must look to patents in the markets they

---

[23] Risch (2011) reports that non-practicing entities file suit an average of 8.3 years after the patent issues, raising the "submarine patent" concern that opportunistic patent-holders deliberately wait until products, businesses, or industries develop around patented technologies. This strategy is not limited to classic trolls; see Mullin (2016).

[24] Phelps and Kline (2009).

[25] RPX S-1, filed January 21, 2011. See: http://www.nasdaq.com/markets/ipos/filing.ashx?filingid=7325067.

cannot do without — principally the US, China, and Europe — if they are to realize economies of scale in the face of global competition. Incomprehensible volumes of patents inevitably serve as thickets against new entrants who lack the resources to evaluate the thicket and assume the risks and consequences of inadvertent infringement.

Research indicating bias in the examination process[26] suggests that national thickets will further disadvantage outsiders — on top of the costs of translation and the difficulties of navigating patents and prior art in a foreign language. Spain has declined to join Europe's pending Unified Patent Court because of the language issue.[27]

With the well-known quality problem in digital patents, it may not make economic sense for anyone to investigate deeply. Costs can mount up quickly. First is the cost of searching for relevant patents. Once patents are identified, they must be evaluated. The retail cost of an attorney's opinion on validity averages more than $13,000 per patent. Then there is the cost of analyzing whether the proposed technology actually infringes the patent (another $13,000–15,000).

Risks are compounded by subjective factors that transcend ordinary economic damages, especially for outsiders. In the US, "willful" infringement can result in up to treble economic damages as a form of punishment. Another is the crafting of remedies to include "deterrence," following the argument that otherwise firms would infringe first and pay only after a judgment.[28] To be sure, uncertainty can be reduced — and is ordinarily reduced in the course of litigation as more information is revealed. But reducing uncertainty through litigation is costly and favors large over small, incumbents over start-ups, and local interests over remote interests. While accusations of bias are often leveled at China, the

---

[26] Webster *et al.* (2014).

[27] See Kluwer Patent Blog (2015). Italy, with similar concerns, joined only after Spain lost its legal challenge to the UPC in the Court of Justice of the European Union.

[28] Of course, this is only the case if the cost of litigation is trivial. As the AIPLA survey shows, it is not trivial, except perhaps in relative terms for the very largest firms.

US system has built-in institutional biases: jury trials,[29] forum shopping by plaintiffs,[30] the International Trade Commission, and the ability of the White House to override USITC decisions on "policy" grounds.[31] All of these were conspicuously on display in the *Apple v. Samsung* litigation. Viewing these aspects of US practice from abroad, it is easy to imagine that the Federal Circuit's recent decision awarding Apple disproportionate damages for design patent infringement is simply another episode of local favoritism.[32]

As noted, value in secondary markets is based primarily on evidence of use, i.e., arbitrage value based on information failure. But patent applications are not published until 18 months after filing, and few companies can monitor the firehose of applications in digital technology. Since boundaries of digital patents are often unclear and prior art may be difficult to find,[33] opportunities for assertion are not readily apparent. Assertion is often delayed until the patent falls into the "right" (or "wrong") hands. (Patent assertion entities do not file suit until an average of 8.3 years after the patent issues.)[34]

Uncertainty is endemic to innovation, but here technological and market uncertainty is compounded by legal uncertainty. Ironically, this multidimensional uncertainty is generated by an intellectual property regime that is intended to minimize risks associated with investments in innovation and to promote public knowledge. And uncertainty can be used as a weapon — for example, to extract settlements from smaller players who simply cannot afford to contest a claim of patent infringement.

---

[29] Moore (2003).

[30] Forum shopping entails both the ability to choose pro-plaintiff venues such as the notorious Eastern District of Texas (favored by trolls) and the ability to litigate before a hometown audience, as Apple was able to do against Samsung.

[31] See Kahin (2013b).

[32] The Federal Circuit interpreted a 19th century statute to require Samsung to pay all profits earned for the product. The US Supreme Court has recently granted certiorari in this case, an indication that the Federal Circuit's decision may be overturned. See: http://www.scotusblog.com/wp-content/uploads/2016/01/Apple-v-Samsung.pdf

[33] Both problems are related to the degree of abstraction. The Supreme Courts has made it clear that abstract ideas are not eligible for patents, but abstraction is relative and hard to define. *Alice v. CLS Bank* (2014) is the most recent formulation.

[34] Risch (2012).

The most notorious example, MPHJ Technology Investments, sent letters to thousands of small businesses intimating that they were infringing a "scan-to-email" patent and offering a license at $1,000 per worker. This kind of behavior does not attract academic interest, but it has attracted public attention — at least 18 states have recently passed legislation against bad-faith patent assertions at a time when federal legislation has stalled.[35]

Large companies with successful products at stake are also vulnerable — not just from trolls but from larger, more established firms. Published accounts are rare because settlement agreements typically preclude the parties from talking, but Silicon Valley attorney Gary Reback vividly recounts an IBM lawyer's threats against Sun (not a small company but a latecomer relative to IBM):

> "OK," he said, "maybe you don't infringe these seven patents. But we have 10,000 U.S. patents. Do you really want us to go back to Armonk [IBM headquarters in New York] and find seven patents you do infringe? Or do you want to make this easy and just pay us $20 million?"[36]

An elaborate account of "patent bullying"[37] in litigation is provided by Ted Sichelman's detailed account of Verizon's attack on Vonage, an innovative voice-over-IP service. In its patent attack on Barnes and Noble, Microsoft demanded that Barnes and Noble sign non-disclosure agreements in order to be informed of the specific patents that it was asserting, in essence shrouding patents in trade secrecy.[38]

7. Global technical standards and FRAND commitments

While patents remain territorial, *global* technical standards are essential to developing new markets for digital technology. Standards build

---

[35] See Gugliuzza (2015).

[36] Reback (2002). Reback is known for his work on behalf of Netscape in motivating the US Department of Justice case against Microsoft.

[37] See: https://papers.ssrn.com/sol3/papers.cfm?abstract_id=1856703

[38] Feldman (2013).

confidence in investment at all points in value chains by enabling interconnection, interoperability, and economies of scale and scope. In some areas, where widespread implementation is desired (Internet, Web, software), standard-setting organizations discourage use of patented technology or require that participants license essential patents royalty-free.[39] In areas where patents and royalties are expected (computers, cellular communications), participants must commit to licensing patents on FRAND terms.[40]

FRAND commitments are contractual and global, applicable everywhere, while judicial interpretations of what "reasonable" means and when injunctions may be appropriate may vary from jurisdiction to jurisdiction.[41] Until recently, standards organizations have left these terms undefined, but the new IEEE policy (cleared with the US Department of Justice and adopted in 2015) states that FRAND is to be calculated for the smallest saleable unit and sets limitations on when injunctive relief can be granted.[42] Debate over the new IEEE rules has pitted some royalty-dependent companies — Ericsson, Nokia, Interdigital, and Qualcomm — against the rest of the industry.

Patents are normally treated as absolute property rights that owners are free to exercise as they wish.[43] Patent offices regulate the application and examination process; beyond that, patent-related behavior is left alone, subject only to private ordering and, to a limited extent, competition law. Historically, standards-setting organizations have been reluctant to impose requirements beyond disclosure based on the personal knowledge of the participant and a commitment to licensing

---

[39]These organizations include IETF, W3C, and OASIS.

[40]RAND (FRAND without the "fair") is used in the US, but the practical effect seems to be the same (so far).

[41]Reasonable could be defined under the contract or interpreted according to the national law of a designated country (France in the case of ETSI).

[42]Contreras (2015).

[43]The claims in patents are said to be equivalent to the metes and bounds that delimit real property boundaries. But unlike real property, which is precisely surveyed and visibly delineated, with two sides to every boundary, patents are defined by words and boundaries that may be overlapping and "fuzzy," especially in software and business method patents.

under FRAND terms.[44] This has left the implementation of FRAND commitments to be worked out between individual companies, typically at a portfolio level. This works reasonably well because many of the larger companies are already cross-licensed with each other. However, disputes flared around the smartphone, which brought many different technologies, and industries, together in one product.

A few recent high-profile cases have sought to clarify the meaning FRAND commitment and the use of injunctive relief — provoking debate and providing impetus for the new IEEE policy. A dispute between Motorola and Microsoft was especially important because the judge provided a detailed analysis of apportionment that embraced the smallest saleable unit as the starting point for calculating royalties, with damages ending up far smaller than what Motorola had asked for.

8. Investor–state dispute settlement

Trade agreements bind countries not companies. The WTO does not permit private firms to initiate complaints, but many bilateral treaties and some regional trade agreements, including NAFTA, allow private companies to initiate extra-legal proceedings against states under investor–state dispute settlement (ISDS) provisions. Originally ISDS was designed to discipline developing economies in order to attract foreign investors. However, it has expanded in practice to shield multinationals from evolving regulatory practice, even in developed economies with advanced legal systems.

The ISDS proceeding is not available to a domestic corporation, unless it is able to game the system through a presence in a foreign country in order to initiate the arbitration — which then takes place outside of national (and international) legal systems. In addition to avoiding the rule of law, ISDS has damaged the case for international trade by elevating the

[44] One notable exception is VITA (http://www.vita.com/), which has strict disclosure requirements and a requirement that contributors of patented technology specify the maximum royalty in advance.

interests of multinationals over the laws and regulations of developed democracies and providing them with an advantage over smaller, domestic competitors. On top of the opacity of treaty negotiations (along with periodic leaks that serve as reminders of the opacity), ISDS has politicized trade liberalization in new ways, and allowed certain negative aspects to reinforce each other.

The political reaction has been most pronounced in environmental cases, but a couple of intellectual property cases have struck an emotional chord: the Eli Lilly case against Canada and the Philip Morris cases against Uruguay and Australia stand out. Using NAFTA, Eli Lilly has attacked the Canadian Supreme Court's interpretation of patent eligibility, implicitly attacking the sovereign-based territoriality of patents. The notorious Philip Morris attacks on plain-paper packaging of cigarettes based on claims of trademark devaluation stand as especially aggressive extensions of the global trading system. They were brought both as ISDS claims and using country surrogates like Honduras under the WTO. The Australia ISDS case came through the backdoor of a most-favored-nation (MFN) clause in the bilateral agreement between Australia and Hong Kong.[45] While unsuccessful, the ironic result has been to damage the brand of ISDS, WTO, and trade agreements generally.

The reaction against ISDS has been especially pronounced in Europe, where it has undercut support for the Transatlantic Trade and Investment Partnership (TTIP) and EU–Canada Comprehensive Economic and Trade Agreement (CETA). The European Commission has acknowledged the "toxicity"[46] of ISDS and addressed it by proposing[47] a new "Investment Court System" with higher standards and increased accountability and transparency. However, this alternative has not been accepted by the US, which views the reforms to ISDS in the TPP as sufficient, while it has been viewed by critics as a mere rebranding of ISDS.

To domestic audiences, the USTR boasts "the United States has only been sued 17 times under any U.S. investment agreement and has

---

[45] Another example of mission creep. MFN clauses were designed to achieve tariff parity, not regulatory parity.

[46] See: http://www.politico.eu/article/isds-the-most-toxic-acronym-in-europe/

[47] See: http://europa.eu/rapid/press-release_IP-15-5651_en.htm

never once lost a case. In some instances, we have even received compensation for having had to defend against a case in the first place".[48] Canada's legal system is no less developed, yet as of 2014, Canada had been sued 35 times under NAFTA alone and had lost (or settled) six times.[49] The imperviousness of the US may well be related to the fact that most of the disputes are handled in private by the International Centre for the Settlement of Investment Disputes (ICSID). ICSID is an agency of the World Bank, which (as flaunted by the ICSID homepage) is based in Washington, D.C., where it is perpetually subject to Congressional oversight.

ISDS, in effect, adds a new supranational regulatory layer to the global economy, although this is obscured by the fact that it enables private enterprises to regulate public agencies. In this regard, it contrasts dramatically with broader deregulatory trends over the past four decades — and particularly digitally enabled developments at the local level. To be sure, the TPP and the proposed Investment Court System are improvements, but they still represent additional layering in situations where the need is not apparent, high legal costs can skew outcomes, and decisions can undercut public sentiment for free trade and globalization. This layering further embeds legacy formulations against new evidence, new technologies, and new business models and practices. Both ISDS and the proposed ICS favor legal superstructure over political and economic discourse, even though this implicitly trumps national rule of law.[50]

The TPP does so in a roundabout way that superficially makes it appear that intellectual property is excepted from ISDS.[51] The Investment chapter provides:

> This Article shall not apply to the issuance of compulsory licences granted in relation to intellectual property rights in accordance with the TRIPS Agreement, or to the revocation, limitation or creation of

[48] USTR (2014).

[49] Freeman (2015).

[50] To be sure, ISDS tribunals are not of one mind in expanding their mission. Some of the arbitrators are wary of expansionist decision-making that may undercut the legitimacy of the tribunals over the long term. See Sornarajah (2015).

[51] Seidenberg (2015).

> intellectual property rights, to the extent that the issuance, revocation, limitation or creation is consistent with Chapter 18 (Intellectual Property) and the TRIPS Agreement. (Article 9.7, paragraph 5)[52]

A close reading reveals that not only is intellectual property included, but the burden is on national governments to show the consistency of their laws and rulings with both TRIPS *and* the TPP intellectual property provisions.

Unbound by precedent and free from oversight, ISDS thus not only constrains the scope and adaptability of policy-making but burdens and embarrasses governments, undermining the authority and legitimacy of the state. The reaction against ISDS in Europe, where it resonates with resurgent nationalism, threatens to derail the TTIP. Since the US hosts the principal ISDS forum (ICSID) and has never lost a dispute, it has largely avoided the ISDS controversy and has resisted the EU effort to create an alternative Investment Court System. For a home audience, the TPP is even pitched as a US creation, with the USTR website proclaiming that the agreement is "made in America." TPP is presented as a counter to the Regional Comprehensive Economic Partnership (RCEP), which includes China, India, Japan, South Korea, and the 10 ASEAN countries. As expressed by President Obama: "When more than 95 percent of our potential customers live outside our borders, we can't let countries like China write the rules of the global economy. We should write those rules…."[53] The trade pact rivalry takes on this larger significance after the launch of the Asian Infrastructure Investment Bank, initiated by China and perceived as competing with the World Bank.[54]

---

[52] See Seidenberg, *supra*. Footnote 19 to this paragraph reads:
"For greater certainty, the Parties recognise that, for the purposes of this Article, the term "revocation" of intellectual property rights includes the cancellation or nullification of those rights, and the term "limitation" of intellectual property rights includes exceptions to those rights."

[53] Statement of the President on the Trans-Pacific Partnership, October 5, 2015. Available at: https://obamawhitehouse.archives.gov/the-press-office/2015/10/05/statement-president-trans-pacific-partnership.

[54] Perlez (2015).

9. Discussion

Nearly 60 years ago, economist of knowledge Fritz Machlup famously wrote:

> If we did not have a patent system, it would be irresponsible, on the basis of our present knowledge of its economic consequences, to recommend instituting one. But since we have had a patent system for a long time, it would be irresponsible, on the basis of our present knowledge, to recommend abolishing it.[55]

Today, the patent system has become far more complex and embedded. It has been modestly internationalized by TRIPS and globalized in practice by global value chains and ICT standards — while remaining territorial in administration, law, and practice. It has been subsidized and exploited in ways that were not apparent when Machlup wrote the foregoing as part of the Senate Committee's intensive examination of the US patent system. Most importantly, the dramatic rise of ICT as a general-purpose technology makes clear that the standard for "irresponsible" depends greatly on the context and methodology.[56]

At the bottom, there are three nearly intractable problems underlying digital patents and patent practice: cost, quality, and proportionality. Policy perspectives vary according to market position, technological/business context, and professional discipline. However, the complexity of the system, secrecy surrounding practice, and institutional weight of patent administration and adjudication mean that legal perspectives dominate. But the legal community has little interest in containing the scale and scope of the system. Many patent professionals resent the role of the US

---

[55] Machlup (1958), echoing his colleague Edith Penrose in her 1951 book, *The Economics of the International Patent System*.

[56] Compare Bessen and Meurer (2008) finding dramatic inter-sector differences for aggregated private benefits and costs; Federal Trade Commission (2003) finding similarly large qualitative differences based on testimonial evidence; Mansfield (1986) finding stark division between the industries he surveyed as to effects of the patent incentive; and Fontana *et al.* (2013) finding limited industry variation (and remarkably, finding that fewer than 10% of award-winning technologies were patented).

Supreme Court in limiting patent-eligible subject matter, and in Europe, the patent community tried unsuccessfully to keep the Court of Justice of the European Union from overseeing the new Unified Patent Court.

The bureaucratic framework and the precedent-oriented inertia of legal reasoning keep political discussion bound to the individual patent and to distinctive, coherent, and marketable inventions such as the telephone or the airplane. This mythology exalts the individual inventor over other less tangible, social aspects of innovation.[57] In a digitally empowered world, the individual patent is ordinarily overshadowed by the operation of portfolios, while portfolio practice is obscured not only by costs of evaluating these many unique legal instruments but also by a dark web of confidential agreements. Despite the premise of public disclosure built into patent law, opacity and secrecy abound.

The basic political economy of patents helps explain why it is so difficult to address the fundamental problems. It is not in the immediate economic interest of the professional community to limit scale and scope — or the many characteristics that give patents private value ("strength"). Aside from occasional hand-wringing, there is virtually no discussion of the cost of patents. The European Commission investigated insurance for patents in the early 2000s under the assumption that Small and Medium Enterprises (SMEs) needed and wanted it in order to assert patents and in the belief that there was a well-functioning insurance market in the US. The project was abandoned after the Commission learned that the principal demand was for defensive insurance, that the patent insurance market in the US was anemic,[58] and that any subsidy scheme would be cumbersome and costly. However, there may be renewed interest in light of the costs that the pending Unified Patent Court may impose on SMEs.[59]

The cost issue arose recently in the US when Intellectual Property Owners (IPO) succeeded in getting a "loser pays" provision into reform

[57] Lemley (2012).

[58] Patent infringement is not covered under standard errors and omissions policies. The very limited market for patent insurance is documented by the Betterley Report every April (see: http://betterley.com/samples/ipims16_nt.pdf).

[59] Amar (2016).

legislation in 2013. While "loser pays" is common in European laws, it would have been a radical departure for civil litigation in the US, where each party is normally responsible for its own attorneys' fees, win or lose. But litigation is especially costly in the US, and the business culture is more litigious. The change would have pushed the winner-take-all character of the patent system to a new extreme, and the uncertainty surrounding patent litigation would have imposed disproportionate risk on small patent holders, such as universities.[60] The provision would have increased total costs because it would naturally entail added fighting over the reasonableness of attorney expenses. The stand-off over implementation of "loser pays" ultimately played a large part in derailing new reform legislation in the US.[61]

The cost problem is intimately related to the quality problem but they are rarely addressed together. Patent quality will remain limited if the USPTO spends only $3,713 per application and relies on college graduates with a B.S. as primary examiners. As a practical matter, this means that the determination of quality gets borne by the private sector. Some scholars have justified this — even the use of simple registration instead of examination — on the grounds that few patents are worth the expense of a rigorous examination at the front end.[62] While this view is not widely accepted, it illuminates the real tradeoffs between certainty and expense that undermine the investment-promoting rationale of the patent system.[63]

---

[60] Under Bayh–Dole, universities must retain ownership of patents generated from federally funded research but may grant exclusive licenses.

[61] To some extent, the concern over cost in the US has been mitigated by the availability of plaintiff representation based on contingency fees, although this does nothing to help small defendants. Successful small plaintiffs get only half of economic damages to which they are entitled, and lawyers are naturally reluctant to take cases when the amount in controversy is low or there is a good chance of losing.

[62] Lemley (2001).

[63] Taken to an extreme, this view advocates for an examination-free registration system, such as the utility models that China and other countries enjoy. The early US patent system was registration-only, but this proved subject to abuse, and examination was introduced in 1836.

Although quality and cost are clearly interdependent, they are framed very differently. Quality has become a perpetual mantra as new administrators take office and promise to do better, but remain dogged by specific evidence of failure. However, the difficulty of measuring quality makes progress elusive. Ideals of precision and predictability — central to the rationale of promoting investment — can only be approached with increasing cost and delay, i.e., through litigation. Increased frequency of assertion by non-practicing entities leads to increasing concern about reputation since no firm wants to appear an easy target.

At the same time, portfolio-level licensing and cross-licensing diminish the quality and cost problem for larger players. The ability to trade non-exclusive rights at scale partially solves the problem of large numbers of patents of indeterminate value through negotiations between peers by experienced professionals. Portfolios buy "freedom to operate" — a term seldom heard outside of patent practice. The implication is that large producing firms are effectively able to buy their way out of the patent system, a luxury not available to small firms. Large firms have reason to cooperate in this way: They are likely to have a shared interest in expanding the market and, as is often the case in the digital sector, have multiple business relationships with each other that they are reluctant to jeopardize.[64] And they have a common interest in protecting their market positions against newcomers.

But the differences in professional interest and perspective resist a common perspective on quality. For the lawyer, the test of quality is whether the patent is upheld by the highest court. For the engineer, the test at the patent level would be inventiveness, but in law this is a binary question of whether the invention is obvious or not.[65] This binary test is keyed to "the person having ordinary skill in the art" — a diminishing standard in a global economy where frontier innovation is led by the

---

[64]This minimized the case for policy intervention based on hold-up or royalty-stacking. It was badly disrupted in the smartphone wars, in part by the dramatic cross-industry convergence of functionality and in part by Steve Jobs' emotional response to Android (all the more remarkable in that Samsung was a major supplier of components to Apple).

[65]Economists, by contrast, talk of the "height of the inventive step" as a variable that can be tweaked along with patent breadth and duration.

top 1%, not by journeymen. Simple behavioral economics says that my patents are better than yours, but an objective engineer might want a patent system that is aligned with her way of working creatively without deferring to lawyers. Ideally, patents should serve as "beacons of light," not as landmines that require the constant vigilance of para-professionals and lawyers. The system should work as effectively to focus and contextualize digital knowledge, much as the Orange Book[66] works for pharmaceuticals.

The low threshold for patentability is yet another "applicant-friendly" subsidy, like the explicit dollar subsidy built into the front of the administrative process. Patent law is riddled with other indirect or implicit subsidies that cannot be quantified readily but are deeply embedded, such the burden placed on the examiner to show that a patent should not be granted, the *ex parte* nature of the examination process, and the burden placed on defendants to show that patent should not have been granted.

Other implicit subsidies include automatic availability of continuations (a uniquely American concession to applicants) that make it impossible for the USPTO to conclusively reject patents.[67] This makes it cumbersome to quantify the allowance rate for comparison with other patent offices, none of which permit automatic continuations. The burden of these implicit subsides is an externality that is borne by other innovators. A landscape full of questionable patents imposes a cognitive burden, legal costs, and financial risks that should be factored into investment decisions but cannot be evaluated with much objectivity or confidence. Surveys, including the detailed Berkeley/Kauffman survey of start-ups, fail to come to grips with how start-ups handle these difficult tradeoffs. Academic research naturally focuses on private benefit, which can be measured to some degree, and neglects risk and uncertainty, which cannot be measured.[68]

One solution would be to improve quality and limit the proliferation of questionable patens by raising the statutory threshold from "ordinary

---

[66] See: https://www.accessdata.fda.gov/scripts/cder/ob/

[67] Lemley and Moore (2004).

[68] As the limited literature on patent insurance shows, it also difficult to rationalize. See Rodrigo (2009).

skill in the art" to that of an "expert." This standard would be less hypothetical since expertise is easier to specify and assess than ordinariness. In short, it would raise patent examination to something close to peer review and the Daubert standard for admitting expert testimony in court.[69] Of course, any substantial raise in the inventive step requirement would deflate the cottage industry around patents and draw fire for making the patent system less accessible.

Fortunately, there is some progress on proportionality. By ending the Federal Circuit's rule of automatic injunctive relief, the Supreme Court has indirectly focused new attention on damages — especially the meaning of "reasonable royalty." Litigation around FRAND commitments puts the apportionment problem in a new context, in which the economic calculus can be made in the context of a complete global market. The calculation of damages is the one point where economics already enters the patent system, and this opening should be exploited further.[70]

Nonetheless, there are contending perspectives within industry, as reflected in the split over the new IEEE patent policy. One side reflects the linear model of innovation, with patents protecting core inventions, which are then refined and fitted to innovative products and services. The other side stresses engineering skills and combinatorial innovation — assembling the varied functionality of technology into products and services in advance of the competition.

In terms of assessing reasonable royalties, this implies choosing between the primacy of patent markets and the primacy of product markets. However, patent markets remain dark and opaque, shrouded in non-disclosure agreements (NDAs). By contrast, product markets are based on the measurable, relatively transparent reality of global markets for recognized products and services. In digital technology, the distance between patents and products is vast, yet proportionality in remedies demands some effort to bridge that distance. Focusing on the small saleable unit as in IEEE policy and recent case law is an important step toward

---

[69] Daubert was the first of a trilogy of Supreme Court cases that set the standard; see: https://en.wikipedia.org/wiki/Daubert_standard

[70] The final recommendation of the 2003 FTC report reads: "Expand consideration of economic learning and competition policy concerns in patent law decisionmaking."

bridging that distance. In part, this requires understanding the ownership structure based on the number of patents and standards within products.[71]

It remains extremely difficult to assess the private benefits and costs of patents, let alone the social benefits and costs. The AIPLA surveys afford a concrete sense of the most palpable private transaction costs — but not the costs associated with assertion prior to or independent of formal litigation. The naïve view focuses on the patent in hand from the perspective of the patent holder — not the possible counterparties to a patent, i.e., the aggregate uncertainty of widely dispersed liabilities.[72] These will be greater when the threshold of inventiveness is low, and high levels of independent inventive activity increase the likelihood of innocent infringement. At the same time, global-scale innovation and patent office examination on the cheap limit private benefits because they increase the odds that further investigation will show the patent to be invalid. This skews the system in favor of large multinationals who can most efficiently navigate patent thickets while reaping direct and indirect tax benefits.

There is clearly demand for government-subsidized entitlements that can be asserted against rivals, cash-rich companies, or soft targets such as small businesses. But the effects of low-quality patents in creating widely dispersed, contingent, largely undocumented liability are unknown, despite the economic stakes. Despite the dense interweaving of public and private functions, accountability is lacking. Despite greater attention from economists and empirically oriented scholars, survey data remains outdated (Carnegie-Mellon), semantically useless (NSF BRDIS), or selective (Berkeley study of start-ups). Studies by the USPTO's Office of Chief

---

[71] See Biddle *et al.* (2010) (finding 251 standards in a laptop). The number of disclosed patents in a standard varies greatly; see: http://www.law.northwestern.edu/research-faculty/searlecenter/innovationeconomics/documents/Baron_Pohlmann_Mapping_Standards.pdf. Conservative practice favors overdisclosure, so many declared patents are not in fact essential. See: http://www.frlicense.com/LTE%20Final%20Report.pdf.

[72] The only effort to measure net social benefits and costs remains that by Bessen and Meurer (2008), who relied on public company records and event studies (effects of patent litigation on market value). They found a net benefit prior to effects of the Federal Circuit and a net cost thereafter, although a continuing net benefit for pharmaceuticals and chemicals.

Economist, such as the report on "intellectual property intensive industries," amount to politically driven cheerleading.[73] The United Kingdom Intellectual Property Office (UKIPO), by contrast, stands out in its systematic agenda setting[74] and willingness to tackle hard problems, such as patent thickets.[75]

There are ways to alleviate these problems: raising the inventive step threshold, requiring the real-time reporting of patent assertions, patent office self-examination when patents are found invalid, a lesser presumption of validity to reflect the nature of ex parte examination, eliminating automatic continuations, etc.[76] But such reforms would reduce the strategic scope of the patent system and would threaten tactical and strategic prerogatives for patent applicants. Politically, they may be impossible.

An alternative is to first build greater accountability into the system. One step that could help diminish the problems of quality, cost, and proportionality would be to subject patents to the same kind of annual *ad valorem* taxes that applies to real and personal property. This tax would be similar to the maintenance fees that are presently paid every few years, except that instead of supporting the full transaction costs of the system, maintenance fees now just subsidize the examination process, encouraging overuse by applicants and grants instead of denials by the PTO. Instead, the tax would underwrite the costs of the judicial infrastructure used to litigate patents — just as fees must do for the new Unified Patent Court in Europe. It could subsidize costs of disputes for

---

[73] Department of Commerce (2012): "We defined patent-intensive industries as ones with above-average patent intensity (patent/job ratio) when comparing all industries." Naturally, aggregating all "above-average" industries results in impressively large numbers, but similar results could be engineered by aggregating "lawyer-intensive" industries — or even "sexual-harassment-intensive" industries.

Another example is the PTO's defensive study of software patent validity based on a handful of district court (and ITC) decisions, criticized in Kahin (2013a).

[74] See: https://www.gov.uk/government/uploads/system/uploads/attachment_data/file/564009/Research_and_evaluation_priorities_2016_2017.pdf

[75] See: https://www.gov.uk/government/publications/a-study-of-patent-thickets

[76] One interesting proposal, which would do much to make markets for bare patents more transparent, is to require disclosure of license fees. See Lemley and Myrhvold (2008). The idea has received little if any traction.

SMEs and universities, who are now disproportionately disadvantaged by the cognitive burden and high and transaction costs. Like maintenance fees, a tax would encourage the abandonment of worthless patents, thereby reducing the clutter and noise in the system. Self-reported value would create presumptions that could be applied to assess damages, transfer pricing, and compensation for public use.[77]

10. Conclusion

For much of history, duties on incoming goods were a principal, if not primary source of government revenue. Borders and location were dominant factors, just as they were for state-funded military ventures. Based on simple asymmetries in value from one place to another, traders arbitraged physical goods, opportunistically combining import and export, going out and coming back. Today, trading takes place within a complex environment of FDI, capital flows, value chains, alliances, licenses, travel, immigration, and networks — in which the movement of intangible resources is nearly invisible.

Digitization has blurred borders in many other ways — and not just at the international level. It has expanded the scope of economic activity — products, services, and business models. Software — infinitely functional, malleable, extensible, and versatile — is "eating the world".[78] It has enabled multisided platforms and algorithmically empowered markets to become proprietary ecosystems that overlap, cooperate, and compete with each other. These gigantic enterprises vie for attention, shaping how people organize their economic and social lives.

But policy oversight remains limited and constrained. Multilateralism has stalled. Policy development is channeled incrementally into trade strategies that draw in a widening range of policy domains. Unlike conventional democratic policy development, trade policy has become an exercise in geopolitical maneuvering and strategic prioritization.

---

[77] Adding another layer to an already complex system adds further to the complexity of the system — just as the unitary patent adds to the complexity of the European patent system. It may make more sense to redesign maintenance fees to accomplish similar objectives.

[78] Andreessen (2011).

The negotiations process is wrapped in a unique form of public secrecy, which in the US is open to industry through the ITAC committees (individual members are listed but without their affiliation/employer). Drafts leak from time to time, drawing attention to the undemocratic secrecy of the process. The drafts are inevitably disclaimed as works in progress, defusing public debate until a final unamendable text is released for approval, which to date has been nearly automatic. Much of the TPP reiterates the terms in bilateral treaties. Much is merely exhortative, aspirational, or expressive of intent to cooperate. Mandatory provisions are buried by the 6,000 pages, much as click-wrap licenses or consumer disclosures are obscured by length and details of limited relevance.[79] This accretive process nevertheless assumes a practically immutable constitutional stature distributed over hundreds of lesser treaties and embracing an expanded ISDS as a global supreme court promising multinational enterprise a global common law of "legitimate expectations."

The digitally expanded scope for private action may outpace conventional policy-making but it also expands opportunities for realigning governance, asset management, and knowledge-sharing between private and public domains. In fact, digitization has brought new territory and new forms of organization that mix private and public attributes, such as standards development organizations, privately owned and operated markets, subsidized knowledge networks, public utility-like Internet services, massive exploitation of public sector information, etc.

While many legal regimes involve a balancing of public and private interests, nowhere are public functions and private interests more entangled than in patent systems. Yet this interaction is shielded from oversight by the systemic complexity and the highly technical nature of subject matter and law. Patents defy the third principle of the Clinton/Gore Administration's Framework for Global Electronic Commerce: "Where governmental involvement is needed, its aim should be to support and enforce a predictable, minimalist, consistent and simple legal environment for commerce."

[79]Recently dramatized by the Norwegian Consumer Council's reading of the terms and conditions for a typical number of apps on a smartphone. See Associated Press (2016).

Ironically, it is in the digital environment where this principle is most violated — where government intervention generates a firehose of legal instruments on the cheap, using an applicant-friendly *ex parte* process that has no parallel in other public agencies. Yet as a consequence, the value of the individual patent is diluted, and advantage lies in numbers, size, and leverage — portfolios, patent funds, and resources available to operate and litigate under uncertainty; market size; ability to gather intelligence and expertise; and the geo-economic clout to manage trade negotiations for national advantage. Because of the high transaction costs and portfolio-level efficiencies, increasing returns provide disproportionate advantage to scale — for states as well as firms.[80]

The munificence of digital technology has created new layers of hundreds of thousands of legal instruments below the measured economy of traded goods, and globalization has opened up broad prospects for how these instruments can be arbitraged and asserted — by specialists, multinational enterprises, and national governments.

References

Amar, L. (2016). "The European Commission's take on the UPC's impact on SMEs," *UPC Blog*, January 19.

Andreessen, M. (2011). "Why Software Is Eating The World," *Wall Street Journal*, August 20.

Associated Press (2016). "See A 32-Hour Webcast of Norwegians Reading the Fine Print," *New York Times*, May 26.

Balsillie, J. (2014). "Intellectual Property Must Be Protected," *Waterloo Region Record*, February 3.

Bessen, J. and M. J. Meurer (2008). *Patent Failure* (Princeton University Press, Princeton, NJ, USA).

Biddle, B., A. White, and S. Woods (2010). "How Many Standards in a Laptop? (And Other Empirical Questions." Available at: https://papers.ssrn.com/sol3/papers.cfm?abstract_id=1619440.

[80]For example, the US condemns state cyberespionage on behalf of specific companies, but not in support of trade negotiations on behalf of domestic industries. See Sanger (2014). How would this policy apply to a small country with a single firm in a trading sector?

Clover, S-J. (2016). "Despite a Year of Conflict, the EPO is Still Rated as Number One for Quality and Service by IAM Readers," *IAM*, May 26.

Contreras, J. L. (2015). "IEEE Amends its Patent (FRAND) Policy," *Patently-O*, February 9.

Cotropia, C., C. Quillen, and O. Webster (2013). "Patent Applications and the Performance of the U.S. Patent and Trademark Office," *Federal Circuit Bar Journal*, 23.

Department of Commerce (2012). "Intellectual Property and the U.S. Economy: Industries in Focus." Available at: https://papers.ssrn.com/sol3/papers.cfm?abstract_id=1619440.

Federal Register (2013). "Setting and Adjusting Patent Fees, Final Rule," January 18. Available at: https://www.federalregister.gov/documents/2013/01/18/2013-00819/setting-and-adjusting-patent-fees

Federal Trade Commission (2003). Chapter 3, "Business Testimony." In *To Promote Innovation: The Proper Balance of Competition and Patent Law and Policy*, Chapter 3. Report by the Federal Trade Commission.

Fontana, R., A. Nuvolari, H. Shimizu, and A. Vezzuli (2013). "Reassessing Patent Propensity," *Research Policy*, October.

Frakes, M. and M. F. Wasserman (2013). "Does Agency Funding Affect Decisionmaking?: An Empirical Assessment of the PTO's Granting Patterns," *Vanderbilt Law Review*, 66.

Freeman, S. (2015). "NAFTA's Chapter 11 Makes Canada Most-Sued Country Under Free Trade Tribunals," *Huffington Post*, January 14.

Kahin, B. (2013a). "Software Patents: Separating Rhetoric from Facts," *Science Progress*, May 15.

Kahin, B. (2013b). "Patently Geopolitical," *IP Watch*, August 26.

Kluwer Patent Blog (2015). "Spain would have been better off inside the Unitary Patent and the Unified Patent Court," *Kluwer Patent Blog*, October 20.

Lemley, M. A. (2001). "Rational Ignorance at the Patent Office," *Northwestern University Law Review*, 95(4).

Lemley, M. A. (2012), "The Myth of the Sole Inventor," *Michigan Law Review*, 110, p. 709.

Lemley, M. A. and K. A. Moore (2004). "Ending Abuse of Patent Continuations," *Boston University Law Review*, 84, p. 63.

Lemley, M., and N. Myrhvold (2008). "How to Make a Patent Market," *Hofstra Law Review*, 36, p. 257.

Machlup, F. (1958). "An Economic Review of the Patent System," Study No. 15 of Committee on Judiciary, Subcommittee on Patents, Trademarks, and Copyrights, 85th Congress, Second Session.

Mansfield, E. (1986). "Patents and Innovation: An Empirical Study," *Management Science*, 32(2).

Moore, K. A. (2003). "Xenophobia in American Courts," *Northwestern University Law Review*, 97, p. 1497.

Mullin, J. (2016). "IBM Sues Groupon over 1990s Patents Related to Prodigy," *Ars Technica*, March 3.

Perlez, J. (2015). "China Creates a World Bank of Its Own, and the U.S. Balks," *New York Times*, December 4.

Phelps, M. and D. Kline (2009). *Burning the Ships: Intellectual Property and the Transformation of Microsoft* (Wiley).

Rich, G. S. (1960). "The Principles of Patentability," *George Washington Law Review*, 28, p. 393, 407; *Journal of Patent Office Society*, 42(2), p. 85.

Risch, M. (2011). "Patent Troll Myths," *Seton Hall Review*, 42, p. 457.

Robbins, C. A. (2009). "Measuring Payments for the Supply and Use of Intellectual Property." In *International Trade in Services and Intagibles in the Era of Globalization* , Chapter 4, M. Reisndorf and M. J. Slaughter (eds.) (University of Chicago Press, Chicago), pp. 139–171.

Rodrigo Fuentes, J. (2009). "Patent Insurance: Towards a More Affordable, Mandatory Scheme?" *Columbia Law Review*, 10, p. 267.

Roin, B. N. (2014). "The Case for Tailoring Patent Awards Based on Time-to-Market," *UCLA Law Review*, 61, p. 672.

Sanger, D. (2014). "With Spy Charges, U.S. Draws a Line That Few Others Recognize," *New York Times*, May 19.

Seidenberg, S. (2015). "TPP Strengthens Controversial IP Arbitration," *IP-Watch,* November 30.

Sornarajah, M. (2015). *Resistance and Change in the International Law on Foreign Investment.*

USTR (2014). "The Facts on Investor-State Dispute Settlement," *Tradewinds*, March.

Van Pottelsberghe, B. (2014). "Industrial and Corporate Change," 20(6), footnote 9.

Webster, E., P. H. Jensen, and A. Palangkaraya (2014), "Patent Examination Outcomes and the National Treatment Principle," *The RAND Journal of Economics*, 45, pp. 449–469. doi: 10.1111/1756-2171.12053.

Part 4

The Critical Role of Technical Standards

Chapter 10

The Impacts of Technical Standards on Global Trade and Economic Efficiency

Gregory Tassey

Economic Policy Research Center
University of Washington
E-mail: gtassey@outlook.com

1. Introduction

The globalization of the technology-based economy offers substantial opportunities in the form of unified markets that are both larger and more diversified. In economic terms, the resulting larger markets enable specialization by individual economies in specific tiers of worldwide supply chains, while still realizing product scale and scope efficiencies. This, in turn, drives higher profits and worker incomes than would be possible in a compartmentalized global economy.

Such efficiency, attained through specialization and competition, can be enhanced by trade agreements, such as the Trans-Pacific Partnership (TPP) and the Transatlantic Trade and Investment Partnership (TTIP). The broad scope of such agreements, both in terms of national economies covered and the scope of economic activity affected, can enable the establishment of competitive industries in a large number of national economies. The result will be substantial gains in overall global economic efficiency and consequent increases in the standard of living for workers.

The bottom line is that the nature of modern technologies is such that ample opportunities exist for many economies to successfully participate in global markets, with each one adopting specific targets of expertise that will pay off in terms of greater economic growth in domestic output, corporate profits, and incomes for workers.[1]

However, the realization of such advantages will only be achieved if: (1) individual economies identify and invest in industries within broader global supply chains where they can exploit comparative advantage; (2) a technical infrastructure, largely based on standards, is developed and implemented to ensure efficient product portfolio development, production, and commercialization for each economy's selected strategy; and (3) the standards infrastructure is uniform across all economies involved in global supply chains. Doing so efficiently maximizes global economic growth and welfare.

Critically important is the fact that the above characterization implies a public–private asset growth model in which industry and government co-operate to develop and use several types of technology infrastructure (Tassey, 2007, 2010, 2014). Doing so is essential to overall R&D, production, and marketing efficiency.

Further, such a model must be based on the understanding that modern technologies are complex systems, so that the resulting domestic investment strategy must provide not only technical infrastructure to support development of the targeted technologies but also the subsequent integration of new products into final product systems.

Therefore, given the high risks of developing specific technologies and commercializing them on a global scale and the high costs of establishing the technology infrastructure needed to enable a domestic industry capable of competing in global markets, it is essential that investment decisions be based on variables that reflect (1) the complex *systems nature*

---

[1] See: https://en.wikipedia.org/wiki/Transatlantic_Trade_and_Investment_Partnership and https://en.wikipedia.org/wiki/Trans-Pacific_Partnership for more detail and reactions by stakeholder groups.

of modern technologies and (2) the required public and private sector roles for providing the supporting *technical infrastructure*.

This chapter provides a conceptual framework of global technology-based industries and associated markets that rationalizes the positive economic welfare impacts of trade and then uses this framework to demonstrate the economic efficiency impacts of modern technology infrastructure, in particular standards, and thereby promotes an understanding of the increased global economic welfare achievable by the implementation of an effective global standards infrastructure.

2. The Systems nature of modern technologies and the requirements for standardization

The fact that technologies are complex systems means that they are composed of numerous hardware and software components. This opens up a substantial number of markets for the various components of such systems. Economies can choose to develop expertise in a subset of these markets and compete effectively — as long as the second requirement, an effective technical infrastructure, is in place to define functional interfaces among the components of the global technology system.

That is, if these interfaces are consistent across the global economy — i.e., effective global standards are in place — national economic strategies can be targeted at components for which domestic expertise exists or can be developed, knowing that system integration can be executed at some other location. Such a situation is particularly advantageous to smaller developing economies that only have the resources to compete in a few components of a much broader technology product system. However, the rapid expansion of global technology-based competition means that even the largest economies must adopt some degree of technology specialization.

More broadly, the economic role of interface standards as described above is only one of a number of roles for standards that are required by technologically advanced industries. In fact, standards are a ubiquitous technical infrastructure. As described in more detail below, the development of individual components within a particular domestic economy also needs global standards, such as materials characterizations, equipment calibrations, and data format specifications, that not only increase productivity but also assure domestic customers and those in other economies that a component will perform as desired within the broader product system technology. Collectively, they significantly affect the "expansion path" of a technology: direction, scope, and speed of its evolution. The US semiconductor industries have over 1,000 standards covering the various categories identified above; without them, these industries could not exist.[2]

Finally, today's complex technology-intensive products are difficult to sell for the simple reason that buyers cannot easily measure/test for performance to see if the product meets their specifications. The difficulties in "product acceptance" testing lead to high transaction costs that, in effect, add to the price of a new innovative product, thereby slowing its market penetration. Standards for testing new products greatly reduce risk by providing believable test data at relatively low cost, thereby accelerating market penetration. In summary, standards affect the efficiency of the entire cycle of technology-based economic activity. Because of increasingly intense competition, their timing as well as their quality are crucial to efficient development, commercialization, and expansion of markets for new technologies. The fact that such technically complex standards are essential for all phases of technology-based economic activity means that the standardization process requires rigorous and intensive public-sector support.

3. The economic roles of standards and their use in global high-tech industries

In this context, standards have significant implications for regional economic trade agreements. Specifically, their structure directly affects the potential economic benefits from such agreements because access to global markets for all competitors requires conformity to a single set of

---

[2]The SEMI International Standards Program lists almost 1,100 standards for semiconductor industries. See: http://www.semi.org/en/sites/semi.org/files/data15/docs/ContentsbyTopic0615.pdf.

standards. Non-conforming economies will find transaction costs associated with doing business in global markets to be prohibitively high.

The fundamental role of multinational trade agreements such as the TPP and the TTIP is to eliminate tariff and non-tariff barriers across substantially all trade categories, which includes goods, services, and investments in productive assets, as underscored throughout this book. The ultimate objective is to integrate national markets into regional ones and thereby realize greater efficiencies by facilitating specialization on the supply side and the realization of economies of scale through access to larger markets. Individual domestic economies will realize greater market opportunities and easier access to critical inputs, thereby promoting innovation, productivity, and overall competitiveness.

However, such globalization of supply chains and related markets poses a threat to domestic industries protected through previously erected trade barriers. Inefficient industries will have to be restructured and domestic workers will have to be reskilled, in order for targeted comparative advantages to be realized. Thus, the removal of tariffs and quotas initially cause shifts in location of various high-tech products and services, as effective relative prices adjust. Such "transition costs" must be managed by each.[3]

As indicated by the above discussion, high-tech industries rely on standards to perform a number of functions at all stages of economic activity. Specifically,

(1) *R&D Efficiency*: R&D requires qualified (standardized) measurement and test methods and critically evaluated science and engineering data bases that enable accurate characterization of materials, as well as data formatting and classification frameworks that

[3]Unfortunately, politicians often appeal to workers' fears of job losses and stagnant incomes resulting from globalization of a variety of industries by rejecting international trade agreements, in effect arguing for protecting non-competitive jobs (at least at current pay levels). This is the antithesis to what is proposed here and elsewhere; namely, an aggressive multi-asset investment strategy to raise productivity and hence competitiveness, which, in turn, will raise real wages (Tassey, 2016a).

allow new technologies to be transferred to subsequent steps in the innovation process.

(2) *Production Efficiency*: Production requires standardized conceptual models and performance metrics to control processes for quality and yield, which are implemented through such functions as equipment calibration and real-time machine performance monitoring and adjustment.

(3) *Variety Reduction*: Where large economies of scale are important, product variety reduction may be necessary to reduce unit costs and thereby achieve acceptable price points.

(4) *Commercialization*: As buyers' assurance of adequate component and system performance before consummating purchases is essential, successful innovation and subsequent market penetration require cost-effective product acceptance standards that reduce transaction costs that can become excessively high when trying to buy or sell complex products.

(5) *System Integration*: The final and perhaps most critical step in producing effectively functioning product systems of hardware and software components requires standardized interfaces to allow competition at the component level and hence achieve maximum innovation, which, in turn, increases the productivity of the final technology system.

As a consequence of this diverse set of economic functions and the rising importance of technology-based economic growth for increasing productivity, growing attention and investment are being directed at the role of technical infrastructure, including standards. However, standards have considerable public-good content in that a standard by definition is commonly used by both competitors and their customers. Thus, although standards are ubiquitous in technology-based industries, their public-good content can greatly constrain their *quality* and the *timing* of availability.

When standards first began to be used more than a century ago, they were almost exclusively product standards (either the entire product or a component). The objective was almost always to reduce variety in order to achieve economies of scale. This focus was compatible with the mass production character of the Second Industrial Revolution.

As product technologies have become more complex (i.e., intrinsically intricate systems of hardware and software), more and more standards have become necessary to specify functional interfaces among components and also processes and procedures across all three stages of economic activity (R&D, production, and commercialization/market transactions). It is important to note that the vast majority of these new functions of standards are non-product in character. Functions (1), (2), (4), and (5) from the above are non-product standards. Function (5) is also significant in that it has a pronounced effect on geographical location because it enables specialization.

3.1. *Product standards*

Most standards apply to an "element" of a product. Each element is characterized by several attributes. The set of these elements collectively characterizes the product. When product innovation occurs, most or even all product attributes are fluid. Over time, competitors enter the market with competing designs. Eventually however, users begin to resist excessive diversity because substitution (switching) costs can be high and economies of scale — and hence lower prices — are not realized. Eventually, therefore, one or more major attributes of the generic product technology achieves dominant market status and becomes a *de facto* standard (Function 3) and drives standardization of other product elements.[4]

Such market dynamics determines the timing of product element standards, and the sources of scale and scope economies determine the dominant product element. The evolutionary process by which a product element becomes dominant is important because standardization of the

[4]An example is the architecture of a microprocessor, which refers to the layout of components within a computer's central processing unit (CPU). A microprocessor's architecture drives the structure and function of the other components of a computer and all software that runs on it. This situation means consumers will incur significant "switching" (re-learning) costs, if the decision is made to adopt an alternative computer technology platform. The growing number of users attracts even more users who can interact with (and learn from) existing users and a growing infrastructure. This process leads to the so-called "lock-in" phenomenon.

design of the entire product is increasingly unlikely in an era of complex system technologies where complexity facilitates competition among suppliers and better serves the increasingly heterogeneous demand of users. In fact, complete standardization would most likely mean "lock-in" to a dominant supplier, which would be a sub-optimal result (David and Greenstein, 1990).[5]

As a precursor to the above result, alternative technologies compete until one version of the dominant product element gains sufficient market share to become the *de facto* standard and all suppliers convert to it. This standardization of a critical component, say the architecture of a microprocessor, drives the structure of the entire computer.

If the innovator is a large firm, its market control can subvert the competitive process by which a product element standard is achieved. Such control can accelerate increasing returns by forcing acceptance of the monopolist's proprietary technology element as the standard.

However, the globalization of high-tech markets with more competitors is making single-firm dominance more difficult. In response, various combinations of vertical and horizontal consortia are promulgating product element standards by consensus, at least within a single large economy or a regional trading block, rather than wait for market dynamics to finally pick the dominant design.

3.2. *Non-product standards and system productivity*

Functions (1), (2), (4), and (5) in Table 1 all leverage economic efficiency at one or more of the major stages of technology-based economic activity. Because they are product-technology neutral, their existence leverages the efficiency of all participants in global markets. Without such standards, the evolution of a technology will be significantly retarded as costs at each of the three stages of technology-based economic activity will be high. Figure 1 depicts the multiple types of non-product standards and the points in the technology development and commercialization process

[5]Occasionally, a single product element standard is imposed on an industry to ensure economies of scale. An example is the specification of the number of lines of resolution for HDTV broadcasts (see: http://en.wikipedia.org/wiki/High-definition_television).

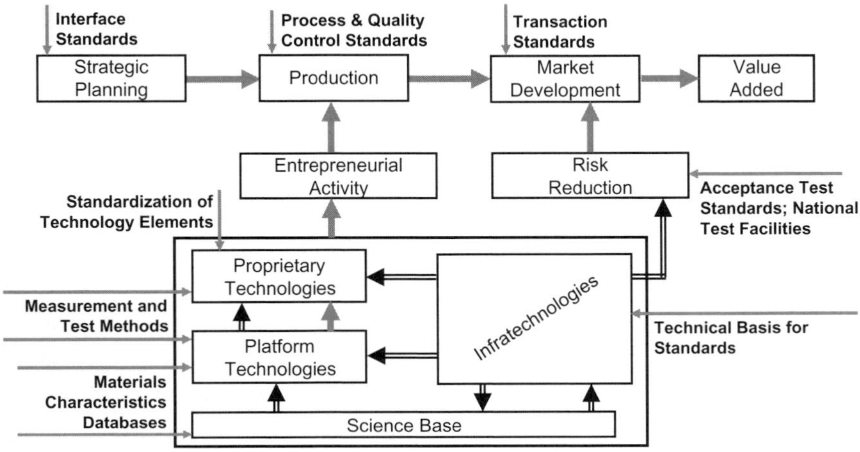

Figure 1. Roles of non-product standards in technology-based industries.

Source: Tassey (2015).

where each type of standard delivers its infrastructure role. The thin solid (blue) arrows indicate the points of impact, while the thick solid (red) arrows show directions of proprietary technical knowledge flows. The double-stem thick arrows indicate the direction of impact of public technical knowledge (often in the form of standards).

4. The interaction of global market structure and standardization

The relentless global expansion of the capability to create, manufacture, and integrate technology into advanced products and services has led to complex "global value chains" that confound their mapping and understanding the opportunities and challenges for individual economies wishing to take advantage of high value-added market opportunities (Dollar *et al.*, 2017).

The complex systems nature of advanced technologies means that a number of economies are typically involved in designing, producing, and assembling hardware and software components that are eventually assembled into final product/service systems by yet additional economies. In this arena, determining national- and industry-level strategies is influenced to a significant degree by the availability of technology infrastructure, including standards.

4.1. *Local versus global product standards for single markets*

As indicated in the previous section, changes in the nature of a technology over its life cycle result in iterative changes in the broader set of product element standards. Such evolutionary change also affects market structure. As a product technology's set of elements evolve, global product standards are hard to achieve because of the changing technological content. As indicated, this can lead to several versions of the product technology co-existing. As a result, sub-markets often evolve around existing versions of a product standard (referred to as "local" standards). Cellular phone, automobile safety, and HDTV standards are examples. Such alternative versions can exist for some time and historically have evolved first within large economies, which are frequently the innovators of new technologies and have large diversified internal markets. Today, however, increasingly intense global competition is shifting the meaning of "local" to economic regions or trading blocks (Tassey, 2015).

On the one hand, the competition among such sub-markets more efficiently sort out the most productive technology alternatives, which eventually maximize economic welfare. On the other hand, economies of scale and scope are not realized to the maximum and customers are still faced with the risk of choosing the wrong local standard (i.e., the risk that it does not become the eventual global standard).

Specifically, the "risk" is the potential to incur significant costs from switching between local standards relative to perceived benefits accruing to users. These switching costs are perpetuated when suppliers with monopolistic control of portions of the market for the generic product (such as a computer) fail to agree on a single standard for a dominant product element (such as a processor architecture and associated operating system).

The argument for global standards is that even if an innovating economy maximizes profits in the short run, such profits will be reduced by the efficiency constraints of local standards. Thus, long-run global profits will be diminished as aggregate growth is reduced by (1) lack of price competition and eventual switching costs or (2) lower productivity from an inability to optimize system design by integrating components from a larger number of vendors in other economies whose market participation would be made possible by standardized interfaces.

With respect to evolution of global market structure, if several local standards exist across the global economy, one can become the global standard through aggressive marketing by the firms in one of the local markets or initial buy-in by one or more large users. The resulting increasing returns can result in a monopoly position for the innovator.

However, while some economies of scale are realized quickly, there is a good chance that this version of the technology is not the optimal one, as the result was not due to broad global market dynamics. Further, if the product happens to be an especially critical component, the monopoly can be extended into related component markets, that is, the component innovator can evolve into a "turnkey" product system supplier.

Without the above types of barriers to a single, competitive global market, the general industry supply chain framework for these interactions is depicted in Figure 2. Platform technologies provide the conceptual basis for specific product innovations, which include product attributes

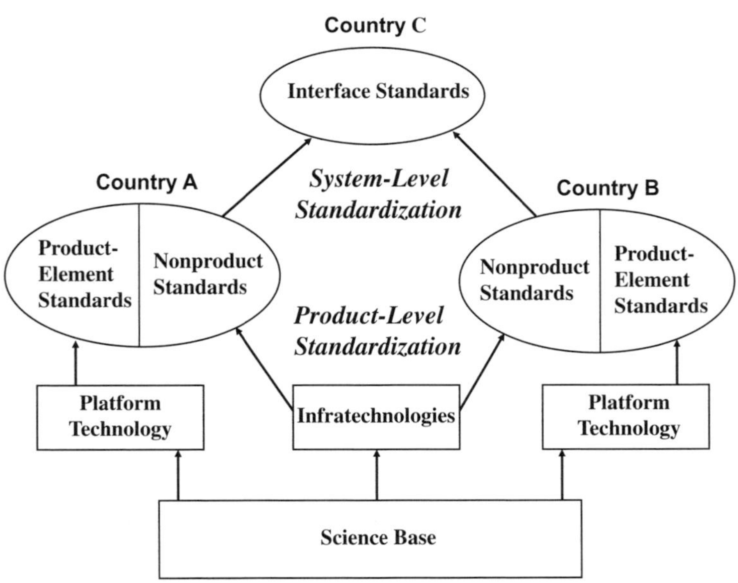

Figure 2.

that eventually become standardized (e.g., HDTV broadcast system specifications such as frame size, scanning system, and frame rate).

As defined above, infratechnologies provide (1) the technical basis for a range of measurement and testing methods, equipment calibration procedures, evaluated science and engineering data, etc., at the component industry level (Countries A & B), which, as standards, enable R&D and production efficiency; (2) interface standards at the product system level to enable efficient system design and integration by Country C, using components from A & B; and (3) efficient market transactions between the system integrator, Country C, and its domestic customers plus those in countries A and B.

One issue raised by this framework is whether system-level standards are simply the result of the aggregation of specific product element standards or they have a structure based on the intrinsic nature of the system technology. This is an important question for international standards organizations that work toward global markets unified through free trade and a supporting unified standards infrastructure.

The reality for international standard setting is that a technology system is determined to a significant extent by the structure of its components, but system design also drives both component structure and the interfaces among components (Jervis, 1997; Meadows, 2008). Thus, an iterative process for setting component and system standards is implied.

More specifically, because several levels of standardization exist within modern technology systems, standards development cannot take place totally from the bottom up. For example, two machines can be made to communicate through interface standards. However, as more machines are added to, say, automate a factory (the "system"), chaos ensues without some overall "system architecture". An architecture provides an integrating force for all product-level standards in the system, driving the structure of the interfaces among the several layers constituting the control hierarchy of the automated factory (Tassey, 2015).

The standard-setting dynamics might be described as recursive, in that technological advances in components affect system design and vice versa. Thus, economies where system design/integration is located will have considerable influence on both product and non-product standards.

Complicating matters is the fact that large firms with market strategies focusing on turnkey systems have an incentive to resist open systems to protect their market shares resulting from their horizontally integrated strategies. The increasing returns that make such monopoly positions possible can be attained by one supplier or prolonged market segmentation by several competing suppliers. Either situation has the potential to constrain economic efficiency (David and Greenstein, 1990).

Yet, some economies will condone or even encourage the belief that local (national) standards will retain more value-added within the domestic economy by reducing foreign competition, and domestic firms might possibly be able to sell to the rest of the world products and associated services based on the local standard. An example is the early development of cellular telephones where US firms were divided between two standards, CDMA and TDMA, which encouraged European firms to push their own standard, GSM. The result, of course, was incompatible digital formats (Lowell, 1998).

Thus, impacts on domestic industry structure from global standardization can be important and have significant effects on the achievement of individual national economic growth objectives. "Open systems" on a global scale allow small and medium companies in all economies to participate in markets for system technologies by supplying components in which they have a competitive advantage. This diversification on the supply side of the market allows system optimization by users (increases productivity) and greater price competition (lowers costs for buyers).[6] Trade pacts facilitate this approach to the detriment of non-participants.

There are also examples of open and closed systems co-existing. The most prominent example is the smartphone. Apple was the innovator and used its early monopoly position to implement a closed system. That is, it

---

[6]In the case of information infrastructures, attaining truly open systems has been extremely difficult. One of the major reasons for this difficulty is the complexity of IT systems technology. Each one of a number of layers in the hierarchical structure of a typical information system must have its own architecture and these architectures must be integrated into a larger network. Today, systems technologies not only have complex hardware and software platforms but also require "portability" rules and protocols to manage information flows.

is the sole provider of phones based on its operating system (OS X) and uses this monopoly to control all applications that run on it.

In contrast, Google's Android is a "quasi-open" operating system that smartphone developers, such as Samsung and LG, can cheaply license to produce their own lines of phones. However, Android companies' phones are not interoperable, that is, applications from one company's phones will not run on another company's phones. Still, this quasi-open system strategy has resulted in a collective global market share that has become much larger than Apple's (Tassey, 2015).

Thus, the interactions over a technology's life cycle between product technology elements and types of standards (Figure 1) are complex. Certainly, the potential benefits of standardization are significant. However, inappropriate use of standards can and does result in negative economic effects, such as restricting global competition by promulgating standards that favor domestically developed technologies.

Further, standards in high-tech industries are often based on intellectual property rights (IPR) of one or several companies, who can then access the IPR in an attempt to maintain monopoly positions and extract monopoly rents for its use in the standard. Finally, as technologies are increasingly complex systems (Figure 2), system design will drive standardization simultaneously, with standardization at the product/component level.

5. Degree and timing of standardization

In complicated system technologies, such as distributed data processing, telecommunications, or factory automation, standardization evolves in reaction to the evolution of both embodied and disembodied technologies. The particular pattern is determined by several factors: the pace of technological change embodied in each component category; disembodied technology development, which determines the overall system architecture and organization; and changes in market structure, which affect the incentives and ability to force the standardization process. The economic impact of the complex set of factors affecting timing of standardization is magnified in global markets by the different rates of technological advance across economies involved in a particular industry and the consequent need for adjustment to existing standards.

As an example of timing effects, the evolution of numerically controlled machine tools would have suffered from early total standardization of data formats because rushing the standardization process would have severely compromised the range of performance attributes potentially available to users over successive generations of machine tool technology. This negative impact would have occurred because standardized data formats are used to compare and assess different vendors' products, so the format for comparison is developed based on current technology platforms. Investment in new platforms that might not be able to be evaluated by existing data formats would therefore be inhibited.[7] Thus, only a degree of standardization was initially optimal in the technology's evolution. The policy message is that complete standardization too early in a technology's life cycle can constrain innovation.

In contrast, delaying the decision of standardization of a critical product technology element can retard private investment and thereby impede realization of consumer and producer surplus gains. This situation creates an incentive to choose a standard early in the technology's life cycle when the impacts identified earlier can be realized.

As technology-based systems become increasingly important and "windows of opportunity" for making successful investments in the associated markets continue to shrink, adjusting standards in accordance with the evolution of the technology is increasingly critical.

A rigid standard, even if it is competitively neutral, will stifle the introduction of new technologies into the system. In contrast, if a standard is updated frequently, then version consistency (forward mobility of existing standardization among current system components) can be a problem.

In data processing and communications networks, for example, efficiency is possible only if standard interfaces are provided on all the communication paths in the network. Such interfaces need to be defined between application programs, data formats, network protocols, printer control codes, human/machine interfaces, and so on. But, these system elements are all evolving at different rates and therefore need updated interfaces at different points in time.

---

[7] The more performance-oriented the standard, the less will be the risk of inhibiting innovation. See: http://news.thomasnet.com/IMT/2001/05/18/the_move_toward/.

The systems nature of the technology-based economy increases the importance of the need for all components of a system — both public and private — to be assessed within the context of their intrinsic performance and how such performance affects the overall efficiency of the system. This is a critical point because it is the system that delivers the ultimate economic benefit. Moreover, a system's elements are all evolving at different rates and thus need updated interfaces at different points in time.

Along another dimension, just as modern technologies have a hierarchical structure, so must the associated set of standards (Figure 2). The evolution of component technologies drives the need for standardization at each level in the product system. Standardization at one level then creates demand for standards at adjacent levels in the system. For example, a numerically controlled machine tool is a system of components such as the controller and numerous sensors. The multiple interface standards involved in a machine tool allow modularization of the product, which permits custom design but prevents overall tool obsolescence. However, these standards must be linked by a common data format, if all components are to function together as an efficient system (Link and Tassey, 1987).

In technology systems, the establishment of a new standard, no matter how well conceived, can be costly to comply with. For example, a new interface standard for electronic product data exchange can require conversion of existing databases, which can be time-consuming and expensive. Also, introducing such a data format standard may solve one problem (data exchange) but create another (security).

Initial standards associated with innovation can be hard to update due to time and cost requirements, coupled with installed-base effects. This can be particularly true for standards such as those providing interoperability as markets grow and the number of participants increases.

In summary, standards can facilitate innovation and subsequent market penetration. However, the content, timing, and flexibility of standards have to be right or new technologies will have difficulty penetrating markets or even reaching the marketplace (Branscomb and Kahin, 1995). An important element of this management imperative is that standards remain relatively fixed across individual *product* life cycles based on a particular technology platform. However, they typically change between generations of a technology and must frequently be completely restructured when radically new technologies emerge.

In terms of overall dynamic efficiency, once a product element standard is set, network externalities are typically realized. However, the marketplace dynamics that result in one firm's version of the technology becoming the standard do not guarantee that this version is the optimal one.

As this lock-in effect ensues, many developers of related products conform to the standard and purchasers of the standardized technology invest substantial resources in learning to absorb and use it as well as complementary technologies and infrastructures. These sunk costs create a reluctance to switch to a new standard and related cluster of technologies, thereby perpetuating sub-optimal growth.

Considerable resources can therefore be allocated to sub-optimal technology, which can extend over long periods of time. An example is DOS — a clearly inferior operating system to alternatives such as Apple Computer's operating system (Mac OS). Nevertheless, it dominated the personal computing market because the Microsoft/IBM combination was able to lock in the PC market.

Nevertheless, although the firm that controls a core product element standard can capture increasing returns to scale as initial market penetration begets dominance, this may be an acceptable price for rapid market growth from an economic welfare perspective, as long as the standardized version of the technology element is at least a good one and sufficient opportunity exists for market forces to eventually replace it as the standard.[8]

Another "second best" solution can occur when established market positions promote evolutionary as opposed to revolutionary migration of customers, who demand backward mobility to existing technology.[9] Mobility is more difficult and expensive, the more radical the new technology. If Microsoft, Apple, and IBM were freed from maintaining

---

[8]Arthur (1996) provides an excellent analysis of the increasing returns-to-scale phenomenon that characterizes many technology-based industries and the consequent motivation for firms to set and control product-element standards.

[9]The "installed-base" and related "installed wisdom" effects can significantly delay adaptation to changing global economic conditions. See Tassey (2007, 2010).

compatibility with existing generations of hardware and software, new generations of systems software would likely evolve at a faster pace.

Such dynamics have obvious implications for global markets. If the domestic industry is the innovator and creates a lock-in effect, other economies may have constrained competitive opportunities over the entire technology life cycle. However, the transition from one technology platform to the next will always present opportunities for competitors to gain market share by being at least contributors to the new technology platform. They thereby benefit from early-cycle increasing returns. In fact, the relatively large public-good content in the early phases of a technology's development at least offers the opportunity for international co-operation and hence a more level playing field as the new technology platform matures.

In this regard, international trade agreements, by opening (i.e., integrating) domestic markets, can facilitate co-operation in a technology's early development phases where the public-good content is greatest. And, of course, the subsequent unified market will drive common standardization and do so in a timelier manner, thereby increasing collective economic welfare, as the new technology pushes out the old one with its lock-in or other inefficient structures.

However, just as several local standards can exist for some time in a single national market, the above evolutionary processes will often proceed at different rates across national markets, leading to inefficiencies in technology product system creation and subsequent life cycle management. International trade agreements, such as the TPP, can promote reconciliation across national economies with respect to timing and content of evolutionary standards management, thereby realizing the scale and scope economies from large global markets.

6. Policy implications

The US accounts for less than 5% of the world's population, which means that 95% of the world's consumers live outside the country. Thus, exports are becoming increasingly vital to support future economic growth. As other economies gain technological expertise, the availability of lower-cost quality inputs to support domestic industries will further increase domestic export growth potential. These supplier industries in other economies will also experience higher growth rates. Thus, if the emerging global economic infrastructure, including standards, is adequately configured, economic welfare can be increased for all participants.

However, the complexity of modern technologies has created the demand for an equally complex supporting technical infrastructure, implemented largely through a variety of standards affecting all stages of technology-based economic activity. Not only must these standards be structured to promote access to the relevant markets for all competitors but also a management system must be in place to facilitate standards development and to assure proper timing of their implementation.

In this way, uniform multinational standards will lead to considerably greater global economic growth. Global standardization of the various functions of standards described in this chapter will greatly increase economic efficiency, not only by opening new markets to domestic companies and the consequent realization of economies of scale and scope but also by increasing the efficiency of domestic resource allocation through the promotion of specialization (comparative advantage) in specific technologies and related markets. At the same time, the availability of efficient system integration across economies will inhibit inefficient turnkey system strategies and thereby yield more effective product and service systems through greater competition.

However, it is also a fact that transitions from local/national to global standards will require new strategic foci by domestic companies and consequently new investment and commercialization strategies. Thus, given the relentlessly growing complexity of modern system technologies, a complete and uniform set of standards for all stages of technology-based economic activity along with implementation assistance are essential to enable this specialization, thereby including more companies from more economies in the global marketplace. Increasing the breadth of participation will enhance innovation and overall economic efficiency, leading to greater economic benefits for the global economy.

Nevertheless, some economies can and will benefit to a greater or lesser extent than others. The alternative policy responses are to either resist making the necessary investments to increase productivity and thereby take advantage of larger and more diversified markets, or to embrace the opportunities presented and implement both trade adjustment assistance programs and stimulate both public and private investment to acquire newly available markets shares.

The policy implication, as stated succinctly in a Brookings Brief review (Meltzer and Bahar, 2017), is to "focus on better assisting people who lose their jobs — whether due to trade, technology, or productivity gains — to find new work. This could include improved education and training, wage insurance to encourage workers to renter the workforce early, and allowances to encourage people to move to where the jobs are".

Unfortunately, the US has limited trade adjustment assistance programs available and it does little long-term planning for the required strategic investments necessary to increase productivity and, thereby, competitiveness.

References

Arthur, B. (1996). "Increasing Returns and the New World of Business." *Harvard Business Review*, **74** (July–August): 100–109.

Branscomb, L. and B. Kahin (1995). "Standard Processes and Objectives for the National Information Infrastructure." In *Standards Policy for Information Infrastructure*, B. Kahin and J. Abbate (eds.) (MIT Press, Cambridge, MA).

David, P. A. and S. Greenstein (1990). "The Economics of Compatibility Standards: An Introduction to Recent Research." *Economics of Innovation and New Technology*, **1**: 3-41.

Dollar, D. *et al.* (2017). "Measuring and Analyzing the Impact of GVCs on Economic Development." International Bank for Reconstruction and Development/The World Bank. Washington, DC. Available at: https://www.brookings.edu/research/measuring-and-analyzing-the-impact-of-gvcs-on-economic-development/.

Jervis, R. (1997). *System Effects: Complexity in Political and Social Life* (Princeton University Press, Princeton, NJ).

Link, A. N. and G. Tassey (1987). *Strategies for Technology-Based Competition* (Lexington Books, Lexington, MA).

Lowell, S. (1998). "Global Standards for the Global Marketplace." SAE Technical Paper 981494. doi:10.4271/981494.

Manyika, J. *et al.* (2015). "Can Productivity Growth Save the Day in an Aging World?" McKinsey Global Institute, January.

Meadows, D. (2008). *Thinking in Systems: A Primer* (Chelsea Green Publishing, White River Junction, VT), pp. 145–165.

Meltzer, J. and D. Bahar (2017). "NAFTA under Trump — The Myths and the Possibilities." *Brookings Up Front*, February 23.

Tassey, G. (2007). *The Technology Imperative* (Edward Elgar, Northampton, MA).

Tassey, G. (2010). "Rationales and Mechanisms for Revitalizing U.S. Manufacturing R&D Strategies." *Journal of Technology Transfer*, **35**: 283–333.

Tassey, G. (2014). "Competing in Advanced Manufacturing: The Need for Improved Growth Models and Policies." *Journal of Economic Perspectives*, **28**(1): 27–48.

Tassey, G. (2015). "The Economic Nature of Knowledge Embodied in Standards for High-Tech Industries." In *The Routledge Handbook of the Economics of Knowledge*, C. Antonelli and A. N. Link (eds.) (Routledge, London/New York).

Tassey, G. (2016). *Needed: A New U.S. Economic Growth Strategy* (Information Technology and Innovation Foundation, Washington, D.C.). Available at: https://itif.org/publications/2016/02/01/why-us-needs-new-tech-driven-growth-strategy.

Chapter 11

Standard-Essential Patents and the Distribution of Gains from Trade for Innovation*

Florian Ramel[†], Maximilian von Laer[†] and Knut Blind[*,†,§]

[†]*Technische Universität Berlin*
Sekr. MAR 2-5, Marchstr. 23, 10587 Berlin, Germany
[‡]*Fraunhofer Institute of Open Communication Systems*
Kaiserin-Augusta-Allee 31, 10589 Berlin, Germany
[§]*Rotterdam School of Management*
Erasmus University
Burgemeester Oudlaan 50, 3062 Rotterdam, Netherlands

Abstract

An earlier study finds that standard-essential patents (SEPs) enhance trade and additionally lead to a phenomenon of global cross-licensing; that is, global value chains (GVCs) arise along those countries in which companies will act as both the licensor and licensee of SEPs. This, however, simultaneously leads to the situation that SEP-induced GVCs are

---

*This chapter was originally prepared as a Think Piece for the *Mega-Regionalism — New Challenges for Trade and Innovation* workshop held at the East-West Center, Honolulu, Hawaii, on January 20–21, 2016.

a relatively closed-off club of companies that profit from the artificial temporary monopoly awarded by patents and their connection to standards. This chapter proposes several paths for future research with the goal to uncover the micro-mechanisms of these findings. The principal questions asked are:

(1) How can intermediate producers (i.e., global network suppliers, for instance, in Korea and China) profit from GVCs in which many SEPs are used?
(2) How is upgrading (i.e., moving up in the GVC toward performing more elaborated tasks) in an SEP-induced GVC possible?
(3) How can a company reach a position that allows it to exert control over a GVC?

1. Introduction

Formal standards provide specifications of technologies. They are developed by industry representatives and are then openly available for use by anyone. In many ways, they fulfill the definition of a public good[1]: The use of a standard is non-rivalrous (i.e., it can be used by an unlimited number of actors simultaneously) and non-excludable (i.e., no one can be barred from using it). Standards therefore serve to diffuse technologies in the markets (Blind, 2004).

Patents also provide specifications of technologies. Yet they work very differently from standards. Patents grant the patent-holder the exclusive right to use their protected technology during the duration of the patent grant. Because patents grant this excludability and are simultaneously rivalrous in their nature (there can only be one patent of a kind, never two identical ones), they can be characterized as private goods. Patents thus give the patent-holder a competitive advantage over their competitors. Nevertheless, patents facilitate the diffusion of technologies

[1]The reader is referred to the debate about the public good character of standards. Standards combine aspects of public as well as private goods. To name only a few, Berg (1989) highlights the external effects that the adoption of a standard by a firm can have on the market and the acceptance of a standard, while Garcia and St. Clair (1992) call standards "impure public goods" (p. 9) as they can be produced privately and serve a private purpose but still improve public welfare.

because patent-holders can license the right to use the protected technology to other market actors (Blind *et al.*, 2006, 2009).

In recent years, a new factor has come into play that combines standards and patents: standard-essential patents (SEPs; Lerner and Tirole, 2015). When a standard implements a patented technology, the standard-setting organization (SSO) needs to ensure that everyone who is going to use the standard will receive a license from the patent-holder in order to be able to do so. The patent then becomes standard-essential and the respective licenses usually adhere to so-called FRAND (Fair, Reasonable, and Non-Discriminatory) terms. SEPs have mainly emerged in the Information and Communication Technology (ICT) sector because standards in this area often implement hundreds or thousands of patented technologies. Examples are the standards used in smartphones, or in technologies such as UMTS or MPEG. Due to the ongoing digitization and interconnectedness, other sectors like automotive or household appliances are also starting to be affected by SEPs (Bekkers *et al.*, 2014). SEPs play a major role in many court cases and "patent wars" between large ICT firms. They have therefore not only aroused huge research interest but also have significant economic impact (e.g., Bekkers *et al.*, 2011, 2014; Blind *et al.*, 2011; Pohlmann *et al.*, 2015).

Another development in the global economies concerns the way production takes place. One product or service is not necessarily produced in a single factory or within one country anymore. There is a clear shift toward production in global value chains (GVCs), that is, a product or service is completed in several production stages in multiple countries (Baldwin, 2013). In the course of production, intermediates are shipped from one country to another. The geographical length of GVCs, the number of stages that are used to produce a single item, and the number of countries involved in a single GVC have all increased over the last two decades (Miroudot and Nordström, 2015).

With a surge in SEP disclosures to SSOs, the first usable time series of SEP data are now becoming available. At the same time, new ways to measure global trade allow to map the actual flow of goods, intermediates, and services in GVCs. This availability of new data allows us to test the influence of SEPs on GVCs for the first time to find out how SEPs affect the layout of global production patterns.

In Section 2, we discuss our empirical results in detail. We find that SEPs enhance trade and additionally lead to a phenomenon of global cross-licensing, that is, GVCs arise along those countries in which companies will act as both licensor and licensee of SEPs. This, however, simultaneously leads to the situation that SEP-induced GVCs are a relatively closed-off club of companies that profit from the artificial temporary monopoly awarded by patents and their connection to standards. In Section 3, we propose further avenues for research, especially by focusing on the microeconomic impacts of SEPs. Section 4 concludes with recommendations for the next steps to be taken.

2. The macroeconomics of SEPs — First insights

Our dataset provides us with a list of the patents that were disclosed to SSOs as being standard-essential from 1990 to 2010.[2] From the data, we construct an SEP stock variable for the years 1995, 2000, 2005, and 2008. We are restricted to those years because of the limited coverage of the matching trade data. The SEP stock is the stock of patents of a country that were disclosed to an SSO as being standard-essential until the end of the given year.[3]

From the descriptive statistics of the SEP stock variable, we see that it has risen sharply during the observed time frame from less than 500 in 1995 to almost 20,000 in 2008, which underlines the notion of a surge in SEP disclosures. In addition, there are enormous differences in the SEP stocks between the different countries: there are some countries that hold no or only a few SEPs whereas other countries hold many. The countries whose SEP stocks are above the third quartile of all SEPs in the sample in 2008 are China, Germany, Japan, South Korea, and the US (in alphabetical order).

In order to assess the GVCs, we use trade in value-added data. These data allow us to see how many inputs (raw materials, intermedi-

[2]The data are kindly provided by IPlytics GmbH.

[3]There are 20 countries in the sample: Austria, Belgium, Brazil, China, Denmark, Finland, France, Germany, Italy, Japan, South Korea, Netherlands, Poland, Russian Federation, South Africa, Spain, Sweden, Switzerland, the UK, and the US.

ates) a country imported, how much value its domestic producers added to these inputs, and how much of the value-added was eventually exported to the next country in the GVC. We restrict the data sample to the ICT sector because SEPs are only really prevalent in this industry (Bekkers *et al.*, 2014).[4]

2.1. *Results from an earlier empirical study*

In order to identify the effects of SEPs on GVCs, we use a gravity model of trade (Anderson, 1979; Anderson and van Wincoop, 2003; Deardorff, 1998; Tinbergen, 1962). This model allows us to estimate the effects of SEPs on trade in value-added while we control for a number of other effects that might affect the trade flows: geographical distance, cultural similarities between the countries, idiosyncratic effects of certain years or countries, or the effects of patents and standards (see Ramel and Blind (2015) for the full model).

We present the stylized trade effects from the estimation for the most relevant variables in the following paragraphs. First, for reference, we report the effects of patents with no reference in standards on trade flows. They enhance exports because of the competitive advantage they grant and decrease imports because patent-holders produce a superior technology so that no imports of other, inferior technologies are necessary. This result is in line with former research (e.g., Anderton, 1999; Blind and Jungmittag, 2005; Greenhalgh *et al.*, 1994).

Although SEPs are patents as well, the estimation results differ in the case of the SEP variable. They show an overall trade enhancing effect, yet in contrast to "regular" patents, the effect is also positive on the importer side. The positive co-efficients for both exporting and importing countries are an expression of cross-licensing between domestic and foreign companies. Cross-licensing originally refers to the "preferred means by which large companies clear blocking patent positions amongst them-

[4]Our trade data covers the ISIC Rev. 3 classes 30–33: Manufacture of office, accounting, and computing machinery; electrical machinery and apparatus n.e.c.; radio, television, and communication equipment and apparatus; medical, precision, and optical instruments; and watches and clocks.

selves" (Shapiro, 2001, p. 129). In the ICT sector, companies are technologically interdependent. Patent thickets, that is, technologies in which many different patents overlap, are very common (Bekkers *et al.*, 2014; Graevenitz *et al.*, 2011, 2013). In addition, ICT innovations are often cumulative, that is, they build upon each other (Bessen and Maskin, 2009). Former research shows that, under these circumstances, cross-licensing is common (Nagaoka and Kwon, 2006; OECD, 2013; WIPO, 2011) and that it is especially typical in the realm of SEP-referencing standards (Bekkers and Martinelli, 2012).

The cross-licensing thus allows the producers to use all technologies they need, including those that require licenses from competitors. SEPs thus play an important role in supporting GVCs, in which intermediates are exchanged equally back and forth in order to combine them into a standardized product. Hence they have a positive effect on trade in both directions.

The close intermingling of producers in the GVCs may, however, also have a downside. SEP-induced GVCs are a relatively closed-off club of companies. Once a standard that implements SEPs is published, this technology is locked in and all SEP holders in the standard enjoy a temporary monopoly. SEPs would then act anti-competitively (Lerner and Tirole, 2015) and become an anti-trust issue like patent pools that are suspected to provide a platform for collusion (Farrell *et al.*, 2007; Lerner and Tirole, 2004). The pools are suspected to create monopolies that complicate the integration of outsiders into production flows that require licenses from SEP holders. Moreover, through the standard, patent pools can be used to lock in a certain market for the involved companies ensuring safe and potentially long-term profits (Choi *et al.*, 2004; Shapiro, 2001). This kind of licensing strengthens the relationships between the involved firms and increases the market power of the incumbents (Choi *et al.*, 2004; Grindley and Teece, 1997).

We try to grasp this issue with the use of a third variable, a multiplicative interaction term between the SEPs and the stock of national standards in a country. National standards are known to constitute barriers to trade because they cause high adaptation costs for foreigners (Blind, 2004). If a country now has a large stock of national standards and at the same time a large stock of SEPs, we assume the protectionist momentum to be so strong that the negative effects for outsiders predominates the positive effects for the incumbents. Testing this empirically, we indeed find the trade effects to be negative, that is, the larger the combined stock of national standards and SEPs in a country, the more trade decreases. This seems to provide first empirical evidence for our assumption.

The patterns that we present lead to further implications and questions that need to take into regard the industry structure of the GVCs.

2.2. *SEP disclosures and the GVC position of countries*

The data we use contain 378,003 patents that were disclosed as standard-essential. It is noteworthy that, as shown in Table 1, the two companies that engage in disclosing SEPs most heavily account for almost half of all the disclosures of 1,155 companies in total. The six most actively disclosing companies account for almost three quarters of all disclosures.

We see that there is indeed a very small group of firms that control the majority of SEPs. This industry structure certainly supports the notion of a club when describing the market actors involved with SEPs. On the one hand, it is easier to reach an agreement about cross-licensing when fewer players are involved in the negotiations, which in turn facilitates the creation of GVCs that rely on SEPs. On the other hand, with so few players involved, these can easily construct high barriers to market entry for outsiders. This leads back to the notion of pools exhibiting cartel-like structures.

An immediate eye-catcher when regarding Table 1 is that the two leading SEP disclosers are non-manufacturing companies. InterDigital and Qualcomm pursue business models that are based on developing, licensing, and trading IPR. Especially with regard to GVCs, the question arises as to whether there is a separation between the IPR-developing and IPR-trading companies, and the manufacturers.

Even though the majority of SEPs are controlled by only a few SEP holders, there are many more market actors such as subsidiaries, independent producers, procurers, wholesalers, retailers, etc., involved in related GVCs. All of these companies are part of the production, service, and distribution network yet none of them hold SEPs. Dedrick *et al.* (2010) analyze the GVCs of the production of iPods and notebook PCs as

Table 1. Descriptive statistics of SEP disclosures by company (total until 2014).

	InterDigital	**Qualcomm**	**LG Electronics**	**NOKIA**	**Ericsson**	**Motorola**	**Remaining 1,149 companies**
SEP disclosures	87,437	84,890	46,657	29,444	19,338	14,556	95,681
% of total disclosures	23.13	22.46	12.34	7.79	5.12	3.85	25.31
			74.69				25.31

Source: IPlytics GmbH (authors' calculations).

an illustrative example of this fact. They find that there are numerous companies in several countries involved in manufacturing these products. Only the enterprises on the very top of the value chain, those that provide the product designs, labels, brands, and overall coordination of the process, hold the SEPs or the licenses.

The fact that SEP holders can influence the layout of a certain GVC but need the support of companies on lower levels of the chain to make the final products generates externalities, for example, manufacturing companies that naturally want to profit from being part of a GVC. In order to support such firms, some countries actively make policy decisions to facilitate entry into GVCs. Emerging economies especially see GVCs as one way to foster their economic development and to engage in global markets. For example, Brazil deliberately tried to strengthen its consumer electronics industry so as to become one of the major global manufacturers in this sector. A decision by governments to get involved in certain GVCs can then, as in Brazil, lead to economic success. Nonetheless, as a mere manufacturer with no innovative activity of its own, it will be difficult to move up in the GVC and to steer and control it eventually (Gereffi and Sturgeon, 2013).

In the next section, we discuss some aspects for future research that emerge from these considerations.

3. Aspects for future research

The principal thoughts that we stress in this section can be summarized as follows:

- There is a strong oligopolistic structure in the SEP-related businesses. Still, the gains from these GVCs are shared between the SEP holders and those companies that do the manufacturing but do not hold SEPs themselves.
- China, for example, as an emerging economy, has managed to become one of the major SEP holders because government action focused on fostering growth in R&D. Other emerging economies, such as Brazil, have positioned themselves as mere manufacturing countries in the

SEP-induced GVCs and struggled to achieve growth of other domestic markets outside of ICT manufacturing.
- The actual microeconomic flows of trade in value-added, together with a disaggregated look at where SEPs are actually being invented, are a promising avenue for further research. Detailed and more disaggregated data will be increasingly available to pursue this track.
- The disclosure of SEPs is concentrated in European, international, and US-based SSOs and consortia. Other countries seem not to be able to attract SEP holders into their standard-setting bodies.
- Investigating who participated in the standard setting of standards that reference SEPs will allow drawing a more nuanced picture of the mechanisms at work.

3.1. *Upgrading via SEPs and the role of government policy*

The empirical facts lead us to a first interesting question for further research: How can intermediate producers profit from GVCs and how is upgrading (i.e., moving up in the GVC toward performing more elaborated tasks) in an SEP-induced GVC possible? In a further step, the question would extend to: How can a company reach a position that allows it to exert control over a GVC?

Generally, adhering to the standards of a GVC is seen as essential for upgrading (Gereffi *et al.*, 2005; Kaplinsky, 2010; Nadvi, 2008). Especially in SEP-induced GVCs, this seems intuitive because the final product is heavily standardized. Only adhering to standards will, however, not result in upgrading. A company that wishes to upgrade must be able to fulfill the requirements that are needed on a higher level in the GVC. These requirements will be reached through success in business and the subsequent possibilities to grow (Nadvi, 2008). Especially when businesses have first access to GVCs because of government actions, it might be difficult to achieve this kind of growth. Whenever government actions strengthen a specific part of the economy, this very part usually grows whereas the surrounding sectors and markets stay at the same level or may even decrease in strength because of the new allocation of resources (Gereffi and Sturgeon, 2013). Therefore, a policy-driven, top-to-bottom approach at first sight appears not be the right path to move domestic companies up in GVCs.

Upgrading in GVCs will rather be achieved whenever a company shifts from being a standard-taker to a standard-setter. Research on developing countries shows successful upgrading when representatives of these countries actually take part in the standard setting at SSOs and influence the standards in a way that the domestic innovations of the respective country become part of the standard (Kang and Motohashi, 2015; Nadvi, 2008). However, as we regard the special case of SEPs, becoming a mere standard-setter still does not suffice to play a major role in a GVC. Referring back to the results from the trade data, we realize that only those companies that actively engage in cross-licensing shape GVCs. Therefore, in order to gain such a position, companies need their own patents as bargaining chips in standard setting. Without these, they are unlikely to receive from other firms the licenses that they need to be able to play a crucial role in the GVCs.

Obtaining own SEPs requires heavy investments in R&D that can generate outputs that will actually be regarded as standard-essential. Generally, this kind of R&D is driven by companies that are already experienced innovators. For this reason, it is especially difficult for companies that are laid out as manufacturers in a GVC to construct and manage R&D departments that produce the required output. Only the growth of a company, as we mentioned above, is likely to result in such R&D. In this regard, Caviggioli *et al.* (2015) show that market entrants disclose SEPs that are of narrower technical specifications than incumbents. That is, in order to enter into the market, companies need to hold patents that cover very specific technologies for the standard. There is empirical evidence in our data sample that it is also possible for emerging countries to become an important player in GVCs. In Section 2, we describe which countries hold the majority of SEPs in 2008 (China, Germany, Japan, Korea, and the US). In 1995, the top five countries were Germany, the UK, Japan, Sweden, and the US. China and South Korea thus managed to move up into this top-five group within 13 years.

Kang *et al.* (2014) investigate this development and point out that it is the result of two separate strategies. The authors show that South Korea relied on technology transfer from leading countries that hold the majority of SEPs, such as the US, and applied these technologies successfully. The country was therefore a typical standard-taker, yet has managed over time to learn from the inflowing innovations and to build its own technological strength with own SEPs. Kim (1997) investigates, for the case of South Korea, how capability formation can take place in developing economies by domestic firms acquiring, assimilating, and improving foreign technologies. The author confirms that South Korea could create growth because it was able to profit from being a developed country in general and from having a number of powerful domestic incumbents such as Samsung and LG.

The case of China is, however, very different. Instead of fostering the import of technologies, local governments can alternatively bring forward the indigenous R&D in their countries with the goal to reach higher levels in the GVCs. This strategy is actively pursued by China, for example, by establishing a sophisticated culture of national standardization (Ernst, 2011a, 2011b). Especially in SEP-related fields, China developed its own technologies and SEPs. As Kang *et al.* (2014) note, this is only possible because of the enormous size of the internal Chinese market. The internal market creates a sufficient amount of demand for Chinese technology products so that domestic standards can survive economically. After years of successful domestic development, the Chinese industry is now savvy enough to be able to produce SEPs of international interest. China, therefore, pursued a successful policy-induced path which was eventually successful, thanks to perseverance and a large domestic market.

Although the cases of South Korea and China are certainly more complicated in reality, these findings show in a nutshell that there are different, very specific ways to success and that one of these ways may actually be a strong top-to-bottom approach. Because of the competition and high entry barriers in SEP-related markets, it proves therefore to be difficult to upgrade in these GVCs without concentrated government actions combined with the right business environment. Still, SEPs seem not to be very prominently placed on the political agendas. While the European Commission and the US Department of Justice have published statements regarding litigation and anti-trust issues (European Commission, 2014, 2015b; U.S. Department of Justice and U.S. Patent & Trademark Office, 2013), we do not know of any government policy concerning the fostering of SEP development at local companies at this point. Given the economics of SEPs, a political program concerning SEPs similar to, for example, the Chinese national standardization strategy, would be a promising step. Such a program should be supported by future research looking into microeconomic time series data that cover the involvement of specific companies in the SEP-induced GVCs and that ideally indicate SEP-related government policy actions as well.

The data we have available at the moment show clearly that the SEP-heaviest market actors are few and are indeed very powerful. Consequently, the question arises if and how a smaller company, even if it succeeds in developing SEPs, can become influential in this market while holding only a few SEPs. A simple solution would be to sell the SEPs off to one of the major players that hold patent portfolios. In this case, the oligopolistic structures would be strengthened. Otherwise, companies have to try to build their own SEP portfolios either by R&D, by strategic acquisitions of other patents, or by taking over other companies. In any of these cases, for smaller companies, the distribution of committee chairs at the SSOs plays a crucial role. If company engineers will be able to go to the standard-setting meetings, their technology is more likely to succeed (Kang and Motohashi, 2015). If now the home country of a smaller firm actively positions herself at the SSOs, as was the case with South Korea and China, this supports both small and large domestic businesses. In a study concerning regular patents, Serrano (2010) shows that, in the ICT sector, 23.9% of the small innovators but only 7.9% of the large innovators trade their patents. If we assume these numbers to be similar for SEPs, we can conclude that larger SEP holders would rather keep their SEPs to themselves and use them, while smaller SEP holders might however be inclined to sell their IPR, possibly because they are too small to survive as IPR holders in their GVCs. To examine if only large SEP holders participate in standard-setting meetings or if there are also smaller firms present, and to look into the trade patterns of SEPs in a comparable way to the aforementioned study, will therefore be a feasible way to find out if SEPs in standards do hamper innovation in small companies due to a closed-door policy or if there are equal opportunities for all market actors.

3.2. *The role of the institutional environment*

The closer examination of the effects of SEPs on trade also raises the question whether the international system of standardization with its main institutions like IEC and ISO is fit for the challenges of the 21st century. Standards set trajectories and lock in technologies for the years to come. While this can have a distorting effect on trade already, the issue of SEPs on the gains from trade debate highlights the need to investigate the effects that institutionalized international standardization might or might not have on the distribution of the gains from trade.

Figure 1 exemplifies how much the constellation in international standardization is changing: China went from chairing just nine ISO-committees in 2004 to 60 in 2014, an almost sevenfold increase within ten years. South Korea also more than doubled the number of committees that it chaired, starting from nine in 2004 to 19 in 2014.

While this data shows the involvement of the countries in standard setting, it might be even more insightful to examine at which SSOs SEP

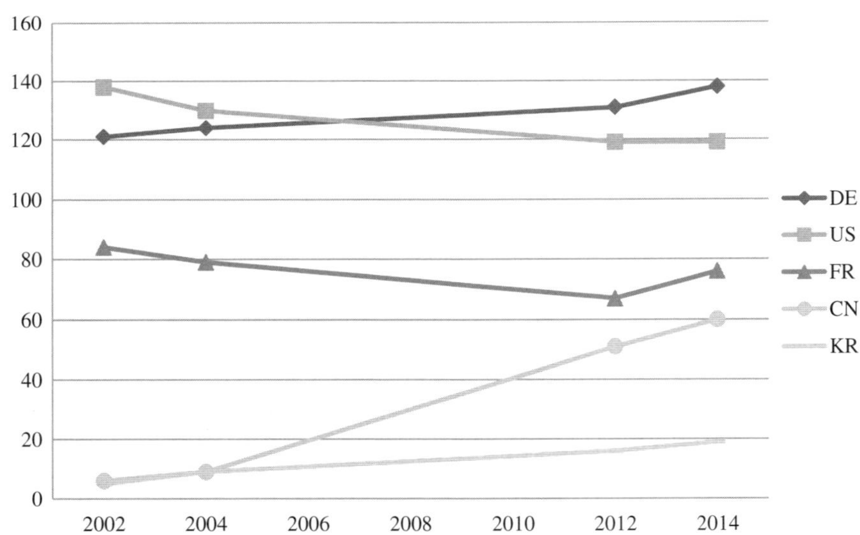

Figure 1. Number of secretariats of ISO (sub-) committees held by the respective country.
Source: Authors' own illustration on the basis of data provided by Expertenkommission Forschung und Innovation (2015).

holders disclose their SEPs. As we see from the data presented in Figure 2, by far the most SEPs are disclosed at the European Telecommunications Standards Institute (ETSI). The other standard-setting bodies are a mix of formal SSOs, such as ISO, ITU, and IEEE, and of a number of standards consortia. Such consortia are private business initiatives or platforms that bring together companies to set standards but that are less bureaucratic than formal SSOs. Especially in the fast-moving ICT business, standards consortia are an important co-ordination mechanism for SEP holders (Baron *et al.*, 2014; Pohlmann, 2014). Regarding the data, it is especially noteworthy that with the exception of ETSI, CEN, and the international SSOs, all of the remaining SSOs and consortia are US-based. There are two possible reasons for this.

One the one hand, SEP holders might go forum shopping for the best platform for disclosure of their SEPs (Lerner and Tirole, 2006): SSOs and consortia offer different terms for SEP disclosure with regard to requirements, remuneration, level of bureaucracy, and with regard to the success of the standard on the market, and SEP holders are able to freely pick the organization they prefer on the market. The European, US-based, and

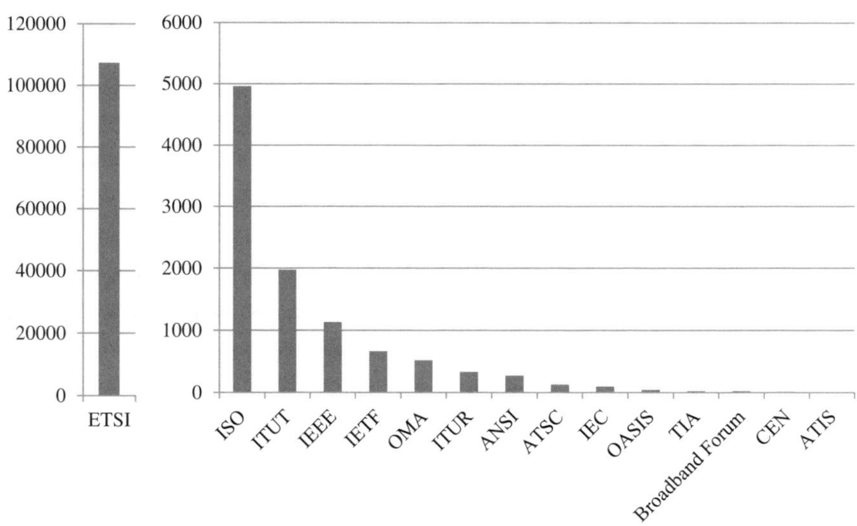

Figure 2. Number of SEP disclosures per SSO (total, 2015 data).

Source: IPlytics GmbH (2015).

international institutions seem to offer the best solutions for SEP holders. Having forum shopping in mind, it might therefore be worthwhile for institutions in other countries to think about the quality of their standard setting and how it fits the needs of SEP holders. Drawing SEP holders into institutions in other countries might also have the benefit of technology inflows that might in turn strengthen the local economies.

The second possible explanation for the US-focus of SEP holders is that the greatest part of the SEP-related innovation takes place at US-based companies. In this case, these firms would choose the SSOs and consortia in the same country to disclose their SEPs and consequently set the global standards. For other countries, this would mean that the fostering of SEP-related innovation would have to go hand in hand with the building of strong standard-setting bodies in the same country so that the innovations will not flow out. Possibly the US concentration is also a result of the IPR regime that is prevalent in the country. The IPR protection in the US is known to be fairly strong; moreover, it is possible to file patents for software which is neither the case in Europe nor in many other parts of the world. A legal environment like this will, of course, attract companies that try to market their IPR in the most economically efficient and safe way. The disclosure of IPR is already a bold step for a company (especially when its whole business model relies on IPR) so that they will be likely to choose institutions in a relatively safe environment.

4. Conclusion

Using a macroeconomic study of trade flows, we are able to uncover the role of SEPs in GVCs of the ICT sector. SEPs prove to enhance trade between countries that are engaged in the ICT-related GVCs due to cross-licensing agreements that benefit both trading partners. As the companies that hold SEPs are a relatively small and closed-off club, SEPs also raise anti-trust issues and might hamper innovations in countries that are not already part of these production patterns. For example, very recently, the European Commission informed one of these firms about ongoing investigations regarding suspected abuse of market power by the means of exclusivity payments and predatory pricing (European Commission, 2015a).

The examples of China and South Korea show that a country can have success that is induced by SEPs under very specific conditions. China, on the one hand, is able to profit from her large local market so that products that rely on a strictly national standardization system can be economically successful. South Korea, on the other hand, relied on technology inflows and on subsequent learning to be able to be innovative by herself.

Despite the success of these two Asian countries, looking deeper into the data of SEP disclosures, we find a concentration in Europe and the US. Assuming forum shopping of the involved companies, we propose that standard-setting bodies need to be laid out attractively with regard to their procedures and also to the local IPR regime in order to become hubs for successful ICT standards.

Generally, we find that the macroeconomic effects of SEPs raise a number of questions that offer interesting paths for future research. We regard it as especially promising to engage in data-driven research on the microeconomic and firm level in order to test some of the propositions we stated in this chapter. As mentioned, most of the standardization bodies as well as consortia are US-based. One characteristic of the US standardization system is the decentralized structure, that is, there are numerous SSOs for different purposes (as opposed to the centralized European system with single national and European institutions). Collecting the SEP disclosure data is therefore difficult as it has to be obtained from each of the hundreds of institutions or from specialized databases. SEP data, when available, generally cover very recent periods.

Regarding trade in value-added data, there are two main sources, one being the OECD-WTO TiVA Database (oe.cd/tiva) we used in the underlying study for this chapter. The other source frequently used by trade economists is the World Input-Output Database (www.wiod.org). Trade data is usually released with a lag of some years, the latest versions of both databases accordingly cover the period 1995–2011. The second challenge with regard to trade data is the availability of sector-specific data. As we only consider the ICT sector, we first need to define the appropriate industry classes and then check this availability of disaggregated data for these sectors. The OECD-WTO TiVA Database, for example, offers data for the sectors "Computer, electronic and optical products" and "Computer and related activities" that probably best approximate the ICT sector, although one might consider to also include "Post and telecommunications." The World Input-Output Database provides a similar arrangement.

In order to give some structured ideas for future research, we lastly propose several research approaches. The six companies mentioned in Table 1 have their headquarters in the US (InterDigital, Qualcomm, Motorola), South Korea (LG Electronics), Finland (Nokia), and Sweden (Ericsson). In an effort to reveal the actual GVC structures, we propose to identify the location of the headquarters of as many companies as possible that have disclosed SEPs. In a second step, following procedures like those of, for example, Timmer *et al.* (2014) or Marel (2015), one can disaggregate the GVC data to identify the actual streams of production in the SEP-related industries. A comparison of the two datasets could provide insights as to where the innovation that leads to SEPs happens versus where the production takes place. Moreover, the GVC data will be able to reveal where the gains from the value-added remain: Are they mainly attributed to the headquarters or does a part of them stay in (or return to) the countries that perform the manufacturing?

Moreover, governmental policy actions could be an interesting strand of research. After the identification of countries that are successful actors in GVCs, be it as manufacturers or as GVC leaders, an analysis of the policy actions that resulted in the success could yield interesting findings.

In case comprehensive firm-level data is available, one is able to follow the flows of SEPs. Interesting questions include who the actual innovators are and how they market their SEPs or potential SEPs. Do small innovators have chances in standard setting to have their inventions included in a standard? Or do they rather sell promising patents to one of the big players and only these have the power in standard setting to make their patents standard-essential? This path of research will allow us to understand the process of cross-licensing in detail. The requirement for very reliable, disaggregated, and sometimes secret data is, however, difficult to meet.

References

Anderson, J. E. (1979). "A Theoretical Foundation for the Gravity Equation." *American Economic Review*, **69**(1): 106–116.

Anderson, J. E. and E. van Wincoop (2003). "Gravity with Gravitas: A Solution to the Border Puzzle." *American Economic Review*, **93**(1): 170–192.

Anderton, B. (1999). "Innovation, Product Quality, Variety, and Trade Performance: An Empirical Analysis of Germany and the UK." *Oxford Economic Papers*, **51**: 152–167.

Baldwin, R. (2013). "Global Supply Chains: Why They Emerged, Why They Matter, and Where They Are Going." In *Global Value Chains in a Changing World*, D. K. Elms and P. Low (eds.).

Baron, J., Y. Ménière, and T. Pohlmann (2014). "Standards, Consortia, and Innovation." *International Journal of Industrial Organization*, **36**: 22–35.

Bekkers, R., L. Birkman, M. Canoy, P. de Bas, W. Lemstra, Y. Ménière *et al.* (2014). *Patents And Standards: A Modern Framework For IPR Based Standardisation* (European Commission, ECSIP Consortium, Rotterdam, Netherlands).

Bekkers, R., R. Bongard, and A. Nuvolari (2011). "An Empirical Study on the Determinants of Essential Patent Claims in Compatibility Standards." *Research Policy*, **40**(7): 1001–1015.

Bekkers, R. and A. Martinelli (2012). "Knowledge Positions in High-Tech Markets: Trajectories, Standards, Strategies and True Innovators." *Technological Forecasting & Social Change*, **79**(7): 1192–1216.

Berg, S. V. (1989). "Technical Standards as Public Goods: Demand Incentives for Cooperative Behaviour." *Public Finance Review*, 17(1): 29–54.

Bessen, J. and E. Maskin (2009). "Sequential Innovation, Patents, and Imitation." *The RAND Journal of Economics*, **40**(4): 611–635.

Blind, K. (2004). *The Economics of Standards: Theory, Evidence, Policy* (Edward Elgar, Cheltenham, UK, Northhampton, MA, USA).

Blind, K., R. Bekkers, Y. Dietrich, E. Iversen, F. Köhler, B. Müller *et al.* (2011). *Study on the Interplay Between Standards and Intellectual Property Rights (IPRs)* (European Commission, Luxemburg).

Blind, K., K. Cremers, and E. Mueller (2009). "The Influence of Strategic Patenting on Companies' Patent Portfolios." *Research Policy*, **38**(2): 428–436.

Blind, K., J. Edler, R. Frietsch, and U. Schmoch (2006). "Motives to Patent: Empirical Evidence from Germany." *Research Policy*, **35**(5): 655–672.

Blind, K., and A. Jungmittag (2005). "Trade and the Impact of Innovations and Standards: The Case of Germany and the UK." *Applied Economics*, **37**(12): 1385–1398.

Caviggioli, F., A. de Marco, F. Rogo, and G. Scellato (2015). "Patenting Strategies and Characteristics of Declared Inventions in the Long Term Evolution Standard." *R&D Management*.

Choi, T. Y., J. Budny, and N. Wank (2004). "Intellectual Property Management: A Knowledge Supply Chain Perspective." *Business Horizons*, **47**(1): 37–44.

Deardorff, A. (1998). "Determinants of Bilateral Trade: Does Gravity Work in a Neoclassical World?" In *The Regionalization of the World Economy*, J. A. Frankel (ed.), (University of Chicago Press, Chicago).

Dedrick, J., K. L. Kraemer, and G. Linden (2010). "Who Profits from Innovation in Global Value Chains?: A Study of the iPod and Notebook PCs." *Industrial and Corporate Change*, 19(1): 81–116.

Ernst, D. (2011a). *Indigenous Innovation and Globalization - The Challenge for China's Standardization Strategy* (Institute on Global Conflict and Cooperation; East-West Center.

Ernst, D. (2011b). "Toward Greater Pragmatism? China's Approach to Innovation and Standardization." SITC Research Briefs, Policy Brief No. 18, Institute on Global Conflict and Cooperation.

European Commission (2014). "Competition Policy Brief: Standard-Essential Patents." Issue 8.

European Commission (2015a). "Antitrust: Commission Sends two Statements of Objections on Exclusivity Payments and Predatory Pricing to Qualcomm." Available at: http://europa.eu/rapid/press-release_IP-15-6271_en.htm (IP/15/6271). Accessed 15 December 2015.

European Commission (2015b). "Report on Competition Policy 2014." Brussels (COM(2015) 247 final).

Expertenkommission Forschung und Innovation (2015). "EFI Report 2015: Gutachten zu Forschung, Innovation und technologischer Leistungsfähigkeit Deutschlands."

Farrell, J., J. Hayes, C. Shapiro, and T. Sullivan (2007). "Standard Setting, Patents and Hold-Up." *Antitrust Law Journal*, **74**(3): 603–670.

Garcia, L. and K. St. Clair (1992). "Global Standards: Building Blocks for the Future." With assistance of S. Emery, J. Alic, S. Remoy, A. Davidson. U.S. Congress, Office of Technology Assessment, Washington, D.C.

Gereffi, G., J. Humphrey, and T. Sturgeon (2005). "The Governance of Global Value Chains." *Review of International Political Economy*, **12**(1): 78–104.

Gereffi, G. and T. Sturgeon (2013). "Global Value Chain-Oriented Industrial Policy: The Role of Emerging Economies." In *Global Value Chains in a Changing World*, D. K. Elms and P. Low (eds.).

Greenhalgh, C., P. Taylor, and R. Wilson (1994). "Innovation and Export Volumes and Prices — A Disaggregated Study." *Oxford Economic Papers*, **46**: 102–134.

Grindley, P. C. and D. J. Teece (1997). "Managing Intellectual Capital: Licensing and Cross-Licensing in Semiconductors and Electronics." *California Management Review*, **39**(2): 1–34.

IPlytics GmbH (2015). "Data Analysis on Standard Essential Patent (SEP) Declarations." Available at: http://www.iplytics.com/download/docs/studies/IPlytics%20SEP%20Analysis.pdf. Accessed 14 December 2015.

Kang, B. D. Huo, and K. Motohashi (2014). "Comparison of Chinese and Korean Companies in ICT Global Standardization: Essential Patent Analysis." *Telecommunications Policy*, **38**(10): 902–913.

Kang, B., and K. Motohashi (2015). "Essential Intellectual Property Rights and Inventors' Involvement in Standardization." *Research Policy*, **44**(2): 483–492.

Kaplinsky, R. (2010). "The Role of Standards in Global Value Chains." World Bank Policy Research Working Paper No. 5396.

Kim, L. (1997). *Imitation to Innovation: The Dynamics of Korea's Technological Learning* (Harvard Business Press, Boston, MA).

Lerner, J. and J. Tirole (2004). "Efficient Patent Pools." *American Economic Review*, **94**(3): 691–711.

Lerner, J. and J. Tirole (2006). "A Model of Forum Shopping." *American Economic Review*, **96**(4): 1091–1113.

Lerner, J. and J. Tirole (2015). "Standard-Essential Patents." *Journal of Political Economy*, **123**(3): 547–586.

Miroudot, S. and H. Nordström (2015). "Made in the World?" European University Institute Working Paper, RSCAS 2015/60.

Nadvi, K. (2008). "Global Standards, Global Governance and the Organization of Global Value Chains." *Journal of Economic Geography*, **8**(3): 323–343.

Nagaoka, S. and H. U. Kwon (2006). "The Incidence of Cross-Licensing: A Theory and New Evidence on the Firm and Contract Level Determinants." *Research Policy*, **35**: 1347–1361.

OECD (2013). *OECD Science, Technology and Industry Scoreboard 2013: Innovation for Growth* (OECD Publishing, Paris).

Pohlmann, T. (2014). "The Evolution of ICT Standards Consortia." *Communications & Strategies*, **95**(3): 17–40.

Pohlmann, T., P. Neuhäusler, and K. Blind (2015). "Standard Essential Patents to Boost Financial Returns." *R&D Management*.

Ramel, F. and K. Blind (2015). "The Influence of Standard Essential Patents on Trade." In *EURAS Proceedings 2015. The Role of Standards in Transatlantic Trade and Regulation*, K. B. Skriver, K. Jakobs, and J. Jerlang (eds.) (Mainz, Aachen), pp. 359–376.

Serrano, C. J. (2010). "The Dynamics of the Transfer and Renewal of Patents." *RAND Journal of Economics*, **41**(4): 686–708.

Shapiro, C. (2001). "Navigating the Patent Thicket: Cross Licenses, Patent Pools, and Standard Setting." In *Innovation Policy and the Economy*, A. B. Jaffe, J. Lerner, and S. Stern (eds.), Volume 1 (MIT Press, Cambridge, MA), pp. 119–150.

Timmer, M. P., A. A. Erumban, B. Los, R. Stehrer, and G. J. de Vries (2014). "Slicing Up Global Value Chains." *Journal of Economic Perspectives*, **28**(2): 99–118.

Tinbergen, J. (1962). *Shaping the World Economy — Suggestions for an International Economic Policy* (The Twentieth Century Fund, New York).

U.S. Department of Justice and U.S. Patent & Trademark Office (2013). Policy Statement on Remedies for Standard-Essential Patents Subject to Voluntary F/RAND Commitments. Available at: http://www.uspto.gov/about/offices/ogc/Final_DOJ-PTO_Policy_Statement_on_FRAND_SEPs_1-8-13.pdf. Accessed 14 December 2015.

van der Marel, E. (2015). "Positioning on the Global Value Chain Map: Where do You Want to Be?" ECIPE Occasional Paper No. 01/2015.

von Graevenitz, G., S. Wagner and D. Harhoff (2011). "How to Measure Patent Thickets — A Novel Approach." *Economics Letters*, **111**(1): 6–9.

von Graevenitz, G., S. Wagner, and D. Harhoff (2013). "Incidence and Growth of Patent Thickets: The Impact of Technological Opportunities and Complexity." *Journal of Industrial Economics*, **61**(3): 521–563.

WIPO (2011). *World Intellectual Property Report: The Changing Face of Innovation*. WIPO Economics and Statistics Series (Geneva, Switzerland).

Chapter 12

Patents, Standards, and Borders: Addressing National Disparities Among Holders of Standard-Essential Patents

Jorge L. Contreras

University of Utah, 201 Presidents Cir,
Salt Lake City, UT 84112, USA

1. Introduction

Today's technology product markets, particularly in the Information and Communications Technology (ICT) sector, are broadly international. Products designed in California may be assembled in China from parts sourced in Germany for sale to consumers in India. The global character of technology markets underscores the importance of technical interoperability standards which include protocols and technologies such as Wi-Fi and Bluetooth (wireless networking), 4G LTE (wireless telecommunications), DVD and Blu-Ray (digital media storage), and MP3/MP4 (digital content encoding). These standards enable products and components manufactured by vendors around the world to operate together without customization or firm-to-firm interaction. Recent studies have shown that standards contribute strongly to economic growth and development (Ernst *et al.*, 2014).

Patents covering technical standards have also assumed increasing importance in global trade, business negotiations, and relationships among firms. This chapter considers the impact of patents on international technical standardization and the production of standards-compliant products. In particular, it assesses the impact that "standards-essential" patents (SEPs) have had on individual firm behavior and intra-firm dynamics. It also evaluates available options to reduce disparities between those firms, primarily from large developed economies, that hold significant portfolios of SEPs, and those firms, primarily from the developing world, that do not.[1]

2. Standards and the international standard-setting landscape

While many health, safety, and environmental standards are developed by governmental agencies, the vast majority of technical interoperability standards originate in the private sector (Ernst, 2012; Biddle *et al.*, 2012). In the US, there is an express governmental preference for privately developed standards over government-developed standards,[2] and elsewhere, this preference has generally been supported

[1]The World Trade Organization (WTO) Agreement on Technical Barriers to Trade pmbl., April 15, 1994, Marrakesh Agreement Establishing the World Trade Organization, Annex 1A, 1868 U.N.T.S. 117 (TBT Agreement) is designed to "ensure that technical regulations and standards … do not create unnecessary obstacles to international trade." *See generally* Bremer (2016). In general, such requirements pertain to the use of nationally-mandated health and safety standards to prevent free trade in products. This body of law, which has prompted a significant academic literature, addresses a different set of concerns than the current chapter, which focuses on the ways in which voluntary technical standards and associated patents may advantage or disadvantage firms from different countries. See generally ITC (2016, pp. 71–75)

[2]Office of Management and Budget (OMB) Circular A-119 (1998). See Bremer (2016).

by the market. Some widely adopted interoperability standards (e.g., Microsoft's .doc and Adobe's PDF electronic document formats) are single-firm proprietary formats (*de facto* standards). Over the past two decades, however, most successful interoperability standards have been developed by groups of firms that collaborate within voluntary associations known as standards-development organizations or standards-setting organizations (SSOs). The resulting standards are often referred to as "voluntary consensus standards," which will be the principal focus of this chapter.

SSOs vary greatly in size and composition. The European Commission (2014) identifies three broad categories of SSO:

(1) Those that are formally recognized by governmental bodies. These include:
    (a) *international* groups (e.g., the International Organization for Standardization (ISO) and the International Telecommunications Union (ITU));
    (b) *regional* groups (e.g., the European Telecommunications Standards Institute (ETSI)); and
    (c) *national* groups (e.g., Germany's Deutsches Institut für Normung (DIN), the Japanese Standards Association (JSA), China's National Institute for Standardization (CNIS), and the Bureau of Indian Standards (BIS)).[3]
(2) "Quasi-formal" groups that are typically large international organizations that share many of the characteristics of formally recognized groups (e.g., the IEEE Standards Association, ASTM International, and the Internet Engineering Task Force (IETF)), and

[3]The American National Standards Institute (ANSI) presents a somewhat unusual case, inasmuch as it is a private organization which is recognized in certain capacities by the U.S. government. ANSI oversees, accredits and establishes policy for national SSOs that wish to develop American National Standards. Among other things, ANSI-accredited SSOs must adopt due process and intellectual property policies that comply with ANSI's "Essential Requirements".

(3) Smaller, privately organized consortia (also known as special-interest groups or fora), including groups such as the Bluetooth SIG, HDMI Forum, and USB Forum, and hundreds of other such groups.[4]

Table 1 lists a number of widely adopted ICT standards and the organizations in which they were developed.

Table 1. Selected ICT standards and where they were developed.

Standard	Description	SSO	EC Class
802.11	Wireless networking	IEEE	2
Bluetooth	Short-range wireless networking	Bluetooth SIG	3
CD	Compact disc (digital media)	n/a[5]	n/a
CDMAone/IS-95	2G wireless telecommunications	Qualcomm/TIA[6]	n/a 2
DVB	Digital video broadcast (Europe)	DVB Forum	1
DVD	Digital media	n/a[7]	n/a
Ethernet	Device networking	IEEE	2
GPS	Global Positioning System	n/a[8]	n/a
GSM	2G wireless telecommunications	ETSI	1
H.264	Audio-video encoding	ITU	1
HDMI	High-definition multimedia interface	HDMI Forum	3
HDTV	High-definition broadcast TV (US)	ATSC	3
HTTP	Hypertext transfer protocol	W3C	2
IP	Internet protocol	IETF	2
LTE	4G wireless telecommunications	ETSI	1

(*Continued*)

[4]Updegrove (2015) catalogs more than 1,000 such groups.

[5]The CD specification was developed primarily by Philips and Sony.

[6]CDMA technology was initially developed by Qualcomm, which then submitted it for adoption to the Telecommunications Industry Association (TIA).

[7]The DVD specification was developed primarily by Philips, Sony, Toshiba, and Panasonic.

[8]The GPS standard was originally developed by the US Department of Defense.

Table 1. (*Continued*)

Standard	Description	SSO	EC Class
MP3/MP4	Audio and video compression	MPEG (ISO/IEC)	1/2
PDF	Portable Document Format	n/a[9]	n/a
SDRAM	Semiconductor memory	JEDEC	2
UMTS	3G wireless telecommunications	ETSI/3GPP	1
USB	Device networking	USB Forum	3
V.90	56k modem	ITU	1
VHS	Video cassette media	n/a[10]	n/a
WWW	Worldwide web	W3C	2
XML	Extensible markup language	W3C	2

3. Firm-level participation in standard-setting

Firm-level participation in SSOs varies according to the type and nature of the SSO. ISO, probably the most prominent Category 1 SSO, allows participation solely on a national basis, so that each member state has a delegation that represents its interests at the SSO. Criteria for participation in a national delegation are determined at the national level. The US representative to ISO, for example, is ANSI (see note 2). Other Category 1 SSOs may limit participation to firms and institutions engaged in business in a particular geographic area. For example, the members of the European Committee for Electrotechnical Standardization (CENELEC) comprise the national electrical standardization committees of each European state. Some Category 1 SSOs, such as ETSI, open membership to all interested parties, but offer different membership categories and benefits to those within the region of focus (Europe, in the case of ETSI).

In contrast, Category 2 SSOs are generally open to all interested parties on an equal basis. Participation depends on firms' interest in the

[9]PDF is a proprietary format developed by Adobe.

[10]The VHS format was developed by Matsushita/JVC.

relevant area of standardization, as well as its ability to bear the not-insignificant personnel, travel, and technology costs associated with SSO participation. It is no surprise that large global technology firms participate in 50 or more different SSOs, with the largest involved in more than 100 SSOs each (Baron and Spulber, 2015). Participation in large international SSOs in the ICT sector has traditionally been international in character, with representation from firms and institutions based in North America, Europe, Oceana, Japan, Korea, and India. Over the last decade, Chinese firms have dramatically increased their participation in international SSOs, in some sectors surpassing participation from all countries other than the US (Ernst, 2011; Contreras, 2014). Despite recent gains by China, SSO participation by firms in less-developed countries, particularly in Latin America and Africa, has remained at low levels.

Category 3 SSOs or consortia are usually formed by small groups of firms interested in developing a specific technology or standard. Often these "founder" or "sponsor" firms hold patents relevant to the technology in question (Biddle *et al.*, 2012). Such founders are typically large multinational firms with substantial patent portfolios but may also include smaller specialized firms focusing on the target technology area and large industry participants that manufacture products or systems that will be dependent on the standard, but which do not themselves hold large technology or patent portfolios (e.g., an electrical power utility that may be dependent on new smart grid communications technologies, but which does not itself develop such technologies).

4. Patents and standards

4.1. *Patenting standards*

As noted earlier, standards are sets of protocols and technical descriptions of product features that enable product interoperability. While standards themselves are not patentable, products that are compliant with the technical requirements of standards (often referred to as standards-compliant products) generally satisfy the statutory requirements for patent protection.

The owners of patents covering standardized technologies (referred to as SEPs) are often the firms and institutions that employ individuals who make particular inventive contributions to standards. Some of these contributions may be made jointly and are owned by multiple firms, but in most cases, firms individually submit technical contributions to the standard-setting process and own the resulting SEPs.

Because standards documents are often quite lengthy and complex, sometimes running to hundreds or thousands of pages, multiple inventive concepts are frequently embodied in the same standard, leading to the possibility of multiple patents covering any given standard. For example, Blind *et al.* (2011) report large numbers of patent families[11] declared to be essential to various standards, including WCDMA (1,000 patent families), 4G LTE (1,000 patent families), MPEG-2 and MPEG-4 (160 patent families), optical disc drive standards (2,200 patent families), and DVB-H (30 patent families)).

Ordinarily, if the vendor of a product that infringes a patent is unable, or does not wish, to obtain a license on the terms offered by the patent holder, that vendor has three choices: to stop selling the infringing product, to design around the patent, or do neither and risk liability as an infringer. With standards-compliant products, however, designing around the patent may be impossible or economically infeasible. Moreover, once a standard is approved and released by an SSO, market participants may make significant investments in plant, equipment, and labor, based on anticipated implementation of the standard in products (a situation often referred to as lock-in) (Shapiro and Varian, 2001). In such cases, the cost of switching from the standardized technology to an alternative technology may be prohibitive, thereby increasing the patent holder's leverage in any ensuing negotiation over licensing rates. This phenomenon has been termed patent "hold-up" and is discussed extensively in the literature (Lemley and Shapiro, 2007; Contreras, 2018).

[11] A patent "family" consists of all individual patents deriving from a single, initial patent application. These may include individual patents in multiple countries, as well as multiple patents in the same country derived from the same initial application (e.g., continuation, continuation-in-part and divisional applications in the US).

As noted earlier, complex technological products may implement dozens, if not hundreds, of standards, each of which may be covered by hundreds or thousands of patents. As such, the aggregation of royalty demands by multiple patent holders could lead to cost-prohibitive burdens on implementing standards-compliant products. This situation is sometimes referred to as "royalty stacking" (Lemley and Shapiro, 2007; Contreras, 2018).

4.2. *SSO patent policies*

Over the past two decades, SSOs have responded to the increasing number of patents covering standardized technologies and the perceived threats of patent hold-up and stacking by adopting a series of policy measures intended to address these concerns. SSO patent policies today fall into two general categories: disclosure policies and licensing policies, and often include elements of both. Disclosure policies typically require participants in the standards development process to disclose SEPs that they hold. Licensing policies typically require that participants grant manufacturers of standardized products licenses under their SEPs on terms that are "Fair, Reasonable and Non-discriminatory" (FRAND) or royalty-free (RF).

These commitments purport to assure manufacturers that they will be able to obtain licenses (which may sometimes involve a payment) to sell standards-compliant products covered by SEPs. Perhaps, in part, because FRAND commitments require relatively little administrative overhead to enact, their use has become widespread among SSOs.[12] Nevertheless, a consistent, practical, and readily enforceable definition of FRAND has proven difficult to achieve. No SSO defines precisely what FRAND means, and many affirmatively disclaim any role in establishing, reviewing, or assessing the reasonableness of FRAND licensing terms. This lack of certainty has contributed to recent litigation over FRAND commitments (Contreras, 2013), and leaves most of the details of licensing arrangements to bilateral negotiations among patent holders and potential licensees.

---

[12]FRAND commitments (or similar commitments to license patents on a royalty-free basis) are required of all SDOs accredited by ANSI.

5. Impact of patents on international participation in standard-setting

5.1. *Patenting by SSO participants*

Over the past two decades, there has been a sharp increase in patenting within certain technology standardization sectors, particularly wireless telecommunications (Bekkers and West, 2009). In addition, a core group of firms in the telecommunications sector accounts for the large majority of patent filings covering ICT standards. These firms include Qualcomm, InterDigital, LG Electronics, Nokia, Samsung, Ericsson, and Motorola (Blind *et al.*, 2011; Baron and Pohlmann, 2015). In addition, Contreras (2014) observes a rapid increase in patenting activity by Huawei in the area of Internet standardization. These statistics suggest that patenting behavior is not concentrated among firms of any one country but is distributed at least among firms based in the major developed economies (US: Qualcomm, InterDigital, and Motorola; South Korea: LG and Samsung; Europe: Nokia and Ericsson; and China, Huawei).[13]

When considering levels of patent acquisition, it is important to note that a firm's home jurisdiction is relatively immaterial to the jurisdictions in which it seeks and obtains patents. That is, a large firm with a global market is likely to seek patents in all major markets, no matter where it is based. Thus, in 2014, the ten firms to which the greatest number of US patents were awarded were: IBM (US), Samsung (South Korea), Canon (Japan), Sony (Japan), Microsoft (US), Toshiba (Japan), Qualcomm (US), Google (US), LG (Korea), and Panasonic (Japan) (USPTO, 2015). It is likely that a comparable distribution exists in most other jurisdictions, with at most a modest "head start" advantage for local firms. Thus, in India, research conducted by the author and the Centre for Internet and Society has found that nearly 100% of Indian patents covering mobile device technologies are owned by foreign companies (Contreras and

[13]Though Japanese firms such as Sony, Toshiba, Sharp, and Panasonic have played major roles in many areas of ICT standardization, particularly consumer electronics and digital media, they are comparatively underrepresented in telecommunications and networking SSOs, due largely to early policies adopted by the Japanese government (Contreras, 2014).

Lakshané, 2017). These are, by and large, the same major international technology firms that are active throughout the world.

These findings suggest that in terms of standard-essential patents (and, most likely, all patents), firms can be classified as either "Haves" or "Have-Nots." The Haves are generally large multinational technology-focused firms based in North America, Europe, and the Asia-Pacific economies.[14] The Have-Nots are all others.[15] It is important to note that not all firms based in these key jurisdictions are Haves. Smaller firms and new market entrants in developed economies are also likely to be Have-Nots. Likewise, not all firms based in developing economies are, or must remain, Have-Nots. A prominent example is China-based Huawei which, in the span of just a few years, rose from insignificance to dominance in the area of internet standardization and related patent holdings (Contreras, 2014). Other large firms in China, India, Brazil, and other emerging economies may also be situated to invest the resources necessary to increase their patent portfolios in this manner. However, it appears that most firms in these jurisdictions are likely to remain as Have-Nots.

5.2. *Patent licensing dynamics*

As noted earlier, most SSOs require that their participants license SEPs to product manufacturers on terms that are either FRAND or royalty-free (RF). Thus, at least as to standardized technologies, patent acquisition and enforcement is unlikely to result in outright exclusion of competitors from a market. However, in markets characterized by FRAND (as opposed to RF) licensing, transactions are not always smooth or equitable, particularly in relation to transactions between Have and Have-Not firms.

The situation often plays out as follows: a standard is developed at an international SSO. Firms that participate in the SSO obtain patents covering the standard throughout the world. The standard then becomes

[14] As Ernst *et al.* (2014, p. 856) observe: "As technologies become more complex and inter-related, advanced countries have pre-empted fundamental technologies by aggressive patenting."

[15] For more comprehensive discussions of patent disparities among firms in developing vs. developed economies, see Ragavan (2016, Chapter 2) and Maskus (2012).

implemented in products that are sold globally. By the time the firms in less-developed countries become aware of the potential for sales of such products in their own countries (possibly with locally attractive features, lower costs, or domestically sourced components), the basic product technologies have already been patented by foreign Have firms. Local Have-Nots must thus seek licenses from foreign Haves in order to manufacture standardized products for their domestic markets. As observed by Ernst *et al.* (2014, p. 854) such firms (which they term "latecomers") "are naturally disadvantaged in the world of international standards as they have not contributed to the 'core technology' on which these standards are based ... Latecomer firms are thus forced to accept standards and pay royalties as decided by the dominant economic players."

The royalties sought by foreign patent-holding firms, while arguably reasonable on an international basis, may be viewed as excessive in local markets. The royalty burden owed to foreign firms can thus be viewed as inequitable by local firms and governments, particularly if foreign Have firms enter the market and compete with or displace local Have-Not firms (Ernst, 2015).[16] The perception of unfairness can be exacerbated when foreign firms actively enforce their patents against local market participants in their domestic markets. This situation has recently occurred in India where, over the past three years, multinational telecommunications vendor Ericsson has brought patent infringement suits against several Indian and Chinese handset vendors serving the domestic Indian market (DIPP, 2016; Contreras and Lakshané, 2017; Ernst, 2015).

6. Potential responses

As noted earlier, participating in international standards development and patenting can have beneficial local effects and can help Have-Not firms to advance their development and growth. Likewise, the aggregation of

---

[16] In addition, the royalty burden on local Have-Not firms is often greater than the burden on other foreign Have firms that hold patents that may be used as bargaining chips in cross-licenses with other Have firms. The result is that Have firms that have entered into cross-licensing networks generally have a low monetary royalty burden as compared to Have-Not firms that lack patents essential to relevant standards.

patents by Have firms in developing countries can have significantly negative effects on local firms and innovation (Ernst *et al.*, 2014, p. 861). Accordingly, a range of responses, both public and private, have been proposed to address the perceived disparity in SEP holdings between foreign Have firms and domestic Have-Not firms in less-developed countries. In many cases, these responses are not mutually exclusive and may co-exist within a country or region. The principal responses are considered below:

6.1. *Embrace the status quo*

Action is required to address a situation only if a problem exists. Some might argue that the current patent imbalance between Have and Have-Not firms is a natural result of market-based global trading. The situation is no different than it is in many other industries, including pharmaceuticals, automotive, and aviation, in which a handful of firms from developed countries dominate the market. In such a market, all firms have the potential to succeed based on superior innovation and technical skill.

This potential is particularly salient in the area of technical standardization, in which SSO participation is, in many cases, open to all interested organizations irrespective of national origin. The success of firms from small countries (e.g., Philips (Netherlands), Nokia (Finland), and Ericsson (Sweden)), and from developing economies (e.g., Huawei and ZTE (China)) demonstrates that the "club" of successful market entrants is not limited to firms from the largest developed economies. Thus, special measures designed to create a greater balance between the interests of Haves and Have-Nots could be counter-productive.

6.2. *Go-it-alone standardization*

In the early 2000s, the Chinese government began to realize that Western firms had dominated the wireless telecommunications standards field, and Chinese firms balked at the high royalty rates charged by these firms (Ernst, 2011; Vialle *et al.*, 2012). In response, the Chinese government embarked on a "go-it-alone" approach to 3G standardization, seeking to "catch up" with Western firms by producing a workable 3G technology that would be patented by Chinese firms (Ernst, 2011). The result was TD-SCDMA, a Chinese standard[17] that was developed by the Chinese Academy of Telecommunications Research (CATT) and its state-owned affiliate Datang in collaboration with German equipment vendor Siemens. Though TD-SCDMA is not generally viewed as a technological or market success, the TD-SCDMA experience appears to have advanced China's goal of building domestic technical expertise, patent leverage, and manufacturing capacity for advanced telecommunications products (Vialle *et al.*, 2012). The cost of China's indigenous standardization program, however, has been high. China is reported to have developed more standards than any other country (Ernst, 2011), yet few of these standards are in use outside of China. Today, China has turned increasingly toward international interoperability standards, with Chinese firms playing significant roles in international SSOs (Contreras, 2014) (see Sec. 6.5).

6.3. *Protectionist legal measures*

When a government perceives that its domestic producers are being disadvantaged by foreign interests, a natural reaction is to implement regulations, and undertake enforcement actions, intended to protect the local industry. Of course, expressly protectionist regulation in the area of intellectual property generally flies in the face of widely adopted international treaty obligations such as the WTO Agreement on Trade-Related Aspects of Intellectual Property Rights (the TRIPS Agreement),[18] as well as more recent bilateral and multilateral trade agreements such as the Trans-Pacific Partnership (TPP).[19] Nevertheless, protectionist measures that impede the activity of foreign patent-holders may be disguised as prohibitions of unfair business practices and anti-competitive

[17] China also submitted TD-SCDMA to the International Telecommunications Union (ITU) for recognition as an international standard, and the ITU approved TD-SCDMA in 2000.

[18] World Trade Organization (1999).

[19] Trans-Pacific Partnership (TPP), Chapter 8 (Technical Barriers to Trade).

behavior and may remain on the books for years before they are successfully challenged.[20]

Another protectionist approach is the targeted enforcement of existing regulations against foreign entities. There has been a spate of recent competition law investigations and enforcement actions against large Western holders of SEPs in China, Korea, and India.[21] For example, in February 2015, China's National Development and Reform Commission (NDRC) fined Qualcomm approximately US$975 million for a host of alleged violations of China's Antimonopoly Law in connection with its licensing of SEPs. The Korean and Taiwanese Fair Trade Commissions have also levied significant fines against Qualcomm. And in India, the Competition Commission of India (CCI) has investigated Ericsson in connection with Ericsson's patent infringement suits against Indian and Chinese manufacturers of mobile phones for the domestic Indian market (DIPP, 2016; Contreras and Lakshané, 2017).

A final way that governments can seek to reduce the dominance of foreign patent-holders in domestic markets is through the imposition of compulsory licensing for particular patents or products. This power, which is permitted under TRIPS in special circumstances, has to-date been exercised primarily in pharmaceutical markets in developing economies (Ragavan, 2016, Chapter 2). Nevertheless, the possibility of compulsory licensing exists in other industries that have a significant impact on health, safety, and welfare of local populations (Ragavan, 2016, Chapter 2; Contreras and McManis, 2014). In response to the dominance of the local Indian mobile devices market by foreign patent-holders, some have proposed the imposition of a compulsory licensing regime in this market, a proposal that is beyond the scope of this chapter (Lakshané, 2015).

6.4. *Increase patenting by local firms*

As the competitive advantage possessed by Have firms derives to a large degree from patents on standardized technology, some have suggested

---

[20]For example, Ernst *et al.* (2014, p. 858) argue that "constraining strategic patenting by owners of essential patents that could block innovation should be considered in a latecomer context as one of the criteria to assess success in standardization."

[21]To some degree, these investigations echo similar investigations by US and European competition law authorities.

that it would benefit local firms to increase their own patenting activity (Ramel and Blind, 2015).[22] Increased patenting by local firms would, it is argued, give Have-Not firms greater bargaining power in licensing negotiations with the existing Have firms. While this conclusion is correct on a theoretical level, it may oversimplify the issue. The acquisition of patents is not itself a productive activity, but a by-product of technological innovation. Thus, unless one seeks to encourage speculative patenting divorced from technical development (a goal that most would agree is undesirable), obtaining patents must be coupled with technological development.[23] To the extent that patents cover technical standards, that technical development usually occurs in connection with participation in an SSO.[24] Thus, to enhance their bargaining position, Have-Not firms should, in the first instance, seek not to increase their patenting activity but their participation in international standardization activities (see Sec. 6.5). If they do, their ability to obtain patents covering their technical contributions should follow.

It is, of course, a separate question whether local governments should facilitate patenting by domestic providers. Doing so in a manner that discriminates against foreign firms would generally run afoul of TRIPS and other treaty obligations.[25] However, as discussed in the next section, governments can help their domestic industry engage more actively in international standardization efforts by funding additional R&D and SSO participation.

---

[22]See generally ITC (2016, p. 80, 211), discussing positive effects of strong patent systems on inbound trade in technology products, foreign direct investment, and licensing transactions.

[23]This point is underscored by Ernst *et al.* (2014), who argue that "IPR protection can only contribute to economic development if it takes place as part of a multi-faceted innovation strategy that seeks to strengthen absorptive and innovative capabilities of firms, and to develop a broad-based innovation infrastructure (including standards)" (p. 858, citations omitted).

[24]While individual firms often develop technologies internally which they then bring to SSOs for standardization, a significant amount of revision, compromise, and development also occurs within the collaborative SSO setting.

[25]These obligations require local patent offices to afford "national treatment" to foreign applicants, treating them on the same basis as local applicants.

6.5. *Increase SSO participation by local firms*

Have-Not firms can realize a range of potential benefits from active participation in international SSOs.[26] First, as observed by Büthe and Mattli (2011, pp. 9, 211–212), firms that embed their proprietary technology into industry standards early during the standardization process can realize significant market gains and strategic advantages. Second, SSO participants can influence the direction of standardized technologies in a manner that favors, or at least takes into consideration, local markets and local technology/patent positions. Involvement in charting the future direction of technology standards can also give firms insight into and advance notice of product development and evolution opportunities. Participation may also give local firms opportunities to export interoperable products beyond the domestic market. It may also afford increased opportunities for patenting in domestic markets and abroad, and can inform foreign firms of the technology and patent assets that local firms have available for licensing.

From a policy standpoint, increased involvement in SSOs would give Have-Not firms opportunities to influence SSO policies and practices, particularly in ways that might facilitate licensing and technology dissemination in developing markets. For example, SSO policies could provide that offering lower royalty rates for deployment of standards-compliant products in developing markets would *not* violate the SSO's requirement of non-discriminatory treatment.[27] Likewise, SSOs could

---

[26] Ernst *et al.* (2014) focus on the development of national standardization policies within developing countries that blend elements of the industry-led US approach and more nationalized approaches adopted in China and Korea (p. 861). This approach has much to recommend it. However, the recommendation of this chapter differs from that of Ernst *et al.* (2014) in that it emphasizes the engagement of Have-Not firms in international standardization activities, rather than the development of specific national standardization policies.

[27] Major research universities around the world have adopted a similar stance in a 2007 document entitled "In the Public Interest: Nine Points to Consider in Licensing University Technology". The "Nine Points" document expressly acknowledges that "responsible licensing includes consideration of the needs of people in developing countries and members of other underserved populations."

mandate reduced-royalty or royalty-free licensing in certain markets or under certain conditions.

Despite the many potential advantages of international SSO participation, with a few exceptions, Have-Not firms have not yet made meaningful and sustained contributions to these organizations. This absence is rendered the more notable by express policies intended to ensure broad participation in such SSOs. For example, participation in international Category I SSOs such as ISO and ITU is often determined on a national basis.[28] The national delegations to bodies such as these present good opportunities for involvement by firms from less-developed countries. Some Category I SSOs such as ETSI, and most Category II SSOs, such as IEEE, ASTM and IETF are, by their own policies, open to participation by all interested organizations. Accordingly, the only barriers to participation in these SSOs, which represent a significant portion of global standardization activity,[29] arise from a lack of technical skill, financial resources and interest among Have-Not firms. These deficiencies are, of course, very real and very serious. However, as discussed below, financial barriers to participation can be overcome, at least in part, through national and philanthropic programs that provide resources for technical training and participation in international SSOs. The example of Chinese firms such as Huawei and ZTE (Contreras, 2014), illustrate that it is possible for local firms, with sufficient determination, governmental support and expenditure of resources, to become significant forces in international standardization.[30]

---

[28] See Sec. 6.1.

[29] Because Category 3 SSOs (consortia) are typically formed by small groups of firms with an existing technology and patent position, it is not realistic to hope that they will be fruitful avenues for greater Have-Not firm participation.

[30] Of course, China recently underwent a phase during which it concentrated significant resources on the development of local standards without heavy foreign patent coverage (see Sec. 6.2, discussing initiatives such as China's TD-SCDMA 3G mobile telephony effort, as well as Ernst (2011), which details several such efforts). While many would argue that these efforts were ultimately of limited success, it is possible that they did serve the goal of preparing Chinese firms to participate in international standardization efforts.

6.6. *Incentivizing increased SSO participation*

6.6.1. *Trade agreements*

Trade agreements, despite their potential to facilitate the involvement of local firms in international SSOs, have, to date, done little in this regard. Though the TPP includes an entire chapter devoted to standards, its goal is ensuring that locally developed standards, generally those relating to health and safety, are open and transparent and do not discriminate against foreign producers.[31] The standards focus of the TPP is thus inward-looking with respect to less-developed countries, ensuring that they allow international firms to enter without standards-based barriers, rather than outbound, or helping them to participate in the broader global standardization community.

In addition, future trade agreements could encourage greater openness to Have-Not participation in nationally based SSOs, require that nationally adopted standards originate from open SSOs, and establish international bodies designed to support Have-Not participation in international SSOs.

6.6.2. *Capacity building*

More important than trade agreements may be international and local capacity-building efforts to support greater international SSO participation by Have-Not firms. As noted by Ernst *et al.* (2014, p. 855), "standardization is a highly knowledge-intensive activity that requires well educated and experienced engineers and other professionals." The training and development of such personnel does not come cheaply and requires significant financial and institutional support (Bremer, 2016). For Have-Not firms lacking the internal resources to fund such training and development, such support must come from external sources. These sources could include grants from local governments, non-governmental organizations (NGOs), and multigovernmental organizations (e.g., the World Intellectual Property Organization, WIPO).

[31] See Trans-Pacific Partnership, Chapter 8 (Technical Barriers to Trade) and ITC (2016, pp. 71–75). This is the TBT issue discussed in Footnote 1.

SSOs themselves could also offer support to Have-Not firms wishing to participate. Such programs already exist within some SSOs. For example, the Internet Society (ISOC), a US/Switzerland-based NGO that oversees the IETF, a major developer of Internet standards, regularly sponsors a number of Fellows from developing countries to participate in IETF meetings and other activities.[32] Other SSOs sponsor participation by consumer advocates and other community representatives (Bremer, 2016). Programs such as these can be underwritten by SSO membership dues as well as fees charged for published standards. With such support programs in place, the steep costs of international SSO participation could be defrayed for Have-Not firms, thus broadening overall participation and promoting broader representation in these critical global organizations.

Another component of governmental and institutional support for standardization is educational. Many developing countries such as India already possess world-class educational institutions in the science and engineering disciplines. However, it is not clear that these institutions uniformly emphasize standards education and training. The need for greater education in the area of standards has been noted even within the US by the National Institute for Standards and Technology (NIST), which has funded efforts at several US universities to promote curriculum and program development relating to standards, and itself offers various training programs relating to standards for US government agencies and the private sector.[33]

Finally, it has been observed that many individuals in developing countries, both in government and the private sector, mistrust international SSOs and transnational bodies in general (Maskus, 2012, pp. 166–167). This mistrust may not be entirely unjustified, as there has traditionally been little representation of developing countries at international SSOs, and the needs and views of developing countries are seldom taken into account. Nevertheless, in order to take part more fully in international product development and standardization, these prejudices

---

[32] See http://www.internetsociety.org/what-we-do/education-and-leadership-programmes/ietf-and-ois-programmes/internet-society-fellowship.

[33] See www.nist.gov.

within developing countries will have to be overcome. Once that happens, assuming that the financial and institutional support described above exists, Have-Not firms will be able to engage more meaningfully at international SSOs, which may in turn begin to change attitudes and assumptions about SSOs within the developing world.

7. Conclusion

Patents on standardized technologies are being issued with increasing frequency, and the majority of these patents are held by large multinational firms based in developed economies. As a result, firms from less-developed economies with sparse patent holdings are disadvantaged in both domestic and foreign markets. While protectionist governmental policies can address these disparities, such measures are potentially contrary to international treaty obligations and generally unsuccessful in the long term. An alternative approach involves greater participation in international SSOs by firms from less-developed economies. This increased participation is likely to benefit such firms both in terms of technology development, strengthening of patent positions, and influence over SSO policies. To facilitate increased participation, both financial and institutional support will be required from local governments, NGOs, multinational organizations, and SSOs themselves. To the extent that participation in international SSOs by firms in developing economies can be increased, it could have a meaningful impact on domestic innovation, job creation, technical capability, and manufacturing output.

Acknowledgments

Earlier versions of this chapter have benefited from presentation and discussion at the *Workshop on Mega-Regionalism: New Challenges for Trade and Innovation* (MCTI) (Honolulu, January 20–21, 2016), the *Conference on Innovation, Intellectual Property, Competition and Standard-Setting in the ICT Sector* sponsored by Jindal Global University (New Delhi, August 20–21, 2016), and helpful comments and suggestions from Ashish Bharadwaj, Dieter Ernst, Brian Kahin, and Mike Plummer.

References

Baron, J. and T. Pohlmann (2015). "Mapping Standards to Patents using Databases of Declared Standard-Essential Patents and Systems of Technological Classification."

Baron, J. and D. F. Spulber (2015). "The Searle Center Database of Technology Standards and Standard Setting Organizations."

Bekkers, R. and J. West (2009). "The Limits to IPR Standardization Policies as Evidenced by Strategic Patenting in UMTS," *Telecommunications Policy*, 33, pp. 80–97.

Biddle, B. *et al.* (2012). "The Expanding Role and Importance of Standards in the Information and Communications Technology Industry," *Jurimetrics*, 52, pp. 177–208.

Blind, K. *et al.* (2011). "Study on the Interplay between Standards and Intellectual Property Rights (IPRs), Final Report." Available at: http://ec.europa.eu/enterprise/policies/european-standards/files/standards_policy/ipr-workshop/ipr_study_final_report_en.pdf.

Bremer, E. S. (2016). "American and European Perspectives on Private Standards in Public Law," *Tulane Law Review*, 91.

Büthe, T. and W. Mattli (2011). *The New Global Rulers: The Privatization of Regulation in the World Economy* (Princeton University Press, Princeton, NJ, USA).

Contreras, J. L. (2013). "Fixing FRAND: A Pseudo-Pool Approach to Standards-Based Patent Licensing," *Antitrust Law Journal,* 79, pp. 47–97.

Contreras, J. L. (2014). "Divergent Patterns of Engagement in Internet Standardization: Japan, Korea and China," *Telecommunications Policy*, 38, pp. 914–932.

Contreras, J. L. (2018). "Patents, Technical Standards and Standard-Setting Organizations: A Survey of the Empirical, Legal and Economics Literature." In *Research Handbook on the Economics of Intellectual Property Law, Vol. II — Analytical Methods*, P. S. Menell and D. Schwartz (eds.) (Edward Elgar, Cheltenham, UK).

Contreras, J. L. and C. R. McManis (2014). "Compulsory Licensing of Intellectual Property: A Viable Policy Lever for Promoting Access to Critical Technologies?" In *TRIPS and Developing Countries — Towards a New IP World Order?* G. Ghidini *et al.* (eds.) (Edward Elgar, Cheltenham, UK).

Contreras, J. L. and R. Lakshané (2017). "Patents and Mobile Devices in India: An Empirical Survey," *Vanderbilt Transnational Law Journal*, 50, pp. 1–44.

Department of Industrial Policy & Promotion (DIPP), Indian Ministry of Commerce & Industry (2016). "Discussion Paper on Standard Essential Patents and their Availability on FRAND Terms." March 1.

Ernst, D. (2015). "Global Strategic Patenting and Innovation — Policy and Research Implications," East-West Center Working Papers: Innovation and Economic Growth Series No. 2. Available at: http://www.eastwestcenter.org/system/tdf/private/iegwp002.pdf?file=1&type=node&id=34977

Ernst, D. (2012). "America's Voluntary Standards System — A Best Practice Model for Innovation Policy?" East-West Center Working Paper No. 128.

Ernst, D. (2011). *Indigenous Innovation and Globalization: The Challenge for China's Standardization Strategy* (East-West Center, Honolulu, HI, USA).

Ernst, D., H. Lee, and J. Kim (2014). "Standards, Innovation and Latecomer Economic Development: Conceptual Issues and Policy Challenges," *Telecommunications Policy*, 38, p. 853.

European Commission — Directorate-General for Enterprise and Industry (EC) (2014). *Patents and Standards: A Modern Framework for IPR-Based Standardization*.

Lakshané, R. (2015). "Letter to Prime Minister Shri Narendra Modi," March 24. Available at: http://cis-india.org/a2k/blogs/open-letter-to-prime-minister-modi.

Lemley, M. A. and C. Shapiro (2007). "Patent Holdup and Royalty Stacking," *Texas Law Review*, 85, pp. 1991–2049.

Maskus, K. E. (2012). *Private Rights and Public Problems: The Global Economics of Intellectual Property in the 21st Century* (Peterson Institute for International Economics, Washington, D.C., USA).

Ragavan, S. (2016). *Patent and Trade Disparities in Developing Countries* (Oxford University Press, Oxford).

Ramel, F. and K. Blind (2015). "The Influence of Standard Essential Patents on Trade." Paper presented at *IEEE-SIIT Conference*, Sunnyvale, CA, USA, October 6.

Shapiro, C. and H. R. Varian (1999). *Information Rules: A Strategic Guide to the Network Economy* (Harvard Business School Press, Boston).

Updegrove, A. (2015). "Standard Setting Organizations and Standards List." Available at: http://www.consortiuminfo.org/links/#.VarhPnjDRD0. Accessed July 15, 2015.

U.S. International Trade Commission (ITC) (2016). "Economic Impact of Trade Agreements Implemented Under Trade Authorities Procedures," Report.

U.S. Patent & Trademark Office (USPTO) (2015). "All Technologies (Utility Patents) Report." Available at: http://www.uspto.gov/web/offices/ac/ido/oeip/taf/all_tech.htm#PartB.

Vialle, P., J. Song, and J. Zhang (2012). "Competing with Dominant Global Standards in a Catching-Up Context. The Case of Mobile Standards in China," *Telecommunications Policy*, 36, p. 832.

Chapter 13

Dual Use of Standardization Strategies: Promoting Regional Integration and/or Global Markets

Konstantinos Karachalios and Karen McCabe

Abstract

This chapter discusses how the instruments of standards and standardization, international trade agreements, and government or public policy could collectively help foster economic growth, enhance prosperity through productivity increase, and drive innovation, without raising new protectionist barriers. These instruments should be used in an effective way to empower innovation across borders for an ecologically and socially sustainable prosperity, as — in the same time — the technological, social, and economic landscapes are undergoing tectonic shifts through:

- the rapid rate of technological development and convergence, particularly in the Information and Communications Technology (ICT) sector;
- increased competition and blurring of boundaries in the digital economy, driven by extensive global connectivity as well as the emergence of global innovation networks and new innovation hubs;

- national and regional policies having moved to coordinated broader digital approaches aiming at integrating social and economic priorities; and
- growing challenges affecting cross-border goods and information flow and their impact on individuals' rights to a minimal respect of their private sphere in a world of data-driven innovation.

This chapter will illustrate a potential path for improvement, taking into consideration the current state-of-play of standards regimes, international and megaregional trade agreements, and government policies paradigms. One example that this chapter will discuss is that open technology platforms, such as standards developed in a globally open and inclusive, collaborative, and consensus manner, that could be voluntarily adopted and be thus applicable globally, should be explicitly considered in megaregional agreements. Temptations to use megaregional dynamics to promote "regional standards" and "export" them globally or use them as a tool to seal the regional market from competition, would create counterforces and could eventually lead to the end of the current globalization era. So, while megaregionals could pose a threat for the openness of future ICT platforms, including open compatibility, and interoperability standards, they offer also an opportunity for the above-mentioned types of standards and processes to be recognized. They could foster an evolution in the definition of what constitutes an "international standard" and "international standards body." The current definitions stem from an era where only a nation-centric standardization model existed, perpetuating thus a concept that predates and is not adequate for the Internet era, and which may lead to national or regional industrial policies with new forms of structural discrimination and protectionism. The chapter will discuss how international and megaregional trade agreements could be used to promote globally open standardization processes and their outcomes, as well as supportive regulations, as means to increase economic activity and social inclusiveness in several regions by enhancing cross-border innovation and trade without — at least formal — discriminations across regions.

1. Introduction

Today, the technological, social, and economic landscapes are experiencing tectonic shifts. The world is witnessing unprecedented technological development and convergence, particularly in the Information and Communications Technology (ICT) sector. There is increased competition and blurring of boundaries in the digital economy, driven by extensive global connectivity, as well as the emergence of global innovation networks and new innovation hubs. National and regional policies have moved to broader digital approaches aiming at integrating social and economic priorities. Moreover, in a world of data-driven innovation, there are growing challenges affecting cross-border goods and information flow and their impact on individuals' rights to a minimal respect of their private sphere. These driving forces are compounded by the transition to a knowledge economy and society, with increasing market inequality and exclusion trends, and by concerns regarding the potential for more frequent financial crisis and destabilization in the world economy. These forces and the rise of megaregional agreements are fueling increasingly complex technological, economic, and socio-political ecosystems that are impacting standardization, public policy, and trade agreement development systems and resulting strategies for the use of these instruments.

In the 21st century, we are witnessing a shift to an economy dominated by the production and consumption of intangibles, such as the trade in know-how and ideas on a global scale. We are also experiencing 21st century regionalism, which underpins this complex trade environment and where regulation economics come into play.

As the world stands on the threshold of significant opportunity for addressing and realizing sustainable development goals[1] also through the use of ICTs, diffusion and development of innovative technologies, and engagement of developing economies in the trade ecosystem, there must be caution regarding protectionist practices using megaregional dynamics to promote "regional standards" and "export" them globally or use them as a tool to seal the regional market from competition, as this can create counterforces and eventually lead to the end of the current globalization era.

A fresh examination of the current instruments and an eye toward new institutional frameworks that help to spark and sustain socio-economic

---

[1] Sustainable Development Goals are an inter-governmentally agreed set of targets relating to international development. See: http://www.un.org/sustainabledevelopment/sustainable-development-goals/

change, productivity growth, and sustained economic development are therefore critical. We must consider how the instruments of standardization, public policy, and trade agreements could evolve coherently, promoting the capacity to innovate and adapt, as well as pluralism, transparency, and openness. Further, as the integration and deepening of global markets are accelerated, we must acknowledge the range of cultural values and contextual circumstances that underpin the search for ways of balancing cooperation with market competition and state sovereignty.

The creation of national and regional spaces for the optimal flow of goods, services, finance, and technology benefit long-term growth. Extending globally, such positive dynamics can improve the allocation of resources and the generation of ideas that are foundational to knowledge societies. However, the use of the current instruments of standards, public policy, and trade agreements and their current frameworks may impose undue constraints, including how international standards are defined and used.

This chapter puts forth considerations that promote the need for a new way of thinking about these instruments so that they foster economic growth and drive innovation, without raising protectionist barriers, and suggest how to use these instruments in a more effective way to empower innovation across borders to achieve sustainability and reduce nationalistic and mercantilist tensions. It purports that it is time for a collective effort to identify new ways to improve the current paradigms and strategic interplay of standards, public policy, and trade agreements.

2. State of play

International commerce in the 21st century involves basically 20th century trade practices overlaid by complex cross-border flows related to international production and innovation networks. It includes almost unlimited trade in intermediate goods, services, ideas, know-how, and capital and strictly regulated relocation of knowledge workers. It encompasses a combination of technology and other productivity-enhancing factors across nations, where one nation's exports can become competitive based more on the easier cross-border movement and combination of several nations' technology, labor, and capital versus on the easier movement of goods

(Baldwin, 2014). The complexities and externalities triggered by this new paradigm of more interconnected exchange, coupled with 21st century trade agreements being keystones in many nations' development strategies, are impacting standardization and public policy systems and instruments. Further, with the substantial decrease in trade barriers over the past few decades, technical barriers to trade (TBT), such as standards, have become increasingly important determinants of international trade flows. Unlike tariffs, standards have the potential to not only decrease but also expand trade (Clougherty and Grajek, 2012).

2.1. *Standards and technical regulations*

The World Trade Organization (WTO),[2] with its focus on the avoidance of unnecessary obstacles to trade, has created provisions and general terms for standardization that are defined by the United Nations (UN) system and by international standardizing bodies. Of particular importance is the WTO agreement on Technical Barriers to Trade (WTO TBT),[3] which addresses standards[4] and technical regulations[5] and compliance to these instruments.

[2]WTO is a global international organization dealing with the rules of trade between nations. At its heart are the WTO agreements, negotiated and signed by the bulk of the world's trading nations and ratified in their parliaments. The goal is to help producers of goods and services, exporters, and importers conduct their business. See: https://www.wto.org.

[3]For full text of the WTO TBT, please see: https://www.wto.org/english/docs_e/legal_e/17-tbt_e.htm.

[4]Standards are defined as the definition of terms; classification of components; delineation of procedures; specification of dimensions, materials, performance, designs, or operations; measurement of quality and quantity in describing materials, processes, products, systems, services, or practices; test methods and sampling procedures; or descriptions of fit and measurements of size or strength.

[5]Technical regulations are defined as those that governments specify are mandatory (legal) requirements that (1) must be met under specific laws and (2) implement general agency objectives. These requirements typically address health, safety, and environmental issues; consumer protection; and packaging and labeling. A regulation may consist of agency-developed technical specifications or requirements or may permit use of particular private sector standards as a means of compliance.

The WTO TBT Agreement recognizes that access to markets can be impeded by the use of technical regulations and standards that can vary from country to country and, if set arbitrarily, can be used or perceived as disguised market protection in the form of non-tariff barriers (NTBs) to trade.[6] In this regime, some standards bodies work to bridge technical regulations with standards by promoting that countries consider the use of international standards as one way of describing how to implement technical regulations. In some standards development processes, such as the International Organization for Standardization (ISO), which is a federation of national standards bodies of over 160 countries, and each member is a body that most broadly represents matters of standardization in its country, standards are developed as voluntary documents where they distil international consensus from a broad base of stakeholder groups. Members propose new standards and participate in their development (where ISO members appoint national delegations to standards committees). When the standard has reached consensus by technical experts, the draft standard progresses to ISO members for final vote or approval. Once an ISO standard is approved, it may be adopted as a national standard by ISO members.

In parallel with this model, a new standards development paradigm has emerged in the past few decades. Unlike the country-driven one, it is, from the very beginning of the process, globally open and market-driven, embodying thus both cooperation and open competition for innovation. Participation is open to all materially interested persons and experts from all around the world; there is no nation-centric (regional or national) membership, and neither a passport nor permission is needed.[7] In short, it is independent of representation of countries' standardization or technical regulation matters or agenda, and the standards produced avoid unnecessary national or regional specifications. Standards bodies working in this

[6]Non-tariff measures (NTMs) are policy measures that have an effect on trade flows. They can be imposed at a border, for example, on imports or exports, or applied in a domestic economy. When NTMs have a protectionist effect, they are referred to as non-tariff barriers.

[7]USTR Michael Froman speech, Brussels 2013. See: http://www.gmfus.org/events/ustr-froman-gives-first-major-address-brussels-transatlantic-trade-agreement-gmf

paradigm reach consensus and approval of standards developed through transparent procedures in which the participants and the broader stakeholder community comment and vote on the standards based solely on technical merit. Upon approval, via respective organization processes, the standards have a virtually global potential, where their ultimate global designation is determined by market adoption.

However, in some cases, for these standards to be recognized as "international" by national regulators and governmental agencies, they must go through a joint development or adoption process with one of the traditional, Geneva-based international standardization organizations. This is by all accounts a redundant and expensive process, tantamount to an unnecessary tax on innovation. Moreover, it creates other issues and complications in the context of intellectual property rights, in-country modifications, and coordination of standards revision or update cycles. This is a strange anachronism in a time where there is ample evidence of examples of voluntary standards that are globally relevant and deployed on international scale that were developed outside of nation-centric processes and that have enabled tremendous innovation and economic and social benefit. These include the collective of standards that form the foundation of the Internet and many of those that have transformed the way the world connects via wireless technology.

This brings to light the challenge of what defines an "international standard," and how standards developed outside the traditional nation-centric model are recognized in the current standardization ecosystem, and by extension, recognized in the public policy and trade agreement ecosystems and their resulting instruments. Unlike in the WTO Agreement of Sanitary and Phytosanitary Measures (SPS Agreement),[8] where international standards are defined as those developed by the Codex Alimentarius Commission, the International Plant Protection Convention, and the World Organization for Animal Health, there are no such definitions of international standards as a result of the WTO TBT agreement, and no organizations have been named as developers of international standards under the agreement. There has been discussion within the

[8]For full text of SPS Agreement, see: https://www.wto.org/english/tratop_e/sps_e/spsagr_e.htm

WTO TBT committee[9] about the need to provide greater clarity around the definition of international standards so that countries, in particular, developing countries, can decide where to channel financial and technical resources to participate in the work of international standardizing bodies. If defined narrowly as those organizations that have nation-centric processes, this will be limiting and negatively impact innovation and market growth. The nation-centric concept of standardization embodied in current WTO rules implies namely often significant government influence, and this may lead to the temptation to pick national or regional winners (Karachalios and McCabe, 2013). There should be thus caution on dialogues drifting in the direction of international standards being defined exclusively as those developed by specifically named organizations, leaving out of the context the thriving bottom-up, globally open collaboration platforms in the field of standardization and their impact on innovation and trade.

As Sean Heather, Vice President, Center for Global Regulatory Cooperation and Executive Director, International Policy and Antitrust Policy notes in his article *TTIP and Flawed Policies on Standards Don't Mix*, "In the context of trade, standards aren't important, but the way a government uses a standard is very important".[10] Although written from the perspective of Transatlantic Trade and Investment Partnership (TTIP)[11] and in response to a CEN/CENELEC[12] paper regarding TTIP and the Europe Single Digital Market,[13] the concepts outlined resonate on a broader scale in regard to recognition of a range of standards that objectively meet regulatory requirements that facilitate trade, and having

---

[9]WTO TBT Committee provides the opportunity of consulting on any matters relating to the operation of the Agreement or the furtherance of its objectives.

[10]See https://www.uschamber.com/issue-brief/ttip-and-flawed-policies-standards-don-t-mix.

[11]TTIP is a proposed trade agreement between the US and the EU to promote economic growth.

[12]European Committee for Standardization (CEN), the European Committee for Electrotechnical Standardization (CENELEC). See: http://www.cencenelec.eu/aboutus/Pages/default.aspx.

[13]The Digital Single Market strategy aims to open up digital opportunities for people and business and enhance Europe's position as a world leader in the digital economy. See: https://ec.europa.eu/digital-agenda/en/digital-single-market.

such standards is not a barrier or risk. Standards, including those developed outside of the nation-centric paradigm, can also meet regulator's requirements, increase competition, and offer a greater choice for consumers (Heather, 2015).

2.2. *Trade agreements and public policy*

Trade and standards are not policy objectives onto themselves. They do, however, become priority as a means to promote economic development. As we have observed, there is a strong interplay between trade agreements and standards, and this extends to public policy. Standards role, specifically in public policy, includes fostering increased innovation and productivity; enhancing competition in open markets through open standards; supporting international trade; and helping reinforce trust among trade actors. Standards can provide support to many kinds of policy actions, including trade and competition policy; investment policy; technical regulation; and consumer, security, and safety and environmental protection. They are powerful instruments of governance because of the effects their use can have on goods, services, and quality of life. These effects are evident whether standards are employed by the private sector or by the public sector. In the private sector, they create market incentives for actors to follow internationally accepted practices by applying competitive pressure (while allowing fair competition) and also encourage innovation and growth by fostering technological development when based on broadly available and open technology platforms. In the public sector, they can enable greater transparency and competition in public procurement and provide essential requirements for industry via their referencing into regulations and laws. In either context, standards can considerably improve efficiency and cost-effectiveness and reduce transaction costs (Hufbauer *et al.*, 2001).

Today, governments continue to be subject to pressures to assist domestic firms and industries, especially in difficult times. As incentives to use traditional trade policies such as tariffs and quotas have become more difficult to use, other types of non-tariff policies, such as standards and technical regulations, have become more attractive. While tariffs constitute an important source of distortions and economic costs, the relative role of tariffs in shaping international trade has declined, due in large part to successful rounds of multilateral tariff reductions in the WTO and its predecessor, the 1947 General Agreement on Tariffs and Trade (GATT), as well as further reductions in tariffs achieved through megaregional agreements (Role of Trade Barriers in SME Internationalization, 2006).

In the consideration of NTBs and why governments may prefer them over tariffs, suggestions in a 1997 OECD[14] Working Paper on Measurement of Non-Tariff Barriers (aka Non-Tariff Measures, NTMs), include salient reasons that hold even today, including institutional constraints built into tariff-related WTO rules in national constitutions and in trade policies; the role of industry or private enterprise in influencing policy; considerations of reaction to the policies of trading partners; and uncertainty about the ways in which different policies may perform (Deardorft and Stern, 1997).

As Pascal Lamy, former WTO Director General, commented in the forward to the *World Trade Report 2012*:

> "NTMs have acquired growing importance as tariffs have come down, whether through multilateral, preferential, or unilateral action. Secondly, a clear trend has emerged over the years in which NTMs are less about shielding producers from import competition and more about the attainment of a broad range of public policy objectives. You could say we are moving from protection to precaution. This tendency is discernible in practically every economy, as concerns over health, safety, environmental quality, and other social imperatives gain prominence. Moreover, issues such as these take on a more central role in policy as economies develop and incomes grow".[15]

3. Response

The significance of technical barriers to trade has increased considerably over the past years, as tariffs steadily decline and governments worldwide

[14]OECD: Organisation for Economic Co-operation and Development. See: http://www.oecd.org
[15]For an example, see http://www.eurotechnology.com/2013/08/05/galapagos-2/.

introduce more regulatory requirements to address health, safety, security, or environmental concerns. With this, public policy concerns find greater expression in trade relations than they did in the past. Experts believe that the expansion of the public policy agenda will mean that NTMs will not follow a path of diminishing relevance like tariffs have done, and NTMs will instead continue to rise (World Trade Report, 2012). Significant drivers of this include the rapid rate of technological development, the introduction of resulting new and emerging business models, the engagement of developing economies in the global trade system, and the need for governments to address their nations' economic health while addressing national and global issues of health, safety, and climate and environmental issues. This collectively has established not only an increasingly complex global trade system but also increased the complexity of the interplay of standards, public policy, and trade agreements. Subsequently, it has also brought to the forefront challenges regarding the capacity to implement trade agreements, particularly those on TBTs, which can constitute a hindrance to trade. Considering emphasis on trade as a means to underpin economic development, especially in developing countries, there is a crucial need to address the issue of standards and technical regulations to allow countries to participate effectively in the multilateral trading system.

3.1. *Standards arena*

When a geographically defined community must solve a community-wide coordination problem whereby all parties can realize mutual gains, but only by making mutually consistent decisions, it can adopt an existing standard or produce a new one. In the traditional standardization ecosystem, the geographic choices are a national standard, developed by national standards organizations, such as the AFNOR; a regional standard, developed by regional standards organizations, such as CEN or ETSI; or standards developed by the Geneva-based international standards organizations ISO, IEC, and ITU. In the context of trade, there are views that national, regional, or international standards are one way of overcoming technical barriers in inter-local or inter-regional commerce caused by differences among technical regulations and standards developed independently and separately by local (national) standards organizations.

However, by all its merits, this concept ignores and excludes standardization platforms and organizations that produce very successful global standards outside of the nation-centric standards development model. The traditional mental model further perpetuates the current limited definition and perception of international standards, primarily as those developed through country-centric processes. There is thus an urgent need to expand the concept of "international standardization" to include globally open standardization processes and standards produced through standards bodies adhering to principles of universally open and collective empowerment to strive to develop standards that are chosen and defined on technical merit, in particular as these standards are made globally accessible to all for implementation and deployment, and their adoption by the market is voluntary. It is ultimately this voluntary market adoption that determines their success, promoting thus economic growth through innovation, free choice, and development of new markets. When a country trades more intensely and is exposed to a variety of standards and standardization processes, its firms are exposed to a wider range of new products and new processes. This is where growth prospects are advanced. Firms that are exposed to world markets and global standards are quicker to adopt best practice techniques, increasing their own profits and also raising the productivity of their work force (Hufbauer *et al.*, 2001). A relevant case in point is illustrated by Sherif (2015). Using the concept of interactive learning spaces, Sherif explores how active participation in SDOs enhanced China's technical expertise and know-how. As there is an immense push worldwide to use technological innovations as an economic base for development, learning, innovation, and technology development operate within a National System of Innovation that defines how the various actors learn to interact, assimilate, develop, and leverage technological and market opportunities (Sherif, 2015).

3.2. *ICT sector: An example*

Standardization in the field of ICT is a complex discipline, the parameters of which are constantly changing as the industry underneath it evolves rapidly. The ICT industry has seen massive evolution in the past decade as the very nature of information use and sharing by customers has been changing at an accelerating pace. The ICT sector has been credited with making economies much more productive. Harmonizing the various forms of standardization by allowing equivalency, in legal as well as in economic settings, could enhance the industry.

Megaregional agreements could theoretically pose a threat for the openness of future ICT platforms, including open compatibility and interoperability standards, in particular if the temptation prevails to use regional integration as a tool to defend the new internal market against imports from "outside." However, as both developing and developed economies are relying on an ICT infrastructure and often global industry players not defined by the nationality of their incorporation or residence for improving quality of life, stimulating economies, and meeting global market needs, these agreements also offer an opportunity for the new generation of global standards and processes, those developed outside of the nation-centric model, to be recognized and implemented, which in turn can help advance the adoption of ICTs on a global scale.

3.3. *Cautionary tale*

Of caution is the use of protectionist practices that may get played out in standards setting processes. Trade barriers, specifically NTBs including TBTs, can sometimes be useful, even vital, but they can miss their mark and simply be an attempt to protect the interests of a given group at the expense of the wider community. TBTs, technical regulations, and standards that set out specific characteristics of a product such as its size, shape, design, functions, and performance, or stipulate the way a product is labeled or packaged before it enters the marketplace, may also include measures of technical procedures for confirming that products comply with the requirements stipulated in regulations and standards. Many of these measures serve legitimate goals of public policy, such as protecting the environment or human health and safety. At the same time, product standards and other TBTs have an important influence on market access and the export performance of businesses. They can be costly and burdensome by design, or effect, create a "Galapagos syndrome",[15] and restrict international trade. Coined to refer to Japanese 3G mobile phones, which developed specialized features and dominated Japan, but were unsuccessful abroad, the Galapagos syndrome illustrates how developing indigenous technologies, standards, or products that do not mesh with the global norm or other nations' requirements can result in it being more expensive for foreign competitors to sell products within a country because they would have to develop versions of a product that incorporates the indigenous technology or complies with the localized standard at added costs. This can advantage domestic producers at the expense of foreign ones. If technologies or products are developed for use in only a particular country's market, it will be difficult to export as well, also hindering cross-border trade.

In addition to these types of compliance procedures that are implicit in trade agreements is also the opportunity for trade agreements to address other regulatory factors, such as how products are produced. For example, trade agreements provide the opportunity to address a range of environmental issues as they strive to find balance between supporting open trade and ensuring governments can respond to growing environmental issues. In doing so, however, protectionist measures may purposefully or inadvertently be introduced. International trade can be a pathway for the transmission of weaker environmental policies from one country to another. For instance, a country's lax environmental standards that fail to internalize the social costs of environmental harms, such as pollution from the production of goods, can provide an unfair competitive advantage to its exports, raising concerns that this will create economic and political pressures in the importing country to also lower their environmental standards (Meltzer, xxxx). An example of how some trade agreements are addressing this challenge can be found in the KORUS Agreement, whereby it is stated that "it is inappropriate to encourage trade or investment by weakening or reducing the protections afforded in its environmental laws".[16]

A related concern that adds to the complexity of the standards, trade, and public policy interplay in the context of increases in trade and overall

[16] KORUS Art. 20.3.2. See https://ustr.gov/trade-agreements/free-trade-agreements/korus-fta/final-text.

economic activity among countries comes from environmentalists. They often point to trade increase as a cause for concern as a greater scale of economic activity likely means increases in transportation, shipping, production, and consumption, which are all pollution-emitting activities. Countering this concern is the position that higher productivity is associated with higher real incomes and greater prosperity, which in turn, can benefit the environment in multiple ways. Latter include the creation of opportunities for investment in R&D in clean technology. Higher real incomes can also generate greater ability and willingness to adopt, enforce, and pay for higher standards of environmental quality (Economic Benefit of U.S. Trade, 2015). But how environmental protection measures get incorporated into trade agreements, with a downstream effect in standards development and technical regulation, is an area for deeper exploration and research. In the EPO's *Scenarios for the Future* "Whose Game?" scenario, a trend where Western businesses caught between the need for increased environmental protection and the pressure of low-cost Asian competition is noted. In this scenario, Western governments respond creatively, and barriers in TAFTA[17] negotiations are eventually defined by social and environmental protocols denoting that only products meeting the highest standards of environmental care and good working practice would be admitted to the US and the EU.[18] Extending this scenario to standards development, where product standards that include measurements and test methods are being considered, one can see how a standard can serve a dual purpose: that of addressing an environmental concern but also potentially raising a protectionist wall. This opens the question of how industry-driven global standards bodies will address protectionist behavior, when some actors may try to influence or manipulate the development of their standards. In addition, it presents the dilemma and opportunity for standards bodies to explore and prepare for new categories of actors that may become engaged, such as privacy and civic right experts, environmental NGOs, etc.

---

[17] Transatlantic Free Trade Agreement (TAFTA), also known as Transatlantic Trade and Investment Partnership (TTIP).

[18] See http://www.epo.org/news-issues/issues/scenarios/download.html.

In addressing open standardization, most actors claim the high ground of openness and acting in the public interest. These claims deserve examination and thought, given the policy decisions that governments, standards organizations, and private parties or enterprise are now being called upon to make in reshaping and redirecting the course of national, regional, and international policies and institutions. Technical standards will continue to help define the future of the information and knowledge society and what and how technologies and markets of the future will be (Schoechle, 2009). There is a need for the study of standardization to include also a framework of analysis of public policy aspects in the global economy and trade systems. The analysis should also include the related issues of technical innovation, access to information, intellectual property, small and medium enterprise needs, and emerging economic development and engagement of developing countries in international economic and trade frameworks.

3.4. *Framework considerations*

The instruments of standards, trade agreements, and public policy are functioning in a new era of 21st century regionalism and rise of megaregional trade agreements, where a significant focal point in negotiations centers on trade facilitation, advancing regulatory best practices, and removal of technical barriers to trade within the target region. The application of product regulations and standards is becoming increasingly contentious as an implicit NTB to trade. There has been increasing use of technical regulations as instruments of commercial policy in the context of multilateral, regional, and global trade as well. NTBs are of special concern to developing countries, which may bear additional costs in meeting mandatory standards. Moreover, many industrial and developing countries express frustration with regulations that vary across their export markets, require duplicative conformity procedures, and are continually revised to exclude imports. Further, in this mix is acknowledgment that among the most important policy tools advanced in regional trade talks are the use of mutual recognition of testing and certification procedures and related efforts to harmonize standards, including settling on one international standard (Maskus and Wilson, 2001).

On October 4, 2015, Ministers of the 12 Trans-Pacific Partnership (TPP) countries[19] announced conclusion of their negotiations, resulting in what some have coined as a "21st century trade agreement" since it focuses on services trade, e-commerce, and intellectual property rights and not just physical goods, and it sets a new standard for global trade while addressing next-generation issues. In short, it has a broad agenda to reflect the complex international trade ecosystem including new trade challenges brought on by the digital economy and the engagement of developing economies in global markets. In the TPP Agreement Chapter Eight on Technical Barriers to Trade, the TPP Parties have agreed on transparent, non-discriminatory rules for developing technical regulations, standards, and conformity assessment procedures and to cooperate to ensure that technical regulations and standards do not create unnecessary barriers to trade. As per the US Trade Representative's *Summary of the Trans-Pacific Partnership Agreement*,[20] specifically on the Technical Barriers to Trade Chapter:

> "TPP Parties agree to rules that will facilitate the acceptance of the results of conformity assessment procedures from the conformity assessment bodies in the other TPP Parties, making it easier for companies to access TPP markets. Under the TPP, Parties are required to allow for the public to comment on proposed technical regulations, standards, and conformity assessment procedures to inform their regulatory processes and to ensure traders understand the rules they will need to follow. They also will ensure a reasonable interval between publication of technical regulations and conformity assessment procedures, and their entry into force, so that businesses have sufficient time to meet the new requirements."

As noted in this chapter, with the lowering of tariff rates, the behind-the-border measures of technical standards come to the forefront as one of the

[19] Australia, Brunei Darussalam, Canada, Chile, Japan, Malaysia, Mexico, New Zealand, Peru, Singapore, US, and Vietnam.

[20] See: https://ustr.gov/about-us/policy-offices/press-office/press-releases/2015/october/summary-trans-pacific-partnership

most significant barriers to trade. The TPP Agreement in the Articles of Chapter Eight[21] call out measures to meet the objectives of facilitating trade, eliminating unnecessary barriers to trade, enhancing transparency, and promoting greater regulatory cooperation and good regulatory practice. As elements of frameworks for an effective interplay of standards, trade agreements, and public policy are explored, examining the TPP Agreement's measures can provide potential insight, notably in the context of reform of the governance of standardization.

In TPP Article 8.3 (Scope), it states:

> "Each Party shall take such reasonable measures, within its authority, to encourage observance by local government bodies on the level directly below that of the central government within its territory which are responsible for the preparation, adoption, and application of technical regulations, standards, and conformity assessment procedures with Articles 8.5 (International Standards, Guides and Recommendations), 8.6 (Conformity Assessment Procedures), 8.7 (Compliance Period for Technical Regulations and Conformity Assessment Procedures), and each of the Annexes to this Chapter."

There are specific measures within the respective Articles of 8.5, 8.6, and 8.7 that acknowledge the important role that international standards, guides, and recommendations can play in supporting greater regulatory alignment, good regulatory practice, and reducing unnecessary barriers to trade (8.5). Article 8.7 (Transparency) specifically calls out that "Each Party allow persons of the other Parties to participate in the development of technical regulations, standards, and conformity assessment procedures by its central government bodies" and that a Party may satisfy this obligation by providing interested Parties a reasonable opportunity to provide comment on the measures it proposes and taking those comments into account in the development of the meas-

[21] The scope of Chapter 8 applies to the preparation, adoption, and application of all technical regulations, standards, and conformity assessment procedures of central government bodies (and, where explicitly provided for technical regulations, standards, and conformity assessment procedures of governments on the level directly below that of the central government) that may affect trade in goods between the Parties, except as provided in paragraphs 3 and 4.

ure. It further denotes that "Each Party is encouraged to consider methods to provide additional transparency in the development of technical regulations, standards, and conformity assessment procedures, including public outreach or consultation."

Chapter Eight Article 8.5 acknowledges the WTO TBT Agreement and in its rules reference Articles 2.4 and 5.4 and Annex 3 of the WTO TBT Agreement "in determining whether an international standard, guide, or recommendation within the meaning of Articles 2 and 5 and Annex 3 of the TBT Agreement exists, each Party shall apply the Decision of the TBT Committee on Principles for the Development of International Standards, Guides and Recommendations With Relation to Articles 2, 5 and Annex 3 of the TBT Agreement (G/TBT/1/Rev.10), issued by the WTO Committee on Technical Barriers to Trade." It further states that the "Parties shall cooperate with each other, where feasible and appropriate, to ensure that international standards, guides, and recommendations that are likely to become a basis for technical regulations and conformity assessment procedures do not create unnecessary obstacles to international trade."

The TPP impacts the development of international standards by building trust and developing habits of cooperation among TPP regulators which can help make future cooperation in the setting of international standards more likely. However, in a spirit of cooperation and collaboration, its provisions in Chapter Eight, notably those in the context of the WTO TBT, do not necessarily address the inclusion of SDOs who develop global standards outside of the nation-centric standards development model. Although the TPP has no stringent commitment to harmonize domestic standards to international standards, a caution is observed in that TPP Parties could seek to harmonize standards among themselves, which can be costly for some TPP countries if such standards diverge from international standards or cover areas not yet addressed by international standards. On the other hand, TPP can lead to improved coordination among public (mandatory) and private (voluntary) standards bodies to cooperate to develop standards. Some argue that TPP could be used to facilitate greater cooperation among domestic standards setting bodies that underpin the development of international standards. But this level of cooperation and coordination, coupled with commercial conditions changing in the main markets abroad and the potential for private or voluntary standards to become mandatory when public standards reference private standards, in short, bringing compliance with private standards within their scope, calls for regulators, policy makers, and standards developers to prepare for such conditions emerging in the markets of TPP economies (Meltzer, 2015).

As a fresh examination of the instruments of standards, public policy, and trade agreements is initiated in the context of forging new frameworks for effective interplay of these instruments, consideration of the TPP Agreement Article 8.11 provision for the formation of a Committee on Technical Barriers to Trade comprised of representatives of each Party to intensify their joint work in the fields of technical regulations, conformity assessment procedures, and standards with a view to facilitating trade between and among the Parties is suggested. The Committee's scope includes considering proposals for new sector-specific or other initiatives; encouraging cooperation between and among the Parties in matters pertaining to Chapter Eight, including the development, review, or modification of technical regulations, standards, and conformity assessment procedures; encouraging cooperation between and among non-governmental bodies in the Parties' territories, as well as cooperation between governmental and non-governmental bodies in the Parties' territories in matters pertaining to Chapter 8; facilitating the identification of technical capacity needs; and encouraging the exchange of information between and among Parties and their relevant non-governmental bodies, where appropriate, on the development of common approaches regarding matters under discussion in non-governmental, regional, plurilateral, and multilateral bodies or systems that develop standards, guides, recommendations, policies, or other procedures relevant to Chapter Eight.

With an increasingly complex trade ecosystem, coupled with rapid technology advancement and the drive for sustainable development around the globe, all of which are impacting the interplay of standards, trade agreements, and public policy, a fresh perspective is warranted on how these instruments and the paradigms in which they are developed can enhance innovation and drive economic growth without inserting unnecessary barriers or protectionist measures. A new integrated framework that takes into account 21st century facts, the anticipated future state of technology advancement, and the entry of developing countries onto the trade scene is needed.

Such a framework has not yet emerged. However, it is useful in the formation of such a framework to take the following into account to inform research agenda and discussion, and to ensure that diverse needs and conditions are reflected and durable and reasonable solutions are proposed:

- Innovation supporting trade in the 21st century depends on many factors, including modern and open standards development systems. Today, many standards bodies are high-stakes forums for techno-political decision-making, going beyond mere technical optimization. Commercial stakes, as well as those related to environmental, safety, and health considerations, but also cultural values such as privacy and ethics, are bringing to the forefront a new era of standardization, which needs to be both global and inclusive. In addition, the standardization processes must have sufficient clear rules, also regarding the inclusion of privately owned technologies in the standards. Finally, they must be sufficiently nimble to effectively address the development and commercial application of rapidly evolving technologies with increasingly shorter life-cycles, outpacing traditional, consensus-based decision-making processes. Only then would formal standardization frameworks have a chance to keep pace with these developments and cope with rapidly evolving global trade landscapes and an extending public policy agenda. Otherwise they will inadvertently be replaced by fast-moving private consortia or fall prey to practices of forced localization and "indigenous innovation" strategies or protectionist measures.
- Viewed in the context of current megaregional agreements in negotiation, an opportunity exists to do more on regulatory cooperation, including with regard to standards when those standards are used to comply with government or technical regulation. Such cooperation would also facilitate market development by promoting diffusion of enabling technologies, could improve production efficiency, foster interoperability, and address needs such as safety, security, and accessibility at a scale beyond national and regional boundaries. Standards developed outside of the traditional country-centric paradigm have a

bigger role to play here and should be formally recognized with a refined definition and open perspective of what constitutes an "international standard." Such an approach would put on equal-standing standards developed in a traditional country-centric mode with the ones that are globally open and inclusive from the beginning, supporting thus the availability and use of more than one standard that can meet technical regulatory objectives. This broader definition would lean on the criteria of the WTO TBT agreement that standards bodies must meet in order to have their standards qualified to be considered international in nature.

- From a research agenda perspective, the dynamics, governance issues, strengths, and vulnerabilities of standards development processes that run on open processes and which are inclusive of all stakeholders is suggested. A point of reference is the creation of the W3C that was formed out of concern and realization that it would be challenging for any standards-setting body to keep up with the rapid pace of technical developments related to the Web, and in response, Tim Berners-Lee developed the W3C to meet this need (Simcoe, 2007, p. 270). Analysis of such processes or systems can potentially provide insight that can benefit new models that take a multipronged, multistakeholder, and collective approach in addressing challenges that arise during the pursuit of new frameworks.
- The forms of standardization serve purpose, especially in the ICT sector. There is the need for stability (provided by the arena of formal standards bodies), coping with rapid change (provided by ephemeral consortia and alliances), specific intellectual property and marketing environments, and the need for robust community involvement (provided by Open Source). To tackle the vast emerging standardization needs for mega systems, such as Internet of Things, 5G, smart cities, and other smart "Xs" (homes, vehicles, connected person, e-health, etc.) the groups within each arena will need to more effectively work together also in order to achieve a public good character of standards through open systems and noting that open standardization is preferred compared to closed proprietary solutions (Schoechle, 2009).
- Due to the proliferation of various non-tariff rules and regulations affecting international trade, trade policy has become complex and

multifaceted. Understanding the uses and implications of these trade policy instruments is essential for the formulation and implementation of effective development strategies, as well as standardization strategies. A better understanding of non-tariff measures will help policy-makers to formulate appropriate policy responses and direct the necessary technical and financial resources to where they are needed (UNCTD, 2012).

- Further, programs to help policy-makers understand and use standards to support public policy initiatives would be beneficial, inclusive of such insight into referencing standards in legislation or regulation, as well as leveraging standards to support public decisions or actions. Engagement in standards fora would allow policy-makers to stay abreast of developments in global standardization they may find useful to support or inform policy goals. Vice versa, standards bodies engaged with public policy can collect feedback on experiences of policy-makers in using standards and gain insight into public policy challenges that could be addressed by the development of global standards.
- Finally, trade policy takes into account of and is adapted to specific development needs and situations of trading partners. Today's era of globalization depends on sound trade policies to reflect markets evolution, establish fair trade practices, and expand the possibilities for international trade. With this, policy-makers should work toward standardization policies facilitating global partnerships that support developing countries in building capacity to implement technical standards, notably as developing countries are beginning to overcome some major obstacles in their efforts to expand trade with industrial countries. As noted in this chapter, traditional trade protection measures such as tariffs and quotas are falling away, but to some extent they are being replaced by domestic technical regulations permitting countries to ban products from entering markets if the products do not meet certain standards. This may be more prevalent in the SPS space, but as standards and public policy broaden their respective agenda, these types of protective measures can carry over to TBTs.

4. Conclusion and recommendations

Standards, public policy, and trade agreements are strategic instruments with similar objectives to create value and aggregate markets, facilitate technology diffusion, promote production efficiency and product interoperability, enhance competition, reduce costs, address safety, and enable the communication of important information among buyers and sellers. These instruments are functioning in the context of political, economic, social, technical, and other forces that are rapidly reshaping the global economy. Megaregional dynamics add to this complexity and have the potential to significantly influence the direction and shape of the emerging regional and global trade architectures.

Increased economic growth and expansion of exports in the 21st century will depend on many factors. These include efficient and modern open standards systems and effectively managing technical barriers to trade. There are indeed proponents for regional efforts as an efficient way to modernize standards systems. To illustrate this, Hufbauer *et al.* (2001) take a regional rather than a national approach in their suggestion to setting up accreditation, testing, and metrology infrastructure to share equipment, experts, and information so as to optimize limited funding and to promote regional bodies as venues to develop common positions in international discussions of the development of standards. From the megaregional perspective, the opportunity for promotion and use of globally open standardization processes and their outcome, as well as supportive regulations, can serve as means to increase economic activity and social inclusiveness in several regions by enhancing cross-border innovation and trade without, at least, formal discriminations across regions.

Governments create technical regulations by mandating technical requirements, which may include procedures for testing, conformity assessment, and ongoing compliance. These requirements may embrace internationally recognized procedures or standards or they may be unique to a country or region. It is with great caution that technical regulations should be considered as they can limit manufacturing flexibility, inhibit innovation, delay the time to reach market, and distort product design. With this, the objectives for technical regulations should be to ensure safe and legal products and not to promote certain technologies (where there are alternatives) or promote "national champions." Technical regulations should never be more trade-restrictive than necessary, and governments should consider alternatives whenever possible. If technical regulations are necessary, governments should fairly consider the costs and time to-market delays associated with implementation and enforcement.

As we look at trade agreements and regulatory coherence, particularly in the recently agreed upon TPP agreement, some positive elements emerge for consideration in formulating improved frameworks. These include easing the conditions and costs of trade between TPP countries while affirming the rights of TTP countries to regulate their economies to promote legitimate policy objectives. According to the USTR, this initiative stems from the proliferation of regulatory barriers and NTBs, which have become a major hurdle for businesses seeking to gain access to foreign markets. Some of the goals are to "improve regulatory practices, eliminate unnecessary barriers, reduce regional divergence in standards, promote transparency, and conduct regulatory processes in a more trade-facilitative manner" (Fergusson and Williams, 2015).

Overall, there are many questions that still need to be unpacked in order to understand how current standards and technical requirements are today interacting with public and government entities, policy frameworks, and public-policy–making processes. This includes an examination of the relationships during standards and standards systems' development as well as of standards implementation and oversight, and in measuring their respective impacts. By developing a clear understanding of these issues, an additional piece of the puzzle of today's challenges may be addressed. A framework that builds upon both existing public- and private-sector–led processes and new multistakeholder initiatives to the benefit of all has the potential to significantly advance governance models and frameworks and the creation of policy and implementation tools that can effectively address the complexity of today's interlinked issues in global trade.

Innovation and economic growth do not come with a unique model. They can be achieved through different combinations of standards, agreements, and public policy, and collectively working toward a new framework to identify ways to improve the current paradigms and strategic interplay of these instruments.

References

Baldwin, R. (2014). "Multilateralising 21st Century Regionalism," *Global Forum on Trade Reconciling Regionalism and Multilateralism in a Post-Bali World*, Organisation for Economic Co-operation and Development, Paris, France, February.

Blind, K. (2013). "The Impact of Standardization and Standards on Innovation — Compendium of Evidence on the Effectiveness of Innovation Policy Intervention." Manchester Institute of Innovation Research, Manchester Business School, University of Manchester.

Clougherty, J. A. and M. Grajek (2012). "Empirical Evidence From ISO 9000 Diffusion." NBER Working Paper Series International Standards and International Trade 18132, June. Available at: http://www.nber.org/papers/w18132.

Elahi, S. (2007). *Scenarios for the Future: How Might IP Regimes Evolve by 2025? What Global Legitimacy Might Such Regimes Have?* (European Patent Office, Munich).

European Patent Office (2007). *Scenarios for the Future* (Head of Controlling Office, Munich).

Executive Office of the President of the United States Council of Economic Advisers (2015). *The Economic Benefits of U.S. Trade* (Washington D.C.).

Fergusson, I. F., M. A. McMinimy, and B. R. Williams (2015). *The Trans-Pacific Partnership Negotiations and Issues for Congress* (Congressional Research Service, Washington, D.C.).

Heather, S. (2015). "TTIP and Flawed Policies on Standards Don't Mix.", July 14. Available at: https://www.uschamber.com/issue-brief/ttip-and-flawed-policies-standards-don-t-mix.

Hufbauer, G. C., B. Kotschwar, and J. S. Wilson (2001). *Trade Policy, Standards, and Development in Central America* (World Bank, Washington, D.C.).

Karachalios, K. and K. McCabe (2014). *Standards, Innovation, and their Role in the Context of the World Trade Organization. E15Initiative* (International Centre for Trade and Sustainable Development (ICTSD) and World Economic Forum, Geneva). Available at: www.e15initiative.org.

Love, P. and R. Lattimore (2009). "Protectionism? Tariffs and Other Barriers to Trade." In *International Trade: Free, Fair and Open?* (OECD Publishing, Paris). http://dx.doi.org/10.1787/9789264060265-5-en.

Maskus, Keith E. and John S. Wilson, (eds.) (2001). *Quantifying the Impact of Technical Barriers to Trade: Can It Be Done?* (University of Michigan, Ann Arbor, MI, USA).

Meltzer, J. P. (2014). "The Trans-Pacific Partnership Agreement, The Environment and Climate Change." In *Trade Liberalisation and International Co-operation: A Legal Analysis of the Trans-Pacific Partnership Agreement*, T. Voon (ed.) (Edward Elgar).

Meltzer, J. P. (2015). "Standards and Regulation in the Trans-Pacific Partnership Agreement: Implications for India." International Institute for Sustainable Development, Manitoba, Canada.

Schoechle, T. (2009). *Standardization and Digital Enclosure: The Privatization of Standards, Knowledge and Policy in the Age of Global Information Technology* (Information Science Reference, Hershey, PA, USA).

Simcoe, T. (2007). "Delay and De Jure Standardization: Exploring the Slowdown in Internet Standards Development." In *Standards and Public Policy,* S. Greenstein and V. Strango (eds.). (Cambridge University Press, New York), p. 270.

Sherif, M. H. (2015). 'ICT Standardisation Strategies and Interactive Learning Spaces — The Case of China," *International Journal of Technology Marketing*, 10(2), pp. 113–136.

Swann, G. M. P. (2009). *The Economics of Innovation: An Introduction* (Edward Elgar, Cheltenham).

United Nations Conference on Trade and Development (2013). *Non-Tariff Measures To Trade: Economic and Policy Issues for Developing Countries Developing Countries in International Trade Studies* (United Nations Publication).

Chapter 14

Beyond Government Control of China's Standardization System — History, Current Status, and Reform Suggestions*

Ping Wang

China National Institute of Standardization, China
pingwang100@qq.com

Zheng Liang

Tsinghua University, China
liangzheng@tsinghua.edu.cn

1. Introduction

This chapter traces the ancient beginnings of standardization in China; the more recent evolution of standardization within the Chinese system, including the formation of governmental management systems in the

*This chapter was originally prepared as a think-piece for the East-West Center NSF Workshop *Mega-Regionalism — New Challenges for Trade and Innovation* (MCTI), held on January 20 and 21, 2016, at Honolulu, Hawaii.

planned economy period; the adjustment of the system after the period of reforms and opening-up; and an overview of the current standardization system, the evolution of the government-dominated standardization strategy, and its achievements and weaknesses. This chapter argues that the top-down, government-centered standardization system, which was instrumental to China's successful catching-up, now needs to adjust to new challenges.

The current standardization management system of the government was primarily formed during the planned economy period and is deeply rooted in Chinese tradition and history. The reform measures and the strategies adopted by the successive Chinese governments in the field of standardization have proven to be generally successful in the country's economic development. The technology path-following strategy and the standard "dual-adoption policy", adopted at the initial stages of the period of reform and opening-up (1980–2000), allowed for a great deal of technology transfer in the process of standardization. Standardization became an effective channel for the technology spillover effect through international trade. In China's period of rapid economic development (2000–2010), the strategy of using standardization to foster and promote indigenous innovation by the government was also very effective, and weaknesses in this system were not obvious during that period. However, after the slowdown in economic growth started (2010–2015), problems caused by the government's role in standardization have become increasingly prominent, including excessive governmental intervention as well as institutional fragmentation and multiple decision-making entities within the government, creating discordance and hindering innovation. The authors argue that government withdrawal, for the most part, from voluntary standardization management is a prerequisite for cultivating self-organized private standardization.

2. Overview of the standardization institution and strategy

2.1. *Historical overview*

Through much of China's history, the government has been involved in standardization management.[1] Qin Shihuang (秦始皇) — The First

[1]The ancient standardization that emerged in China is different from modern industry standardization and is sometimes described as nascent or embryonic standardization.

Emperor (259 BC–210 BC) of Qin Dynasty (秦朝), pioneered the use of the emperor's authority in the country to unify the written characters (or scripts, which had varied greatly from area to area during the Warring States Period before Qin Dynasty), carriage axle length (to allow the royal carts to use the same tracing ruts or grooves on roads), as well as weights and measures. Currency became standardized as a circular copper coin with a square hole in the middle. Unified written characters and carriage axles allowed the Emperor's decrees to reach and be understood throughout the country. The unified weights and measures and coinage became an important infrastructure for economic development of the Empire (Wang, 2008). All the later dynasties of imperial governments from the Sui Dynasty (隋朝, 581–618) onward maintained a department (工部) for royal architecture construction, water conservancy, and opening up uncultivated land for growing food grain, etc. (Zhao *et al.*, 2002). In addition to royal project management, the officers also had responsibility for drafting project management guidance and preparing construction budgets and technical standards and specifications. Historically, they played a role of standardization officials to a certain extent, compiling important monographs similar to technical regulations and specifications, such as *Jun Qi Fa Shi* (军器法式), *Bing Qi Zong Yao* (兵器总要), *Ying Zao Fa Shi* (营造法式), and *Gong Cheng Zuo Fa Ze Lie* (工程做法则列).[2]

Around the time of the collapse of the Qing Dynasty (清朝), approximately at the beginning of the 1900s, modern industries and technologies such as architecture, engineering, railway, textile, steel, shipbuilding, telegraphy, and telecommunication began to be introduced into China from the West. Driven by the international "rationalization and standardization," in 1931, an Industrial Standard Committee (ISC) was set up by the Ministry of Industry (the predecessor of the Ministry of Economy) of the Republic of China, (the Kuomintang government), covering the areas of textile, automotive, medicine, machinery, metallurgy, and electric

---

[2] *Jun Qi Fa Shi* (军器法式) and *Bing Qi Zong Yao* (兵器总要) are specifications for arms manufacturing and powder formulation from Northern Song Dynasty (960–1126); *Ying Zao Fa Shi* (营造法式) is an architectural monograph including technical specifications for construction, also from Northern Song Dynasty; *Gong Cheng Zuo Fa Ze Lie* (工程做法则列) is a technical specification for the palace architecture from Qing Dynasty (1734). (See Li *et al.*, 2001, p. 26; Wang, 2009.)

power. The Standard Law was promulgated in 1946. In 1947, the Bureau of Weights and Measures and the ISC merged and became the Central Bureau of Standards, establishing a standardization management system lead by the government from that point onward (Zhong, 2012).

2.2. *The planned economy period (1949–1964)*

From the establishment of the People's Republic of China to the Cultural Revolution (1949–1964), China formed a standardization management system that was applicable to the planned economy system and classified according to sectors.[3] Though standardization management was under the unified leadership of the Central Government, each department and local government under the State Council took responsibility for its own work. The standards system included national standards, ministerial standards (transferred to sector standards later), and local standards (Deng *et al.*, 1986, p. 36). Such standards were prepared by government bodies, and enterprises had no right to set technical standards independently. The function of standardization, from the very beginning, was a means for the government to organize large-scale production in the planned economy. Thus, all standards at that time were mandatory, and enterprises were required to comply with them. The government was responsible for establishing standardization projects and organizing technical experts to prepare and publish standards. Some government-affiliated scientific research institutions had become the main force for preparing standards. By 1963, 32 research institutes and design organizations affiliated with the government had been designated as the core organizations of national standardization (later changed to technical focal points for standardization) (Deng *et al.*, 1986, p. 385).

---

[3] In this chapter, a sector department of government refers to a department that is in charge of a special sector (industry). Then we have phrases such as sector standard (行业标准) and sector standardization (行业标准化), which mean the standard issued by it and the standardization organized by it, respectively. For the meaning of sector, see: "What is the Difference Between an Industry and a Sector?" Available at: http://www.investopedia.com/ask/answers/05/industrysector.asp.

2.3. *The reform and opening-up (1978–2000)*

Along with the acceleration of China's reform and opening-up, the government began to implement a "planned commodity economy" and loosen its control on enterprises. Thus, joint ventures and capital flows became possible. In this context, the State Council issued the "Standardization Law of the People's Republic of China" in 1989. Though the government still maintained the management system established in the planned economy period, it made three major improvements.[4] First, standards began to be classified into mandatory ones and recommendatory ones. The relevant requirement was: "Those for safeguarding human health and ensuring personal and property safety and those for compulsory execution as prescribed by the laws and administrative regulations shall be compulsory, while others shall be recommendatory." Second, it was specified that a technical committee (TC) for standardization shall be established for developing standards in every technical field. Third, enterprises might independently set standards for themselves. Though this system design was still one of standardization under government management, it made important adjustments to the standardization system. Committee standardization was adopted to get close to internationally common practices. Meanwhile, mandatory standards in China were greatly reduced, and enterprises began to enjoy certain autonomy in their own standardization.

After the reform and opening-up, China rejoined international standardization organizations (ISO, IEC, and ITU) (Li *et al.*, 2001, p. 189, 198, 204), developing extensive and friendly communications with standardization organizations of developed countries, and establishing favorable standard exchanging relations. Meanwhile, the China State Bureau of Technical Supervision (CSBTS) established the China Standard Information Centre in 1987. A total of 18 departments (ministries) of the State Council and 23 provinces, autonomous regions, and municipalities established standards information research institutes

---

[4] See: *Law of the People's Republic of China on Standardization* (中华人民共和国标准化法). Available at: http://www.npc.gov.cn/wxzl/gongbao/1988-12/29/content_1481259.htm

(Li *et al.*, 2001, p. 593).[5] The country therefore formed a large standards information network, transferring Western standards information to TCs for standardization and to enterprises and research institutes throughout the country.[6]

In 1993, the former State Bureau of Technical Supervision began to promote a "dual-adoption policy" in the field of standardization — namely that international standards (such as ISO, IEC, and ITU[7] standards) and foreign advanced standards (such as ASTM, IEEE, DIN, and JISC standards)[8] should be mainly adopted during the preparation of various Chinese standards. This was proposed under historical conditions, where there was a large gap between China's overall technical level and competitiveness and those of developed countries. The "dual-adoption policy" was also adopted as an important component of China's technology path-following strategy, in order to introduce foreign advanced technology.

2.4. *The strategy of standards supporting innovation after entering the 21st century (2000–present)*

China had already experienced an important stage of technology and capital accumulation when stepping into the 21st century. Its industry experienced a gradual integration into the global economy, innovation capability was strengthened gradually, and the technical standards strategy was significantly changed accordingly. The "dual-adoption policy"

[5] China National Institute of Standardization (CNIS), China Electronics Standardization Institute (CESI), China Metallurgical Information and Standardization Institute (CMISI), Communications Standards Research Institute (CSRI), Standard Quota Institute of MOHURD, Beijing Institute Standardization (BIS), Shanghai Institute of Quality and Standardization (SIS), Shenzhen Institute of Standards and technology (SIST), etc., all developed starting from the 1980s.

[6] This network is still running now and, with greater expansion of IT services and the Internet, has become a very effective distribution platform for standards information.

[7] ISO: International Standardization Organization; IEC: International Electrotechnical Commission; ITU: International Telecommunication Union.

[8] ASTM: American Society for Testing Materials; IEEE: The Institute of Electrical and Electronics Engineers; DIN: German Institute of Standardization; JISC: Japanese Industrial Standards Committee.

was eventually diluted. The government began to propose an ambitious standardization strategy to support so-called "independent innovation." The enhancement of economic strength had successively promoted domestic industry's technological innovation and enthusiasm toward standardization. Domestic standardization became increasingly active. TCs in various fields actively organized domestic experts to participate in international standardization, changing from passive tracking and learning to substantive participation in the preparation of international standards, and sought more opportunities to turn domestically produced technologies into international standards. The academic community has noticed and analyzed a number of cases of standards supporting independent innovation, including EVD, TD-SCDMA, TD-LTE, AVS, IGRS, GT800, EPA, UOF, and WAPI.[9] These examples reflect China's efforts to develop independent innovation technologies into national standards and international standards, with its capability to participate in international standardization also showing great improvement in a short time.

However, the government experienced great pressure from abroad when developing the policy of standards supporting independent innovation. In 2013, the Standardization Administration of China (SAC) and the State Intellectual Property Office (SIPO) jointly issued the "Interim Management Provisions for National Standards Incorporated with Patents".[10] The process of developing this document intensively embodied the government's attempts to support independent innovation properly at the policy level, but it encountered strong opposition from multinational corporations (MNCs). The SAC's original intention was to not only incorporate useful aspects of the existing patent policies of international standardization organizations but also find a balance between encouraging

[9]EVD is a standard for digital video players; TD-SCDMA and TD-LTE for mobile telephony; AVS for audio–video encoding/decoding; IGRS for home networking; GT800 for digital trunking; EPA for Ethernet communication tools for the plant automation; UOF for Chinese Office Software Document Format Specification; WAPI for symmetric encryption algorithm in WLAN.

[10]Notice of SAC and SIPO to Release Announcement for "Management Temporary Provisions for National Standards Incorporated with Patents" (国家标准委 国家知识产权局关于发布《国家标准涉及专利的管理规定(暂行)》的公告). Available at: http://www.law-lib.com/law/law_view.asp?id=442588.

indigenous innovation and properly protecting intellectual property (Wang and Lee, 2014, p. 937). However, the consultative draft of this policy, published in 2009, has been criticized by MNCs as too strict, and it was argued that the specified license fees, being far lower than usual (Willingmyre, 2009), would damage the interests of foreign enterprises. The SAC has also received more than 2,000 pages of comments from the US (Ernst, 2011, p. 61). Thus, China was required to conform more closely to international norms to protect intellectual property. This demonstrates the difficulties latecomers may have in the global economy in supporting independent innovation through ambitious domestic policy. Although in certain cases, supporting independent innovation can be a powerful leverage to create significant positive change, even gaining support from MNCs, it can also create dilemmas for governments while designing general/common policies supporting independent innovation at the macro-level (Wang and Lee, 2014, p. 937, 939).

2.5. *The current standardization institution*

China's current standardization system is fully managed by the government, including the SAC (under AQSIQ[11]) in charge of national standardization, the bureaus or ministries under the State Council in charge of sector/ministerial standardization, as well as the local governments in charge of the standardization at the local level.

The SAC mainly relies on the TCs for standardization in each field to organize standardization actives at the national level. The SAC has maintained 537 TCs for standardization (sub-committees are not included).[12] Most of these TCs are established jointly with sector departments of the government. Some secretariats are located in the technical focal points for standardization affiliated to SAC or sector departments.

Sector/ministerial standardization involves various technical fields. There are 64 sector (industry/ministerial) standard codes.[13] In the planned

[11] AQSIQ: Administration of Quality Supervision, Inspection and Quarantine.
[12] SAC, "Query for Professional Technical Committee." Available at: http://www.sac.gov.cn/SACSearch/search?channelid=61613.
[13] SAC, "Management Rules for Sector Standards (行业标准管理办法), 1990." Available at: http://www.law-lib.com/law/law_view.asp?id=6886.

economy era, each code corresponded to a field of sector standardization in which a government department was in charge of the management. After the reform and opening-up, several sector departments have been combined or closed. The management responsibilities of many corresponding sector/ministerial standardization have been transferred to the Ministry of Industry and Information Technology (MIIT), giving MIIT jurisdiction over a number of sector/ministerial standards, including 18 sector standard codes. However, some departments responsible for sector (industry) affairs are still independently in charge of the management of sector/ministerial standardization, including the Tourist Administration, the Ministry of Housing and Urban–Rural Development, the National Railways Administration, and the Civil Aviation Administration of China, among others.

Local standardization involves 22 provinces, five autonomous regions, and four municipalities. For each local government, the local Technical Supervision Bureau is in charge of the standardization for the region, while specific standardization generally relies on the local Institute of Standardization (like BIS, SIS, SIST, etc.) and local Associations for Standardization.

After the period of reform and opening-up, all levels of government have promoted the establishment of many standardization associations. For example, there is the China Association for Standardization (CAS)[14] at the national level, and standardization associations under each sector department, such as the China Electronics Standardization Association (CESA), the China Communications Standards Association (CCSA), and the China Association for Engineering Construction Standardization (CECS), among others. All provinces and municipalities at the local level also have their own standardization associations. These associations are generally not fully independent social organizations and have a very close relationship with the government. Most of them do not formulate their own association standards (association standards are not recognized by the Standardization Law), but mainly help the government to promote the standardization at all levels and do some publicity and training work of standardization.

[14]CAS was established in 1979, see: Li *et al.* (2001, p. 591).

It is worth noting that technical focal points for standardization in charge of the secretariat of TCs for standardization play a very important role in standardization in China. Some of them are research institutions of standardization or technical institutions affiliated to different levels of government (such as CNIS, CESI, SIS, BIS, SITS, etc.), and some of them are standardization associations (such as CAS, CCSA, CESA, etc.). In actuality, on one hand, as the think tank of government for carrying out standardization, they provide consultation and suggests schemes on standardization organization management, policy, industrial priorities, etc., for government; on the other hand, they are in charge of organizing the standardization activities on behalf of the government, hosting the secretariat of standardization committee, planning standardization priorities, providing "public service platforms"[15] for different stakeholders, and participating in the international standardization of such organizations as the ISO and IEC on China's behalf.

After entering the 21st century, the Chinese government began to realize that there has been a contradiction between maintaining a government-controlled standardization system and the pursuit of a market economy, and that a more pragmatic approach needs to be adopted. Thus, the Chinese government has encouraged the establishment of industrial innovation alliances outside of government management in hopes of breeding private sector standardization.

3. Discussion

China's standardization system has been shaped significantly by its history and culture. Unlike the US, where the national government has historically been uninterested in standardization (Murphy and Yates, 2009, p. 9), the central government of China has always managed standardization (see Sec. 2.1). Many academics have noted that the Chinese standardization management is characterized by government leadership of the process, and in particular, the standardization system of China is similar to that of some Asian countries, such as Japan, Korea, Singapore, India, etc. (Liu and Jayakar, 2012; Gao *et al.*, 2014). The current status

[15] For instance, the role of CESI in the case of IRGS, see: Ernst (2011, p. 89).

of standardization management of Asian governments has a very close relationship with the special political culture in the historical development of Asia.

3.1. *Positive role at the preliminary stage of reform (1980–2000)*

Standards are an effective carrier of knowledge and technical solutions. International standards are the technological infrastructure of economic development and have made important contributions to the establishment of the global market (Murphy and Yates, 2009, pp. 20–22). Developed countries continually converted the technologies they developed during their processes of industrialization into international standards, which accumulated over several decades. For latecomers entering the global economy when their economies and technology are relatively backward and their industrial standardization capabilities are relatively low, like China's were in the preliminary stages of reform and opening-up (1980–2000), adopting international standards as the policy of rapidly transferring the Western advanced technologies to home is a smart choice.

The Chinese government adopted policies of marketization in order to attract technology and foreign investment and thus promote economic development; adopted a learning-based strategy for scientific and technological development to greatly promote technical progress; introduced TCs as a means of setting standards to get close to the internationally common practices; and admitted the principle of "openness, transparency, and consensus" to try to get more stakeholders into the TCs to prepare standards together and gradually cultivate industrial standardization capability. Such a government-led standardization policy of TCs was very successful at the time. The "dual-adoption policy" (see Sec. 2.3) gave full play to one important role of standards, that of technology diffusion. Advanced Western technology was rapidly transferred to every field of relatively underdeveloped China at that time.

There are two reasons for general technical progress: one is innovation, the other is the technology spillover effect and "learning-by-doing" effect in the trade process, with the "learning-by-doing" type of technical progress mostly obtained from technology spillovers (Zhang, 2000).

Quoting Reischauer (1992): "The development of world civilization is composed of one-tenth originality and nine-tenths transplant." The industrial, science, and technology policies of the Chinese government greatly promoted the technology spillover effect and played a very important role in the scientific and technological progress of China. It was obvious that the government's policies on standardization became one of the important channels of the technology spillover effect in this process. During the approximately 12 years from enacting the Standardization Law in 1988 to the end of the last century, China completed a transformation from being a third-world country with a backward economy and technology and very low standardization capability to the "world's manufacturing plant" (labor-intensive industry structure). According to the data released by the National Bureau of Statistics, the gross domestic product (GDP) stood at a mere 1.51 trillion RMB in 1988 but reached 9.98 trillion RMB in 2000.[16]

3.2. *Promoting or hindering innovation?*

3.2.1. *Fostering and promoting innovation period (2000–2010)*

China entered into the WTO at the beginning of the 21st century and committed to following the rules of the WTO Agreement on Technical Barriers to Trade (TBTs). From then on, China's economy has experienced a new round of rapid development.[17] Because its economic strength has improved to a certain extent, questions about how the country's industrial structure should be transformed and how to further promote scientific and technological progress became very important. The academic community argued that it was difficult for imitation products to be competitive, and the discontinuity of the technological trajectory brought innovation space to China (Liu, 1997).[18] China was in

[16] See the website of National Bureau of Statistics. Available at: http://data.stats.gov.cn/easyquery.htm?cn=C01.

[17] By the end of 2014, China's GDP had reached ￥63.6 trillion RMB; *Ibid.*

[18] Liu (1997) argues that technology development in the same industry generally has two different directions: one is in a brand new trajectory, another one is in an old trajectory, e.g., petrol engine and electronic engine technology in the automotive industry. Everyone might be at the same starting point at a new trajectory of this discontinuity of technology development.

a sharp adjustment period of industry and capital accumulation, hence the relative comparative advantages of capital- and technology-intensive industries were greatly improved. Combining high and new technology with the labor force in China through technical innovation was favorable for the gradual transition from labor-intensive production into capital- and technology-intensive production (Li and Meng, 2006). Entering the WTO was favorable to the transformation of China's industrial structure; and the government's policy is a further important step that promoted the development of technology-intensive industry, including the high and new technology industry (Guo, 2003).

3.2.1.1. Standardization strategy

The government began to strongly support indigenous innovation technology during this period, and the standardization strategy transformation was also started. All levels of government, including different departments (such as SAC, NDRC, MOST, MIIT, MOFCOM,[19] etc.) of the central government, local governments, etc., had established a standardization strategy to support indigenous innovation in their own jurisdictions (jointly or separately) (Wang *et al.*, 2010). Since the country's opening-up to the international economy, substantial changes have occurred in China's standardization strategy, its institution, and its management practices. (Ernst, 2011, p. 21)

The standardization strategy established by the Chinese government is essentially different from that established by private standardization organizations. The strategy of a private standardization organization mainly focuses on its role in economic and industrial development and proposing the strategy about how to exert the standardization function with respect to new areas of industry and industry requirements. In addition, private standardization organizations are limited in resources that can be mobilized to implement these strategies. For the government, the standardization strategy priorities are designed with respect to the government's broader policy goals (e.g., scientific and technological development

---

[19] SAC: Standardization Administration of China; NDRC: National Development and Reform Commission; MOST: Ministry of Science and Technology; MOFCO: Ministry of Commerce.

policy, energy conservation and environmental protection policies, acquisition policy, etc.), and the method of mobilization of the resources at government's disposal (government fiscal funds, state-owned enterprises, etc.) may be considered from the aspect of key industry development that is promoted by the government.[20]

Overall, China's strategy from 2000 to 2010 was successful. The positive impetus of standardization dominated by the government and TCs in supporting indigenous innovation strategy was integral to the rapid development process of the Chinese economy. The innovation and standardization capabilities of domestic industry were in the early stages of growth, so the government's standardization strategy obviously played a positive role. During this period, China also has experienced the trial-and-error process,[21] which sometimes appeared to be a lack of coordination across multiple decision-making bodies of government. The negative effects of this, however, were minimal in the process of economic growth. Dieter Ernst stated that:

> "...compared to the extended history of standardization of Europe, the United States, and Japan, China is a relative latecomer. The speed of

[20] For instance, MOST encourages its technology development projects to translate their achievements into national and international standards; NDRC's strategy of energy conservation and environmental protection has incorporated the development of relevant standards into the planning and development roadmap, resulting in important new energy and environmental standards; NDRC, MOF (Ministry of Finance), and other government agencies have formally utilized technical standards in their procurement project biddings. Another example would be that SAC jointly promotes with other government agencies for industrial standardization pilot and demonstration programs or zones, like the Shenzhen Hi-tech Standardization Pilot and Demonstration Zone, as well as some agricultural standardization pilot and demonstration programs in some provinces of China. Local governments may even support local enterprises and R&D institutions in undertaking national and international standards projects through subsidies. In 2009, MIIT licensed three 3G technologies to CMCC (China Mobile Communication Corporation), China Telecom, and China Unicom, including the TD-SCDMA standard incorporated with Chinese indigenous innovation technology, CDMA2000 standard based on American technology, and WCDMA standard based on EU technology.

[21] Ernst (2011, p. 17) believes that, "China is not the only country where standards fail. And standards may also fail when the government does not play a role."

> learning and institutional adjustments has been impressive: In a very short time, major standards projects have been initiated in the strategically important ICT industry. ...Chinese leadership to use standardization as an enabling platform for indigenous innovation... (is) not only 'virtually unparalleled among developing countries (Kennedy *et al.*, 2008, p. 21)', its scope and depth certainly also goes much further than the standardization strategies in the United States, Europe, and Japan." (Ernst, 2011, p. 102)

From actual cases of innovation supported with standards and the analysis of academia, each case shows different positions in different technical fields and different government's decision-making levels. For example, Suttmeier *et al.* (2006) think that, "the relative importance of techno-nationalism and techno-globalism seems to vary considerably across a spectrum, with WAPI exemplifying the former and AVS and perhaps IGRS the latter." Ernst (2011, p. 24) pointed out that, "There is no doubt that the MLP (Medium- and Long-term Plan) contains techno-nationalist notions of self-evidence. This reflects the initial objective of Chinese policymakers to reduce China's dependence on foreign companies' intellectual property and relevant high patent licensing fees."

The authors think that the combination of techno-globalism and techno-nationalism (or the so-called "neo-techno-nationalism" of Suttmeier) was dominant generally for both government decision-making and specific standardization practice, and narrow techno-nationalism was limited to just a few instances. The diversity of actual decision-making was derived from a different understanding of indigenous innovation and integration into the global economy among the government's multiple decision-making bodies, and also from the specific conditions decision-makers faced within different technological/industry fields. Chinese enterprises faced enormous pressure from MNCs and Western countries in pursuing indigenous innovation, in part as a result of the unfavorable position of their own innovation capability, the winner-take-all competitive strategy of transnational corporations, the actuality of large-scale Foreign Direct Investment (FDI) and patent thicket, and insufficient ability to engage in dialogue with advanced enterprises from developed countries in standardization organizations. As Ernst (2011, p. 48) described,

"Companies from the emerging economies are thrown into this game without much preparation. The new players are thus experiencing with new approaches to standardization." Through analysis of most standardization cases, it can be seen that both the government decision-making and industries' practices were actively integrated into the international innovation system and standardization system. Although cases differ, for the most part, these practices aimed to have the Chinese industry actively participate in global competition. When the government was trying to integrate domestically invented technology into national and international standards, this was done in compliance with international norms. Only in the case of information network security, has the government attempted to ensure the predominant position of indigenous innovation technology in China.

3.2.1.2. Industrial alliance

As previously mentioned, promoting industrial alliance in the process of standardization has become one of the components of China's standardization strategy. Some of the relatively successful alliances include IRGS, AVS, TD, and CSA, among others. Many domestic experts have studied these industrial technology alliances and think that they have changed situations where a company can only struggle alone and faces intense competition, enabling the effective integration of industrial resources to form a complete industrial chain. Further, they believe these alliances have started the transformation from that of standards development completely under the government control to the coexistence of government and private sector standardization. This is of great significance to China in the breakthrough of market economy transition and the improvement of industrial competitiveness (Wu and Yan, 2011; Wang and Liang, 2013).

China's standardization system has witnessed a historic change. Although some domestic experts argue that these alliances do not have real legal status yet, their operation mechanism of standardization is immature, their patent management and intellectual property politics are not perfect (Liang and Zhu, 2015), and China's testing and certification system still does not recognize the alliance standard as a formal one which can be used in testing and certification, the change has opened the gates of standardization system reform. The standardization system as that which is completely under government management has been broken, and a real sense of private standardization has emerged. "Instead of an exclusive reliance on a top-down approach to standardization led by government, China is now witnessing the emergence of diversity, and often hybrid standardization projects that mix and match elements of industry-led and market oriented approaches with the traditional reliance on government" (Ernst, 2011. p. 93).

3.2.2. *The economic slowdown and institutional drawbacks (2010–present)*

In 2011, China became the world's second-largest economy, with its GDP reaching 48.4 trillion RMB. However, after 2010, economic development started to slow down: the GDP growth rate was 10.63% in 2010 and dropped to 7.27% in 2014.[22] At the same time, drawbacks of a standardization system under government control started to appear. Actually, many domestic and foreign scholars have analyzed the drawback of the Chinese standardization system and the need for reform from various perspectives and given some reform suggestions.

As previously mentioned, the Chinese government is not monolithic but consists of multiple decision-making bodies in actuality. Every department (ministry or administration bureau) under the State Council has its offices (departments) at the levels of the provinces and municipalities, districts under provinces, and the hundreds of counties. Therefore, the government system is fragmented as "top-down-channels-cut-by-horizontal-levels".[23] Standardization under government management is the same, which is a very dispersed system (see Sec. 2.5). The difficulty in coordination among government departments has been a long-standing issue. With the gradual enhancement of economic strength and the improvement of innovation capability, the negative effects gradually appear. Kennedy *et al.*

---

[22] See the website of National Bureau of Statistics. Available at: http://data.stats.gov.cn/ks.htm?cn=C01&zb=A0501.

[23] Suttmeier *et al.* (2006, p. 37) call it as institutional fragmentation. Kennedy *et al.* (2008, p. 16) regard it as stove-piping problem. The Chinese saying is "条块分割."

(2008, p. 16) states that "China's new Ministry of Industry and Informatization is intended to solve some of these stove-piping problems and bring greater coherence to the national effort to develop the ICT industry and diffuse IT throughout the society." Ernst (2011, p. 28) argues that "In fact, China's standards system is overly complex and displays signs of fragmentation. Ambiguity is a fundamental source of such fragmentation...Equally important sources are inter-agency rivalries and turf battles among different ministries and their respective stakeholders."

Actually, the multiple decision-making bodies of the government broke the harmonious system assumed in the planned economy period after the rapid development of a market economy. In actuality, the sector/ministerial standards and local standards tend to be stronger than national standards in some cases as the sector and local governments have more resources and "administrative management measures." The Standardization Administration of China (SAC) is only a subordinate organization in the eyes of those ministerial (sector) organizations. The competent departments of sector government have the interests and demands of large state-owned enterprises under their purview, and the local governments have the interests and demands of the local market and local large taxpayers. As such, the national standards are no longer sacred. Therefore, the standards established by different management bodies of government can be in competition. Where the national standards prepared by SAC are inconsistent with sector/ministerial and local standards, the sector government and local government can go their own way according to their jurisdiction. Such cases aggravate confusion about standardization management, which is disastrous for voluntary standardization and far from making the market play a decisive role in the allocation of resources, as called for in the State Council's Notice for releasing the Scheme of Deepening the Reform of Standardization.[24] In order to solve this confusion, there is no better solution under the current system than solving the standardization discordance between ministries and commissions by the State Council. This is also the direct reason for establishing the

[24] State Council's Notice for releasing the Scheme of Deepening the Reform of Standardization (2015, #13). Available at: http://www.gov.cn/zhengce/content/2015-03/26/content_9557.htm

"Coordinated Promotion of Joint Inter-ministerial Meeting System for Standardization" by the State Council in June 2015.[25]

When investigating the SSL or LED[26] industry (2012–2014), the authors found that the existence of multiple decision-making bodies of government not only caused contradiction among standards issued by government agencies affiliated with different ministries and those at different levels but also brought adverse effects on the development of private standardization. The importance of the LED industry in energy conservation created a condition where all levels of government supported the establishment of an LED standard alliance. A dozen similar industrial alliances appeared in the LED field in recent years. Every alliance has its own government background and wants to do standardization work separately. Issues of not recognizing each other and fighting for authority and interests has appeared among them, even leading to shutting the door on the participants of other alliances regardless of the "openness, transparency and consensus" principle of standardization when holding standardization meetings. During investigation, the authors also found some enterprises and testing organizations reporting that some local governments promoted alliance standardization that was actually local protectionism and that other local governments intended to take measures in retaliation.

The authors think that the primary cause of the excessive intervention of all levels of government in the process of promoting alliance standardization is that the current standardization system allows all levels of government to manage voluntary standardization, forming multiple bodies of government standardization management. All levels of government regard the promotion of voluntary standardization as an "important measure of administration." This is an important cause of various discordance phenomena generated in standardization. The negative effects of the system's drawback have begun to grow, so they obstruct the impetus to standardization which supports indigenous innovation.

[25]The State Council's Official Reply to Agree the Establishment of Coordinating and Promoting Joint Inter-ministerial Meeting System for Standardization, June 1, 2015. Available at: http://www.gov.cn/gongbao/content/2015/content_2883233.htm.

[26]SSL: Solid State Lighting; LED: Light-emitting Diode.

3.3. *Deregulation and social self-governance*

Most academic research indicates that the current standardization management system of the government has caused the monopolization of standardization resource by administrative power and the restriction of market operating system, restraining the enterprise enthusiasm for innovation (Wang, 2003; Xue, 2006). Reform of the standardization system is required to adapt to the diversified needs of the modern social interest, and the diversity of benefit requires open public governance. However, the self-organized social groups in the industry shall be the subject of standardization activities in market economy and can play an important role that is irreplaceable by the government and enterprise.

Russell (2014) holds that current standardization has been developed into a more open "private regimes" era, namely an "open standard" era. According to Russell (2014), "There emerged further agreement around the paramount importance of well-defined procedures to guarantee public participation in the production of standards and liberal terms of access to allow public use of standardized technologies." While assessing the US standardization system, Ernst (2013) holds that "The potential advantages of decentralized self-government are well established in theories of innovation and organization.…The vision of local self-government finds ample support in the 'collective action' governance theory developed by Elinor Ostrom" (Ernst, 2013).

The theory of Elinor Ostrom, the winner of 2009 Nobel Prize in Economics, mainly analyzes the common-pool resources (CPRs) of self-organized systems, like marine fishery, public pasture, etc., proposing the analytical framework and eight design principles (Ostrom, 1990). Simcoe (2013) holds that inherent symmetry exists between the commons problems studied by Ostrom and the anti-commons problems in technical standards, thus the analysis method of Ostrom may be extended to studying self-governance in standards setting organizations (SSO). The result of Simcoe's (2013) study describes that "many SSOs could go further to define clear boundaries and mechanisms for monitoring and enforcing licensing commitments...and articulating a clear set of principles for interpreting FRAND[27] commitments in the event they must be enforced."

[27]FRAND: Fair, Reasonable, and Non-Discriminatory.

Our specific question is: what are the prerequisites for self-organized social subject generated in Chinese industry to develop private standardization? Brunsson *et al.* (2000) hold that "When states 'deregulate', abolishing some of their directives, ideas similar to those inherent in the directives may turn up in standards." Therefore, government deregulation is the first condition for implementing the reform of the standardization system. The authors hold that, for China, "deregulation" is not only to allow social groups to develop standardization but also to make the government withdraw from voluntary standardization management, particularly from the management of sector and local recommendatory standards. The current state of multientities of government standardization management needs to be changed in order to truly create space for voluntary standardization.

One core requirement of modern governance theory is the sufficient expression of stakeholder demand and sufficient participation of governance process; "governance malfunction" would occur where the stakeholders fail to reach an agreement, thus a set of sophisticated system arrangement is needed for guidance and coordination (Li, 2003), embodying the "technology democracy" based on the principle of "openness, transparency and consensus" in standardization (Garcia *et al.*, 2005). The confused condition that exists in current standardization management is caused by the deviation from the government function in standardization activities.

The government needs to develop a favorable operating environment for voluntary standardization, establish corresponding supporting policies, and provide financial support for industrial innovation projects, particularly for the R&D projects with common industrial technology or demand (e.g., inspection and detection), in order to cultivate industrial innovation capability rather than organizing the preparation of voluntary standards directly. There is another more important aspect of standardization governance for the government, namely, to achieve the preparation, supervision, and implementation of the mandatory standards (technical regulation) system in relevant fields so as to solve problems in areas including security of the whole society, human health, environmental protection, and resource conservation. This exactly complies with the idea of reform, of "better exerting the government function while making the market play a decisive role in resource allocation."

If the government can deregulate and leave development space for private standardization in the industry, then the rest are tasks for the industry itself. Many people in China hold that China is not a self-organized society and it will be very difficult for it to develop such powerful private standardization organization like those in developed countries. However, the authors do not agree with this view. Through the CSA investigation in recent years, the authors reach the opposite conclusion. Even though the degree of government deregulation is still very limited,[28] and the CSA alliance encountered various difficulties during operation, it still had great successes. An important standard used for LED luminous module interface (CSA 016) was prepared only in one year and one month from March 2012 to April 2013,[29] which solved problems causing bottlenecks in the LED industry. The enterprise participation enthusiasm was unprecedented. Member enterprises not only actively participated in the design of the technical scheme but also supported the necessary experiments and validations. This type of cooperation and support was rarely seen in previous standardization processes that were led by the government during the planned economy period. This standard rapidly entered into industry implementation stage after issuance, and after approximately one year, more than 30 enterprises had begun to produce street-light modules complying with the standard. (Wang *et al.*, 2014).

The case of CSA demonstrates that the industry is imbued with great enthusiasm for standardization developed by self-governance; in cases where the industry's innovation capability is developed to a certain level, so long as the government can relinquish control, industry enthusiasm will find appropriate organizational form and quickly satisfy the industry's standards demands, and all the costs in the course of the standardization will be borne by the stakeholders themselves, which means that the voluntary standardization will become a self-financed process. Of course,

[28] Private standardization, including alliance standardization, still lacks a legal position in the Standardization Law — although alliance standardization is allowed, it is only recognized as an experimental or pilot project. Alliances in many provinces cannot be registered with the government, therefore, they may be regarded as illegal entities.

[29] CSA Standard CSA016-2014: Interface Requirements for Application of LED Lighting Street/Tunnel Light Consisting of LED Module with Heatsink and Separated Control Gear.

such rapid reaction is an instinct of enterprises within an environment of market competition and will allow for trial and error and competition within the industry. This is a process for nourishing and improving industrial standardization capability and conducting self-optimization as well as for optimizing the resource allocation through market mechanisms for industrial standardization.

The breeding of a self-governance capability for industrial standardization needs a process. If the government can relinquish it, some formerly endogenous associations for standardization from the government administrative system will accelerate the transition, and the industry may also generate new associations for standardization, as well as various enterprise standard alliances and consortia. All these organizations shall experience the marketization process, and industry innovation will become more active. However, the government cannot refuse to relinquish control on the grounds of lower self-governance capabilities of an industry, but it must give the self-organized standardization of industry a process to grow in strength.

4. Policy suggestions and unsolved issues

China cannot copy the standardization modes of developed countries like the US and Europe; it shall consider the historical and cultural development of China. The desire of reform of the voluntary standardization system proposed by the authors is a scheme combining the great promotion of private voluntary standardization with the original system in China:

(1) The government shall first totally quit from the sector/ministerial and local voluntary standardization management to leave space for cultivating the self-governance capability of standardization in industry. The government shall construct a civil standardization operating environment and build supporting policies to ensure the legitimacy of private standardization organizations and shall allow diverse forms of civil standardization organizations.

(2) National voluntary standardization shall be in charge of the administration of a single government department to change the multiple

decision-making bodies of standardization into a single decision-making body. However, a Standardization Commission with decision-making power should be established besides the governmental standardization administration and shall be mainly composed of representatives from industry, consumers, scientific research institutions, and government itself, as a means to introduce market selection mechanism in voluntary standardization management. The advantage of this structure is that national voluntary standardization is jointly governed by the government and industry, ensuring communication between industry and government and coordinating industry and government decision-making, where possible.

(3) Strongly cultivate and develop industry consensus such that standardization will be led by self-organized social groups in the market and taking stakeholders, like enterprises, testing organizations, and consumers, into account in the standardization processes. Increasing private standardization in industry will supplement the solid foundation of national standardization. Otherwise, national standardization will lack industry and market foundations.

(4) The practice of sponsoring voluntary national standard preparation with government financing should be changed. Voluntary standards shall be completely formed with self-financing, and a small amount of government funding may be used, but mainly for supporting the participation of experts in international standardization.

(5) The reform of the voluntary standardization system requires simultaneous marketization of conformity of assessment and testing organizations.

A central proposition of the chapter is that the reform of China's standardization system requires the vigor of self-organized standardization that is stimulated from within the industries themselves. If China can generate private standardization organizations equal to those in the US, such as IEEE and ASTM, or in Europe, such as CEN and CENELEC, after several years of development, the standardization capability of the industry then will be truly improved. This type of standardization can truly support industry innovation.

In March 2015, the State Council issued a notice of "Plan for Furthering the Standardization Reforms"[30] (Reform Plan for short). There, relevant requirements were proposed: "integrating and simplifying the mandatory standards... gradually integrating the current national, sector/ministerial and local compulsory standards into national compulsory standards; ...cultivating and developing civil group standards; encouraging the society, institution, chamber of commerce, league and other social organizations and industrial technology alliances possessing corresponding capability to set standards jointly with relevant market entities meeting the demands of market and innovation"; "recommended standards prepared under government leadership shall be limited within public welfare."

From the Reform Plan, we found that the State Council has a clear cognition on the reform issue and has adopted a more bottom-up approach to substantively reform the standardization system. Standardization will be open to all the civil organizations, and private standardization will go public totally. However, it is a pity that this plan fails to remove sector government and local government completely from voluntary standardization management. The current government state of multiple decision-making bodies for voluntary standardization management has not been broken. Moreover, it is hard to clearly limit the scope of "public welfare" stated in the Reform Plan. The reform seems still to be at the halfway mark.

Mandatory standardization system reform is not discussed in this chapter. However, the authors hold that the practice to totally integrate mandatory standards at different levels into only national ones, as stated in the Reform Plan, may make the sector and local government more unwilling to give up the management authority on voluntary standards. Mandatory standards (technical regulations) are an important measure with which the sector government and local government perform their market regulatory responsibilities. Therefore, the correct practice shall be to follow compulsory standards (technical regulations) that shall keep certain leeway for sector government and local government to find their proper position and exert

[30] State Council's Notice for releasing the Scheme of Deepening the Reform of Standardization (2015, #13). Available at: http://www.gov.cn/zhengce/content/2015-03/26/content_9557.htm

their maximum efficiency. Moreover, compulsory standards shall be transitioned into technical regulations to change the current mixed state of a mandatory standard system and a voluntary standard system. The establishment of a favorable and applicable technical regulation system and administrative supervision system, and a mechanism for coordination between a technical regulation system and voluntary standard system, remains the most important but tough task for the government.

It may be foreseen that the standardization system reform of the government faces great challenges. The existing research on administrative management and bureaucratic organization indicates that every government department has an intrinsic motivation of "authority maximization" and is unwilling to reduce its management authority and function without external pressure (Parsons, 1947). As specific to the standardization in China, both the sector-competent department and local government are equally unwilling to withdraw from the management of voluntary standardization. Especially in this case, they still have to face the situation that their management authority on compulsory standardization will be weakened according to the Reform Plan. It is conceivable that the management authority of the sector government and local government on voluntary standardization still remains (though limited) in the "Reform Plan" because the Central Government has had difficulties in trying to persuade those officials to withdraw from the management of voluntary standardization. Therefore, the biggest challenge of standardization system reform comes from the government itself.

Finally, in the future development process, the key to the continued success of the standardization strategy to support innovation and economic development of the country will be whether the government can overcome its own obstacles to the reform of the standardization system and to solve the problems caused by the excessive intervention of government and the institutional fragmentation in a relatively short time.

Acknowledgments

This work was supported by the General Project of National Natural Science Foundation of China (NSFC) (No. 7137 3137), as well as the Contingency Project of NSFC (No. 7144 1025).

References

Brunsson, N. *et al.* (2000). *A World of Standards* (Oxford University Press, Oxford), p. 14.

Deng, L. *et al.* (1986). *Contemporary China's Standardization* (当代中国的标准化) (China Social Sciences Press, Beijing).

Ernst, D. (2011). Indigenous Innovation and Globalization: The Challenge for China's Standardization Strategy, IGCC and EWC. Available at: http://www.eastwestcenter.org/publications/indigenous-innovation-and-globalization-challenge-chinas-standardization-strategy.

Ernst, D. (2013). "America's Voluntary Standards System: A 'Best Practice' Model for Asian Innovation Polices?" *Policy Studies*, (66), p. 15.

Gao, P., J. Yu, and K. Lyytinen (2014). "Government in Standardization in the Catching-up Context: Case of China's Mobile System," *Telecommunications Policy*, 38(2), pp. 200–209.

Garcia, D. L., B. L. Leickly, and S. Willey (2005). "Public and Private Interests in Standard Setting: Conflict or Convergence." Georgetown University, Washington D.C.

Guo, K. (2003). "Study on Industrial Structure Changing Trend and Policy of China after Entering WTO (加入WTO 后我国工业结构的变动趋势及政策研究)," *Industry Economy Research*, (2), pp. 1–13.

Kennedy, S. *et al.* (2008). "Standards, Stakeholders, and Innovation: China Evolving Role in the Global Knowledge Economy." National Bureau of Asian Research (NBR), Special Report #15, September, Seattle.

Liang, X. and C. Zhu (2015). "Analysis and Suggestions on Barriers to the Development of Industrial Alliance Standardization (我国产业联盟标准化发展的障碍分析与研究建议)," *Standards Science*, (4), pp. 16–19.

Li, C. *et al.* (2001). *Quality Standardization Metrology Encyclopedia* (世界标准化计量百科全书) (Encyclopedia of China Publishing House, Beijing).

Li, C. (2003). "Theoretical Basis of Co-governance by Stakeholders and Its Practice," *Management World*, 16(4), pp. 4–87.

Li, X., and Y. Meng (2006). "Making Use Comparative Advantage Theory, Conceiving Industry Development of China after Entering WTO (利用比较优势理论，构想我国加入 WTO 后的产业发展)," *Journal of Guangxi University (Philosophy and Social Science Version)*, (4), pp. 6–9.

Liu, C. and K. Jayakar (2012). "The Evolution of Telecommunications Policy-making: Comparative Analysis of China and India," *Telecommunications Policy*, 36(1), pp. 13–28.

Liu, X. (1997). "Technical Trajectory and Indigenous Innovation (技术轨道和自主创新)," *China Science and Technology Forum*, (2), pp. 32–35.

Murphy, C. N. and J. Yates (2009). *The International Standardization Organization (ISO): Global Governance Through Voluntary Consensus* (Routledge, Oxon).

Ostrom, E. (1990). *Governing the Commons: The Evolution of Institutions for Collective Action* (Cambridge University Press, UK).

Reischauer, E. O. (1992). 近代日本新观, (translated by卞崇道) (Sanlian Publishing House, Shanghai).

Russell, A. (2014). *Open Standards and the Digital Age: History, Ideology, and Networks* (Cambridge University Press), p. 279.

Simcoe, T. (2013). "Governing the Anti-commons: Institutional Design for Standard Setting Organizations." Boston University and NBER, July, p. 23.

Suttmeier, R. P. *et al.* (2006). "Standards of Power? Technology, Institutions, and Politics in the Development of China's National Standards Strategy," (6), p. 31.

Parsons, T. (ed.) (1964). *The Theory of Social and Economic Organization*, trans. A. M. Henderson and T. Parsons (The Free Press of Glencoe, New York; orig. pub. Oxford University Press, 1947), pp. 87–423; now Part One, Chs. I–III.

Wang, P. (2003). "Standardization Management Institution: Problems and Countermeasures," *World Standardization and Quality Management*, (3), pp. 8–10.

Wang, P. (2008). "Standardization Historical Overview," *China Standardization (English Version)*, (5), pp. 27–30.

Wang P. (2009). "The Contribution of *Ying Zao Fa Shi* by Li Jie in Song Dynasty to Ascent Architecture and Construction Standardization," *Standard Science*, (01), pp. 13–17.

Wang, P., Y. Wang, and J. Hill (2010). "Standardization Strategy of China--Achievements and Challenges." East-West Center Working Paper (Economic Series), No. 107, January, pp. 7–11.

Wang, P. and Z. Liang (2013). "Study on Evolution of Standardization in Associations and Alliances of China," *China Standardization*, (8), pp. 59–62.

Wang, P., J. Kwak, and H. Lee (2014). "The Latecomer Strategy for Global ICT Standardization: Indigenous Innovation and its Dilemma," *Telecommunications Policy*, 38.

Wang, P., J. Ruan, and W. Gao (2014). "Innovative Practice of China's LED Lighting Industry Alliance Standardization," *China Standardization (Overseas Edition)*, (9/10), pp. 52–57.

Wu, G. and P. Yan (2011). "Innovation Implications of Industrial Technology Strategic Alliances (产业技术创新战略联盟的创新意义)," *Journal of Zhaoxing University*, 31(7), pp. 81–84.

Willingmyre, G. (2009). "Take Two — China's Proposed Regulations for Patent-involving National Standards". Intellectual Property Watch. Available at: http://ip-watch.org/2009/12/21/take-two-china%E2%80%99s-proposed-regulations-for-patent-involving-national-standards/.

Xue, X. (2006). "National Standardization Management System Should Have an Overall Reform: Take An Example of the Standardization Management System of the United States." China's Reform, 13 October (Section 6).

Zhang, X. (2000). "Entered WTO in Favor of Promoting Technology Innovation of China," *Zhejiang Social Science*, (1), pp. 47–50.

Zhao, Y. *et al.* (2002). *An Ancient History of China* (中国古代史) (Higher Education Press, Beijing).

Zhong, S. (2012). *Modern China's "Standard and Standardization" and China's Modernization* (近代中国的 "标准与标准化" 与中国现代化), *China Standardization*, (1), pp. 85–91.

Part 5

Trade and Innovation Policy — New Perspectives

Chapter 15

Innovation, Trade, and Intellectual Property Rights: Implications for Trade Negotiations

Carlos A. Primo Braga

Fundação Dom Cabral, Brazil

Abstract

Innovation is often considered the cornerstone of modern economic growth. In the same vein, international trade has traditionally been characterized as the handmaiden of growth. Yet, innovation and trade share a complex relationship. This chapter discusses how innovation-leaders (e.g., the US), innovation-followers (e.g., Brazil), and globalizers (e.g., Vietnam) see the role of trade policies in affecting their innovation strategies. Attention is devoted in particular to the "marriage of convenience" between trade and intellectual property rights (IPRs) laws as negotiated in the context of trade agreements. The experiences of negotiating IPRs protection via multilateral (e.g., TRIPS/WTO) and preferential (e.g., TPP) agreements will be contrasted, and the argument will be made that, even among like-minded countries, these negotiations are inevitably characterized by significant controversy. This chapter also discusses to what extent the increased emphasis on innovation in terms of growth strategies is making the goal of promoting higher standards of IPRs protection easier to advance via trade agreements.

1. Introduction

Innovation is not easy. As Machiavelli once noted:

> *Innovation makes enemies of all those who prospered under the old regime, and only lukewarm support is forthcoming from those who would prosper under the new. Their support is indifferent partly from fear and partly because they are generally incredulous, never really trusting new things unless they have tested them by experience.*[1]

As described in Ashton (2015), the process of creation and innovation has many links, and the incentives that drive these activities are complex and multifaceted. In an era of accelerating technological change, the search for the right policies to promote innovation has become an important goal for policy makers around the world.[2] As the author and other colleagues have noted, the "unique nature of innovative activity and the growing interconnectedness of the world economy call for … greater attention to the interplay of openness and technological innovation not only in OECD countries, but also in developing economies".[3]

Innovation is typically defined as "the implementation of a new or significantly improved product (good or service), or process, a new marketing method, or a new organizational method in business practices, workplace organization or external relations".[4] This broad definition underscores the fact that, from an economic perspective, the

[1] Machiavelli (1992, p. 17).

[2] There is an ongoing spirited debate about the socio-economic impact of the ongoing ICT-driven revolution between those that find it less impactful than earlier technological revolutions (e.g., Gordon, 2014) and those that think otherwise (e.g., Mokyr, 2014). As discussed in Dobbs *et al.* (2015, p. 33), however, it is clear that "the period between historic [technological] breakthroughs has been decreasing by orders of magnitude."

[3] See Chandra *et al.* (2009, p. 12).

[4] See OECD and Eurostat (2005, p. 46).

focus is on not only R&D-related "new to the world" goods and services but also the introduction of new products/processes in the context of a country or a firm.

Some analysts would rather differentiate innovation in terms of its role in promoting "horizontal (or extensive) progress" vs. "vertical (or intensive) progress." Thiel and Masters (2014), e.g., characterize horizontal progress as synonymous with "globalization," i.e., progress that focuses on copying/replicating things that are available in the frontier economies. Vertical progress, in turn, is characterized as "doing something nobody else has ever done. From this perspective, if you take one typewriter and build 100, you have made horizontal progress. If you have a typewriter and build a word processor, you have made vertical progress".[5] That is to say, vertical progress is driven by entirely novel breakthroughs.

In this chapter, we adopt the broader definition of innovation mentioned earlier, but we find Thiel and Masters' (2014) approach useful in helping establish a taxonomy of economies along a continuum that runs from globalizers (e.g., Bangladesh, Vietnam), to innovation-followers (e.g., Brazil, China), to innovation-leaders (e.g., Switzerland and the US). This taxonomy is just a device to tentatively classify where different countries position themselves in their innovation journey. It helps nonetheless in framing the debate about trade and innovation.

In analyzing this interaction, this chapter adopts a simple framework to capture the main drivers of innovation:

$$\text{Innovation} = f\,(\text{passion, knowledge, incentives})$$

In other words, innovative activities are influenced by individual passion (to solve problems, create works of art, etc.), the ability to access existing knowledge and to invest in new knowledge (R&D), and the impact of incentives in promoting or impairing innovation. Trade activities and trade policies are particularly relevant in impacting the knowledge channel and framing incentives for innovation. The same can be said about intellectual property rights (IPRs) protection. In what follows, this chapter briefly describes the relationship between trade, IPRs, and innovation. This is followed by an analysis of the "marriage of convenience"

[5] See Thiel and Masters (2014, p. 7).

between trade and IPRs laws in the context of trade agreements at multilateral and preferential levels. The chapter concludes with a discussion of areas for future research.

2. Trade and IPRs policies: The relevance for innovation

The relationship between trade and innovation "...is a two-way process. On the one hand, trade liberalization and investment flows contribute to technology diffusion and innovation. As Romer (1994) argues, trade restrictions reduce the supply of intermediate goods to an economy, hampering productivity and technology diffusion. Needless to say, absorptive capacity in the recipient country also plays a key role in this process. On the other hand, strengthening national innovation capabilities improves a country's ability to engage in and benefit from the international trading system".[6]

There are, however, different perspectives on the role of trade and competition in terms of their impact on innovation and productivity growth. As is discussed elsewhere by the author, for "those that believe that markets operate efficiently and economic agents have limited market power, static efficiency should guide resource allocation. In this view of the world, trade liberalization will improve productivity, fostering economic growth... A different perspective, however, can be derived from a view of the world that emphasizes learning-by-doing and the importance of market imperfections".[7] From this perspective, the scope for inward-looking development strategies (and government intervention) seems more attractive and the emphasis is on dynamic gains that could be fostered by the related rents associated with knowledge capture by domestic actors.

These different strategies and related set of policies will impact innovation via different channels. In countries that follow an outward-oriented development strategy — not to be confounded with a *laissez-faire*

[6] See Primo Braga and Meléndez-Ortiz (2013).
[7] See Primo Braga (2010, pp. 127–128).

strategy — access to knowledge embedded in capital goods imports will be facilitated and incentives for investment in R&D will increase as competitive pressures will encourage investment in innovation as a strategic response to escape competition from new entrants in the industry.

At the same time, it is important to recognize that these results are context-dependent.[8] Accordingly, more competition may also deter investments in innovation by domestic firms when such firms are far away from the international technological frontier. Moreover, it can hinder the development of domestic markets, as argued by those that believe on the benefits of import-substitution industrialization and the potential of government-led industrial policy in advancing effective development strategies. All these arguments have also to be evaluated with respect to the quality of the institutions managing such interventions in any given country. Not surprisingly, as discussed in Krueger (1974), in countries with weak institutions, government interventions can easily be abused for rent-seeking and directly unproductive activities. Moreover, most empirical analyses underscore the importance of trade openness for sustained productivity growth.[9] In other words, market openness not only increases the options in terms of price/quality for key inputs for innovation (e.g., capital goods) but also expands markets from the perspective of innovators by counteracting the anti-export bias of trade restrictions.

The protection of IPRs, in turn, is often presented as a key lever for innovation in market economies. The history of the evolution of IPRs protection around the world, however, is quite convoluted and the origins of these instruments were not always driven by innovation considerations. Actually, patents were originally used in Venice in the Middle Ages to "manage" technology transfers, while copyright law was a mechanism to control the publishing industry at its origins.[10] The rationalization of the protection of IPRs as an incentive for innovation is a more recent development.

As described in Primo Braga and Fink (1996), the modern economic rationale "for the protection of IPRs is often framed in terms of

---

[8]For further details on neo-Schumpeterian growth theory, see Howitt (2009).

[9]See, e.g., De Long and Summers (1991), Romer (1994), and Keller (2009).

[10]See Primo Braga *et al.* (2000).

Arrow's seminal work concerning the incomplete 'appropriability' of knowledge.[11] IPRs can be understood as second-best solutions to the problems created by the 'public good' nature of knowledge. To the extent that they enhance 'appropriability,' IPRs are expected to foster investment in research and development ('R&D') and knowledge creation. They create, however, a static distortion as they constrain the current consumption of knowledge, by enhancing the market power of title holders. In short, IPRs involve a 'bargain' between the producers of knowledge and society, which is mediated by the government.[12] The above rationale is typically used to explain the economics of patent and copyright laws. With respect to trademarks and industrial designs, the basis for protection is more often framed in terms of incentives for investments in reputation (quality) rather than innovation per se. Trade secrets, in turn, are rationalized as a necessary supplement to the patent system. Their main positive role is to foster innovations that do not comply with the strict requirements for 'Patentability' of products and processes."

The long-term trend with respect to IPRs protection in developed economies has been in the direction of the strengthening of these rights. There is, however, no clear theoretical presumption that a movement toward stronger standards of protection will always be welfare-enhancing.[13] Moreover, as some analysts argue, there is no conclusive evidence that strengthening, for example, patent regimes will lead to more innovation. Some analysts would even argue that the only consistent result found in the literature is that "strengthening the patent regime increases patenting!"[14] Needless to say, there are other analyses that dispute this

[11] See Arrow (1962).

[12] See Machlup (1958).

[13] Even among countries that are characterized as innovation leaders, there is recognition that IPRs could be abused in detriment of fair competition and consumers' interests. This is illustrated by the case of standard-essential patents (SEPs), where holders of such patents try to abuse their market power by constraining willing licensees with injunctions. See, for example, European Commission (2014).

[14] For a review of these issues, see Boldrin and Levine (2013).

perspective and underscore the importance of IPRs in terms of innovation incentives and outcomes.[15]

The political economy of this process is driven by interest groups associated with industries interested in increasing their temporary "monopoly rents." In terms of industries, for example, pharma and chemicals are typically among the sectors that are more dependent on IPRs protection, the argument being that the high fixed costs of development of new drugs/compounds requires strong IPRs protection for adequate *ex ante* R&D investments.

At the level of countries, not surprisingly, support for stronger protection of IPRs is typically higher among countries that are innovation leaders — although some "pockets" of heterodox thinking may thrive as illustrated, for example, by the open-source movement in the software industry. The relationship between the development and levels of protection to IPRs, however, is not monotonic in the sense that, as discussed in Primo Braga (1990), once an economy reaches a certain level of development, the pressure from local entrepreneurs to gain access to foreign technology often translates into a weakening of the standards of IP protection (e.g., acceptance of broader use of compulsory licensing) and expansion of imitation activities that will be characterized as "piracy" by foreign IPRs-holders.

Hence, there is a tension between the development strategies followed by different countries over time in terms of their attitude to market-led development (and willingness to integrate with the global economy via an outward-oriented trade policy) and their approach to IPRs protection. In a simplified manner, one could characterize the interaction between trade and IPRs protection in terms of a 2×2 matrix that captures different combinations of the policies in question (see Table 1).

A description of the journey of countries in terms of their development strategies and IPRs protection can be summarized as follows: (1) in early stages of industrial development, countries typically embrace outward-oriented policies (relying, e.g., on exports of natural resources) and have weak

---

[15] See, for example, WIPO (2015).

Table 1.

Trade/Strength of IPR protection	Weak	Strong
Outward-oriented	A	C
Inward-oriented	B	D

protection of IPRs — stage A in Table 1[16]; (2) as development proceeds and local manufacturing evolves, the demand for trade protection tends to increase, and often countries begin to adopt import-substitution industrialization (ISI) policies, amid a weak IPRs environment — stage B; (3) depending on the size of the domestic market, the role of foreign direct investment (FDI) in the economy, the level of distortions associated with the anti-export bias of the ISI-phase, and the pressure of external economic partners (e.g., innovation leaders such as the US and the EU countries), countries gradually migrate either to stage C (embracing modern globalization) or stage D (in which greater protection to IPRs is provided, but the focus of the economy remains on the domestic market).

It is important to recognize that the schematic description summarized in Table 1 is becoming less typical in a world in which global value chains (GVCs) are increasingly influential and the benefits of specialization more evident. This is particularly the case for smaller economies. In this context, the journey can be more direct from A to C with the emphasis remaining always on the external market.

3. The "marriage of convenience"

Why should trade agreements, which typically focus on the rules for progressive trade liberalization, address the protection of IPRs? Actually, there is a distinguished group of trade economists that have consistently argued against the incorporation of trade-related aspects of IPRs in trade agreements.[17] It remains true, however, that IPRs do affect trade flows

[16]It is worth noting that countries at this stage often display strong IPRs laws on their "books," reflecting colonial "inheritances," but have no capacity to implement/enforce the laws. See Primo Braga *et al.* (2000).

[17]See, for example, Bhagwati (2002) and Panagariya (1999).

and, as a consequence, IPRs-holders do have an interest in constraining "piracy" at international level.

In a nutshell, the argument for the relevance of IPRs for international trade can be presented as follows: firms/innovators may be deterred from exporting their products/services into a foreign market if "pirates" can diminish the profitability of the exporting activities via imitation. It is important, however, to recognize that the impact of strengthening IPRs in terms of bilateral trade flows between an innovation-leader and a hypothetical developing (imitation-prone) country is, in the end, an empirical question. After all, such a policy change will have both a market expansion (by displacing "pirates") and a market contraction effect (in view of the enhanced market power of the innovator).[18]

3.1. *The multilateral front*

The history of the "marriage of convenience" between trade policies and IPRs protection has been analyzed extensively in the literature. IPRs are territorial (i.e., the rights are awarded and enforced at national level) by nature, and attempts to promote harmonization and coordination across countries can be traced back to the 19th century. International conventions (e.g., the Paris Convention, 1883; the Berne Convention, 1886) in this area typically adopted national treatment provisions as the basic standard for international harmonization. As international trade in knowledge products and FDI flows expanded significantly in the post–World War II era, conflicts between innovators (at the level of countries and enterprises) and imitators began to increase.

Already in the 1970s, the US began to push for the adoption of an Anti-Counterfeiting Code at the level of the GATT. This effort — that was launched at the final stages of the Tokyo Round (1973–1979) of multilateral trade negotiations — was driven by the lobbying of trademark-holding companies, which were trying to limit counterfeited products in international trade. This attempt did not succeed, but it signaled the way of the future for innovation-leading nations — in particular, the US.

[18] See Fink and Primo Braga (2005) for an econometric analysis of the impact of IPRs on trade flows.

When the Eighth Round of multilateral trade negotiations (the Uruguay Round) under the GATT started in 1986, the strategy was refined to go beyond anti-counterfeiting with a view to establish minimum standards of protection and enforcement across a broad array of IPRs instruments. The appeal of this approach was to connect the strengthening of IPRs protection to the broader trade agenda and to provide access to the dispute settlement mechanism of the multilateral trade system. Most developing countries, in turn, preferred the World Intellectual Property Organization (WIPO) as the institutional locus for IPRs' discussions. The lack of effective enforcement powers in the WIPO conventions, however, is often presented as the reason behind the US efforts in favor of a GATT-related solution.[19]

This time, the strategy succeeded, leading to the Trade-Related Aspects of Intellectual Property Rights (TRIPS) Agreement. In a nutshell, the creation of the WTO was made possible by a bargain that involved the inclusion of two new agreements under the multilateral system of trade governance (GATS and TRIPS), a priority for industrialized nation negotiators, in exchange for new disciplines for agricultural trade and the dismantling of the system of quotas that governed textile and clothing trade (the MFA), priorities for most developing countries.

The relative success of those advocating stronger protection of IPRs at the global level via the negotiation of trade agreements, however, did create reactions from affected trade partners. This was particularly the case with respect to the impact of TRIPS on access to medicines amid public health crises, as underscored by the HIV/AIDS epidemic. This reaction, in turn, led to some adjustments at multilateral level, as illustrated by the Doha Ministerial Declaration (2001), the related Declaration on the TRIPS agreement and public health, and subsequent decisions/waivers on compulsory licensing with special emphasis on the needs of least developed countries.

3.2. *Preferential trade agreements and IPRs*

Industry groups from innovation-led countries, however, continued to lobby for the inclusion of IPRs chapters in trade agreements, focusing on

[19] See Kastenmeier and Beier (1989) and Primo Braga (1990, 1996).

preferential trade negotiations. As discussed in detail by Fink (2012), the new generation of Preferential Trade Agreements (PTAs) negotiated by the US — starting with NAFTA — typically included "TRIPS plus" provisions. The EU also followed a similar track. Moreover, IPRs provisions became standard in bilateral investment treaties entered both by the US and the EU with other nations.

The latest major development in this area was the Trans-Pacific Partnership (TPP) agreement negotiated by 12 countries in the Pacific Rim. Although one could argue that the TPP reflected mainly a geopolitical move by the US to complement the Obama Administration's strategic "Pivot to Asia" and to counteract the growing economic influence of China in the region, there is no doubt that the agreement had significant global economic implications, encompassing nations that account for roughly 40% of the world GDP. 20 TPP was seen as a landmark agreement that would herald a new era in terms of trade governance. TPP-inspired rules may also have a significant impact on the organization of GVCs, as they would introduce more liberal rules-of-origin (the "accumulation of origin" concept) facilitating the operation of GVCs across its members.[20]

The TPP agreement heralds a new era in terms of trade governance. It can be argued, for example, that at least in the medium-term, the megaregionals are eroding the relevance of the multilateral trade system and the WTO. TPP may also have a significant impact on the organization of GVCs, as it will introduce more liberal rules-of-origin (the "accumulation of origin" concept) facilitating the operation of GVCs across its members.

Not surprisingly, one of the most controversial chapters of the agreement concerns its IPRs provisions. The US put emphasis on longer terms of copyright protection, regulatory changes that would effectively translate into longer patent terms and constrain the entry of generic drugs into these markets, as well as additional rules for biologic medicines (pharmaceutical products developed from living organisms), including minimum standards for data protection.

---

[20]For a discussion of the role of geopolitics in influencing trade strategies, see Dieter (2014) and Primo Braga (2015).

The final terms of the TPP agreement did not deliver on all the demands of the US negotiators. Still, several of these "TRIPS plus" measures were initially adopted. Some noteworthy measures adopted in the original TPP Agreement include trademark terms of protection of no less than ten years (TRIPS requirement is of seven years) and the removal of barriers for the protection of sound marks; a minimum copyright term of protection of at least 70 years (TRIPS minimum standard is 50 years) and stronger copyright enforcement (including the possibility of criminal prosecution against acts of removal of rights management information and the requirement that TPP countries be signatories of WIPO "Internet treaties"); requirement of enforceable legal means for the protection of trade secrets (TRIPS does not specify these means); protection of undisclosed test data submitted for marketing approvals (at least ten years in the case of agricultural chemicals and five to eight years in the case of pharmaceuticals; TRIPS does not have such a requirement); the TPP being the first trade agreement to explicitly protect new pharmaceutical products that are or contain a biologic[21]; and adjustment for patent office delays in the granting of patents that will promote harmonization of patent-granting practices among TPP Parties.

It is worth noting that some of these provisions go beyond the "TRIPS plus" aspects that the US had already negotiated on a bilateral basis in the context of its FTA treaties with countries such as Australia, Chile, and Peru. In short, TPP — if the original text were ratified — would provide for higher standards of IPRs protection that better reflect the existing US laws.[22] Inevitably, some of these provisions generated significant controversy even among like-minded countries.

[21] In the US, a biological product is defined as a "virus, therapeutic serum, toxin, antitoxin, vaccine, blood, blood component or derivative, allergenic product, protein (except any chemically synthesized polypeptide), or analogous product, or arsphenamine or derivative of arsphenamine (or any other trivalent organic arsenic compound), applicable to the prevention, treatment, or cure of a disease or condition of human beings." See Bio (2013) for further details.

[22] One major exception — reflecting resistance from other TPP parties — was the term of regulatory data protection for biologics. In the US, this term is of at least 12 years from the date of approval. Article 18.52 of the TPP provides only for an eight-year term of protection. For further details, see ITAC-15 (2015).

3.3. *Implications for a new research agenda*

(1) TPP — the most concrete example of the new megaregionals — has adopted several "TRIPS plus" provisions. From the US perspective, the Obama administration presented the TPP rules as the new "gold standard" for the promotion of innovation via stronger protection of IPRs. Ironically, the perception that the TPP did not address practices and policies that were deemed to adversely affect the balance of trade between the US and TPP members led President Trump to withdraw the US from the agreement as one of the first decisions of the new US administration in 2017. The relevant question that remains is: Would these "TRIPS plus" measures deliver a higher level of innovation among TPP members or would they simply translate into an exercise of rent transfer in favor of firms that already dominate the global innovation ecosystem?

(2) How best to measure the innovation impact of TPP on its members? Should one focus on input indicators (e.g., levels of R&D), output indicators (e.g., expansion of IP-intensive sectors in terms of production and exports), or ancillary activities (e.g., FDI in knowledge-intensive sectors)?

(3) The TPP offers opportunities for controlled experiments contrasting groups of countries with similar levels of development and ecosystems of innovation inside and outside the "club." Will the TPP path drive innovation in a much more effective manner in countries like Chile, Mexico, and Peru *vis-à-vis* Argentina and Brazil; or in Malaysia and Vietnam *vis-à-vis* Indonesia, the Philippines, and Thailand? Or will it "export" to other countries the flaws of the US IPRs system with its emphasis on litigation (as illustrated by the growing role of non-practicing entities (NPEs) — i.e., entities that focus on licensing and litigation of IPRs rather than production and innovation) and strategic behavior to block the introduction of generic drugs?[23]

(4) Will the TPP model, with its emphasis on outward-orientation and stronger protection for IPRs, impact the political economy of

[23] For an excellent discussion of NPEs and strategic manipulation to delay the introduction of generic drugs in the US, see Feldman (2016).

strategies for development? In other words, will countries transition from A to C (see Table 1) in a much faster pace than was the case in the past in an attempt to qualify as eventual new members of megaregionals?

(5) Will TPP developing members become more attractive not only for the location of nodes of GVCs (i.e., integrating into GVCs) but also in terms of upgrading opportunities over time? Upgrading can involve enhancing domestic value-added (i.e., increasing the "thickness" of the domestic economic links of the node), capturing more nodes of the GVC network (i.e., upgrading within GVCs), and diversifying the number/sophistication of GVCs operating in the country (i.e., upgrading across GVCs). Firms participating in these process benefit from adopting higher international production standards, engaging in process upgrading (e.g., by managing support services required for the GVC operation and the preservation of brands, including the adoption of best practices in CSR), and functional upgrading (e.g., by performing more sophisticated engineering and R&D functions). In other words, will TPP rules — including new standards of protection for IPRs — prove to be more effective/relevant for the current phase of economic globalization than existing multilateral disciplines?

(6) Will the TPP further contribute to the erosion of the relevance of the WTO with respect to debate of new trade agenda issues (e.g., GVCs, innovation, digital economy, the relationship between trade and investment)? Will it, for example, impact the relevance of the "juridical branch" of the WTO (its dispute settlement procedures) *vis-à-vis* new trade issues?

(7) Will TPP standards increasingly influence multilateral rules or is TPP going to open a new chapter of fragmentation in terms of international trade rules with significant discrimination against outsiders? What will be the role of the IP-related chapter in this process (e.g., by explicitly stating that state-owned enterprises cannot be excluded from IP enforcement rules, does it create a barrier to entry for countries such as China to join the TPP in the future?)?

(8) Will TPP rules dominate the terms of the debate in framing the IPRs chapter of the ongoing negotiations between the US and the European Union on the Transatlantic Trade and Investment Partnership (TTIP)? Or will the debate be reopened in areas like protection of undisclosed test data or with respect to the TPP-style rules implications for data privacy standards (a potential rerun of the debate on the Anti-Counterfeiting Trade Agreement that was rejected by the European Parliament)?

4. Concluding Remarks

Evolving trade rules both at the multilateral and preferential "fronts" can have a significant impact on the incentives for innovation at national and global levels. They are introducing new disciplines for the implementation of national innovation policies. Concerns about their impact in terms of policy space for developing countries have dominated the debate in the multilateral arena, e.g., how these disciplines will affect policies that support "national champions." At the same time, new rules being developed in the context of megapreferentials are introducing higher standards of IPRs protection closer to American practices. On November 11, 2017, eleven of the original TPP twelve signatories announced that they would be willing to go ahead with a new agreement — the so-called Comprehensive and Progressive Agreement for Trans-Pacific Partnership (CPTPP) — that builds upon the TPP text. There was also a decision to suspend 20 sensitive provisions of the original agreement, on topics such as express shipments, investor rights, and IPRs. Actually, 11 of the suspended provisions are related to IPRs, reflecting the controversial character of some of the rules of the original TPP as discussed above. The suspension of these provisions, rather than simply eliminating them, suggest that the CPTPP partners want to keep the door open for an eventual return of the United States to the agreement in the future.

The effectiveness of these new rules and their implications for the health of the multilateral trade system are important issues not only for the role of innovation in the world economy but also in terms of the future of globalization. This chapter identifies a few themes — focusing on the implications of "TRIPS plus" disciplines in preferential agreements — that merit additional investigation.

Acknowledgments

This paper was partially written while the author was Professor, International Political Economy, IMD, Switzerland. It was prepared as a contribution to the workshop on *Mega-Regionalism: New Challenges for Trade and Innovation*, organized by the East-West Center and SAIS Europe, The Johns Hopkins University. Comments from Dieter Ernst and Michael Plummer are gratefully acknowledged. Any remaining errors are the fault of the author.

References

Ashton, K. (2015). *How to Fly a Horse: The Secret History of Creation, Invention, and Discovery* (Anchor).

Arrow, K. J. (1962). "Economic Welfare and the Allocation of Resources for Invention." In *The Rate and Direction of Inventive Activity: Economic and Social Factors*, p. 609.

Bhagwati, J. (2002). "Patents and the Poor: Including Intellectual Property Protection in WTO Rules Has Harmed the Developing World," *Financial Times*.

Bio (2013). "The Trans-Pacific Partnership and Innovation in the Bioeconomy: The Need for 12 Years of Data Protection for Biologics." White Paper.

Boldrin, M. and K. L. David (2013). "The Case against Patents," *Journal of Economic Perspectives*, 27(1), pp. 3–22.

Chandra, V., D. Erocal, C. A. B. Primo, and P. C. Padoan (eds.) (2009). *Innovation and Growth: Chasing a Moving Frontier* (OECD, Paris).

De Long, J. B. and H. S. Lawrence (1991). "Equipment Investment and Economic Growth," *The Quarterly Journal of Economics*, 106(2), pp. 445–502.

Dieter, H. (2014). *The Return of Geopolitics: Trade Policy in the Era of TTIP and TPP*, Dialogue on Globalization Series (Friedrich-Ebert-Stiftung, Berlin).

Dobbs, R., J. Manyika, and J. Woetzel (2015). *No Ordinary Disruption: The Four Global Forces Breaking All the Trends* (Public Affairs, Philadelphia).

European Commission (2014). "Standard-essential Patents." Competition Policy Brief, June.

Feldman, R. (2016). "Learning from Past Mistakes — The US Patent System and International Trade Agreements." Paper presented at the *New Challenges for Trade and Innovation Workshop*, East-West Center, Honolulu, January.

Fink, C. and C. A. Primo Braga (2005). "How Stronger Protection of Intellectual Property Rights Affects International Trade Flows." In *Intellectual Property and Development: Lessons from Recent Economic Research*, C. Fink and K. Maskus (eds.) (World Bank, Washington, D. C.).

Fink, C. (2012). "Intellectual Property Rights." In *Preferential Trade Agreement Policies for Development: A Handbook*, in J. P. Chauffour and J. C. Muir (eds.) (World Bank, Washington, D.C.).

Gordon, R. J. (2014). "The Turtle's Progress: Secular Stagnation Meets the Headwinds." In *Secular Stagnation: Facts, Causes and Cures*, C. Teulings and R. Baldwin (eds.) (CEPR Press, London; VoxEU.org eBook).

Howitt, P. (2009). "Competition, Innovation, and Growth: Theory, Evidence and Policy Challenges." In *Innovation and Growth: Chasing a Moving Frontier*, Chandra, V., D. Erocal, C. A. Primo Braga, and P. C. Padoan (eds.) (OECD, Paris).

ITAC-15 (2015). "The Trans-Pacific Partnership Agreement," Report of the Industry Trade Advisory Committee on Intellectual Property Rights (ITAC-15), December.

Kastenmeier, R. W. and D. Beier (1989). "International Trade and Intellectual Property: Promise, Risks, and Reality," *Vanderbilt Journal of Transnational Law*, 22(2), pp. 285–307.

Keller, W. (2009). "International Trade, Foreign Direct Investment, and Technology Spillovers." NBER Working Paper 15442, October.

Krueger, A. (1974). "The Political Economy of the Rent-Seeking Society," *American Economic Review*, 64(3), pp. 291–303.

Machievelli, N. (1992). *The Prince*, 2nd Edn. (W.W. Norton, London and New York).

Machlup, F. (1958). "An Economic Review of the Patent System." Subcommittee on Patents, Trademarks and Copyrights, Committee on the Judiciary, U.S. Senate, Study No. 15, U.S. Government Printing Office, Washington, D.C.

Mokyr, J. (2014). "Secular Stagnation? Not in Your life." In *Secular Stagnation: Facts, Causes and Cures*, C. Teulings and R. Baldwin (eds.) (CEPR Press, London; VoxEU.org eBook).

OECD and Eurostat (2005). *Oslo Manual: Guidelines for Collecting and Interpreting Innovation Data*, 3rd Edn. (OECD Publishing, Paris).

Panagariya, A. (1999). "TRIPS and the WTO: An Uneasy Marriage."

Primo Braga, C. A. (1990). Chapters 3, 5 and 7 in *Strengthening Protection of Intellectual Property in Developing Countries: A Survey of Literature*, W. E. Siebeck (ed.) (The World Bank, Washington, D.C.).

Primo Braga, C. A. (1996). "Trade-Related Intellectual Property Issues." In *The Uruguay Round and the Developing Economics*, W. Martin and L. A. Winters (eds.) (Cambridge University Press, Cambridge).

Primo Braga, C. A. and C. Fink. (1996). "The Economic Justification for the Grant of Intellectual Property Rights: Patterns of Convergence and Conflict — The Implications of the New Regime for Global Competition Policy," *Chicago-Kent Law Review*, 72 (December), pp. 439–461.

Primo Braga, C. A., C. Fink, and C. Sepulveda (2000). "Intellectual Property Rights and Economic Development." World Bank Discussion Paper No. 412, World Bank, Washington, D.C.

Primo Braga, C. A. (2010). "Trade and Poverty: An Old Debate Rekindled." In *Peace and Prosperity Through World Trade*, J. P. Lehmann and F. Lehmann (eds.) (Cambridge University Press, Cambridge).

Primo Braga, C. A. and R. Meléndez-Ortiz (2013). "The Innovation Imperative and the WTO," IMD Tomorrow's Challenges, October.

Primo Braga, C. A. (2015). "The Economic and Geopolitical Dimensions." In *New Realities: Business Dynamics at the Frontiers of Globalization*, C. A. Primo Braga and J. P. Lehmann (eds.) (IMD, Lausanne).

Romer, P. (1994). "New Goods, Old Theory and the Welfare Cost of Trade Restrictions," *Journal of Development Economics*, 43(1), pp. 5–38.

Thiel, P. and B. Masters (2014). *Zero to One: Notes on Startups or How to Build the Future* (Crown Publishing Company, New York).

WIPO (2015). *World Intellectual Property Report: Breakthrough Innovation and Economic Growth* (WIPO, Geneva).

Chapter 16

The Information Technology Agreement, Manufacturing, and Innovation — China's and India's Contrasting Experiences*

Dieter Ernst

East-West Center, Honolulu, HI, USA
Centre for International Governance Innovation, Waterloo, ON, Canada

Abstract

The Information Technology Agreement (ITA), a plurilateral trade agreement, seeks to accelerate and deepen the reduction of trade barriers for the critically important Information and Communication Technology (ICT) industry. ITA thus provides an interesting case study of how tariff reduction within an industry-specific trade agreement might affect gains from trade for manufacturing and innovation. From a global welfare perspective, trade expansion could reinforce the

---

*This chapter draws on a paper presented at the agenda-setting workshop, funded by the National Science Foundation, *Mega-Regionalism — New Challenges for Trade and Innovation (MCTI)*, at the East-West Center, Honolulu, January 20 and 21, 2016. Earlier versions have been published as Ernst (2016a), Ernst (2014a), and Ernst (2013).

diffusion of innovation. Developing countries and especially emerging economies could thus reap significant gains for industrial manufacturing and innovation. But whether this really happens is an empirical question. This chapter compares the contrasting experiences of China and India throughout the history of ITA, from the conclusion of the original ITA-1 agreement in 1996 up to the extended ITA-2 agreement on December 16, 2015. India joined ITA early from a position of weakness in electronics manufacturing. This gap in technological development explains why tariff reductions increased IT imports but failed to stimulate domestic electronic manufacturing and innovation. By contrast, China joined ITA six years after India from a position of strength as the global export factory in electronics manufacturing. Through a continuous upgrading of its industry, China was able to reap the gains from trade. Two specific questions are addressed: First, why is it that China's electronics industry has benefited substantially from ITA, while in India the gains from trade liberalization have been overshadowed by major costs that are eroding domestic electronic manufacturing and innovation? And, second, to what degree are domestic economic structures and global network integration useful to explain these different experiences and the very different approaches of India and China to the ITA-2 negotiations about expanding the product lists covered by this agreement? The analysis focuses on the role played by differences in the stage of development of their ICT industry, but also more generally in their growth models and regulations, and their resources and capabilities. In addition, differences are considered in the integration of both countries into the global corporate networks of production (GPNs) and global innovation networks (GINs) of the ICT industry. By inserting these domestic determinants into the analysis of trade and innovation, new insights are expected for the study of trade and innovation within plurilateral and megaregional trade agreements. The chapter concludes with suggestions for policy and further research.

1. Will trade accelerate industrial innovation?

This chapter examines the Information Technology Agreement (ITA), a plurilateral trade agreement that seeks to accelerate and deepen the reduction of trade barriers for the critically important Information and Communication Technology (ICT) industry.

In contrast to multilateral WTO agreements, where all WTO members are party to the agreement, a plurilateral agreement implies that WTO member countries have a choice to agree to new rules on a voluntary basis. Their main purpose is to respond "to the changing needs of industries with agility by …[a]llowing flexibility in the choice of participants".[1] ITA, for all practical purposes, is about reducing all taxes and tariffs on ICT products, without any agreement thus far on non-tariff barriers (NTBs) and other "behind-the-border" regulations. This sets ITA apart from Regional Trade Agreements (RTAs), which increasingly contain provisions on intellectual property and other "behind-the-border" regulations.[2] Even more so does ITA differ from prospective megaregional trade agreements like the Trans-Pacific Partnership Agreement (TPP). ITA thus provides an interesting case study of how tariff reduction within an industry-specific trade agreement might affect gains from trade for manufacturing and innovation.

From a global welfare perspective, trade expansion could reinforce the diffusion of innovation.[3] Developing countries and especially emerging economies could thus reap significant gains for industrial manufacturing and innovation. But whether this really happens is an empirical question that is at the center of this chapter. ITA participants differ substantially not only in the stage of development of their ICT industry but also more generally in their growth models and regulations, and their resources and capabilities. They also differ in how they are integrated into GPNs and GINs present in the ICT industry.[4] These four parameters encompass what might be called *domestic determinants* of the gains from ITA for industrial manufacturing and innovation.[5] Such divergent parameters matter in particular for the analysis of the ITA's impact on developing countries.

As discussed in a related chapter,[6] manufacturing and innovation are closely intertwined, especially in a technology-intensive industry such as

[1] Nakatomi (2013).
[2] Valdes and McCann (2014) and Mumford (2014).
[3] As argued in Curtis (2013).
[4] See Chapters 3, 4, 17 and 18.
[5] For details, see Ernst (2014b).
[6] Ernst (2016b).

electronics. Of critical importance are innovations in generic industrial technologies that allow for new ways of manufacturing existing products (e.g., with new materials or by using smart robots or three-dimensional (3D) printing) as well as for manufacturing new products derived from new ideas, discoveries, and inventions (e.g., nano-scale semiconductors or implanted sensors). This implies that a country can only aspire to become successful in industrial innovation if it has developed a strong manufacturing industry. Even the most sophisticated research and development (R&D) capabilities are of little use to generate innovations if a country lacks large-scale advanced manufacturing facilities, which can draw on an integrated domestic industrial value chain.

The argument is developed by comparing the contrasting experiences of China and India throughout the history of ITA, from the conclusion of the original ITA-1 agreement in 1996 up to the extended ITA-2 agreement on December 16, 2015. The analysis focuses on the role played by differences in the stage of development of their ICT industry but also more generally in their growth models and regulations, and their resources and capabilities. In addition, differences in the integration of both countries into GPNs and GINs are considered. By inserting these domestic determinants into the analysis of trade and innovation, new insights are expected for the study of trade and innovation within plurilateral, regional, and megaregional trade agreements.

Two specific questions are addressed:

- First, why is it that China's electronics industry has benefited substantially from ITA, while in India the gains from trade liberalization have been overshadowed by major costs that are eroding domestic electronic manufacturing and innovation?
- Second, to what degree are domestic economic structures and global network integration useful to explain these different experiences and the very different approaches of India and China to the ITA-2 negotiations about expanding the product lists covered by this agreement?

Section 2 reviews conflicting perceptions of the distribution of trade liberalization gains from ITA-1. Section 3 examines how domestic determinants might explain the contrasting experiences of India and China with ITA-1.The analysis compares the electronics manufacturing industries of

China and India and their different policy approaches. The very different experiences of both countries with the first round of the ITA-1 are illustrated with data on exports, imports, and domestic production, and indicators of progress in innovation capacity.

Section 4 traces negotiations to expand the product scope, culminating in the ITA-2 agreement. The analysis contrasts the different approaches of China and India to the ITA-2 negotiations and examines to what degree domestic determinants might explain these divergent trade negotiation strategies. The chapter concludes with suggestions for policy and further research and insights that might be gained for the analysis of megaregionalism.

2. The information technology agreement — who benefits?

Concluded in 1996, the original ITA agreement (ITA-1) went into effect in April 1997 with 29 World Trade Organization (WTO) member-countries. ITA-1 focused exclusively on tariff reduction, establishing zero tariffs for 217 electronics products.[7] The main product groups covered were computers, semiconductors, semiconductor manufacturing and test equipment, telecommunications equipment, software, and scientific instruments.[8]

ITA provides "most favored nation" (MFN) treatment to all WTO Members, even if those countries have not joined the agreement. The US government was a major driving force behind the establishment of ITA-1. The agreement's agenda was shaped by a core group of developed countries who accounted for nearly all the original signatories, with Indonesia and Turkey the only developing countries formally adopting the Declaration. By 2012, ITA-1 had 78 WTO Members — 36 are non–Organisation for Economic Co-operation and Development (OECD) member countries, and 35 of them are developing countries.[9] The latter

[7] Soon after ITA became effective in April 1997, participants commenced a schedule of phased duty reductions, with all duties slated for elimination by 2000.

[8] Not covered were mainly consumer electronics products, including CRT TV sets, video cameras, and photocopiers.

[9] For a list of participants and their date of participation, see WTO (2012).

include significant players in the electronics industry (China, Taiwan, Malaysia, Thailand, and Vietnam), and other countries, such as India, Egypt, Indonesia, Philippines, and Turkey, which have the potential to become major players.[10]

ITA does not allow any exception to the covered products. This sets it apart from other plurilateral trade agreements, such as the WTO-Government Procurement Agreement (GPA), which allows exceptions by way of offsets (e.g., defence offsets). The only relaxation is identifying certain specified products as sensitive so that they qualify for a phased-in implementation period for extended implementation. India, which signed ITA-1 in 1997, has requested and received such an extension.[11] When China joined ITA-1 in April 2003, it requested and received a three-year phase-out for import-sensitive items.[12]

ITA-1 has enabled a substantial increase in the trade in electronics products that are covered by the agreement. A study by the US International Trade Commission finds that aggressive tariff liberalization facilitated growth in ITA trade from \$1.2 trillion to \$4.0 trillion in 2010.[13] Opinions differ, however, on the distribution of trade liberalization gains. A widely held perception in the US is that developing countries can reap substantial benefits. For instance, Ezell (2012) argues that ITA-1 is likely to

> "benefit developing countries in three principal ways: (1) reducing tariffs on a broader range of ICT products encourages greater adoption of ICT products that play a key role in spurring economic growth; (2) lower prices realized by reducing tariffs on ICTs increases the productivity of all other industries in a developing economy; and (3) by lowering the price of a key input, the ITA has undergirded development of the burgeoning

---

[10]WTO members that have not joined ITA include Mexico, Brazil, Tunisia, South Africa, Argentina, and Chile. Of these countries, only Mexico is a leading exporter and importer of electronics products. Its government, however, has decided that non-membership better fits its economic interests.

[11]India and several other developing countries, including Costa Rica, Indonesia, South Korea, and Taiwan, implemented extended duty staging to 2005 on a product-by-product basis as permitted by the ITA Declaration.

[12]See China's Accession Protocol at http://www.wto.org/english/tratop_e/inftec_e/itscheds_e.htm .

[13]Anderson and Mohs (2010).

> ICT software and services industries in many developing countries such as India, Indonesia, Malaysia, and the Philippines."

Unfortunately, this argument neglects the fundamental differences among ITA-1 participants in their capabilities for manufacturing and innovation, as well as in their position in GPNs and GINs. As argued in this chapter, these structural differences go a long way to explain why ITA participants differ in their capacity to reap the potential gains from trade liberalization. In addition, as stated by the US International Trade Commission, "The paucity of conclusive research on the impact of the ITA on global trade attests to the difficulties in empirically measuring the effects of the ITA and signals that … considerable discussion and analysis are still needed to determine the magnitude of the ITA's impact on IT [information technology] trade and technology diffusion" (Anderson and Mohs, 2010, p. 41).

Closer to reality are industry insiders and US government officials who argue that leading US multinational corporations (MNCs) "benefit disproportionally" from ITA-enabled trade liberalization.[14] ITA-1 has indeed provided significant benefits for leading US vendors of ICT products. For example, take semiconductors, an important product covered by ITA-1. While before 1997 the US share of the worldwide market in semiconductors hovered around 40%, since the signature of the ITA-1 agreement, it has moved up to around 50%.[15] From 2005 to 2009, semiconductors were the number one product export from the US on an aggregate basis, with exports totaling US$48 billion (US$10 billion more than automobile exports, in second place). And in 2011, US semiconductor producers had global sales of US$152 billion, representing more than half the global semiconductor market.[16]

---

[14] [16] Quote from presentation by Greg Slater, Intel's director for Trade and Competition Policy, at an event hosted by the Information Technology and Innovation Foundation (ITIF) on *Boosting Exports, Jobs and Economic Growth by Expanding the ITA*, Washington, DC, 15 March 2013. At the same event, Miriam E. Sapiro, Deputy US Trade Representative, argued that "ITA has been one of the most successful agreements ever undertaken in the multilateral trading system," as it had boosted US IT exports.

[15] [17] Semiconductor Industry Association (2010).

[16] USITC, hearing transcript, November 8, 2012, p. 39, 41 (testimony of Ian Steff, Semiconductor Industry Association).

ITA-1 also provided a big push to the expansion of GPNs/GINs of US ICT companies. The US International Trade Commission finds that ITA-1 boosted Foreign Direct Investment (FDI) in China, especially by leading US MNCs, which "had a major role in China's accelerating ITA exports, as … [these companies] sought to reduce costs by directly adding capacity in China. Once China joined the WTO, products exported from China were guaranteed MFN access to other countries, providing strong incentives for multinational corporations to establish production and assembly operations in China" (Anderson and Mohs, 2010, p. 17).

3. What explains the contrasting experience of India and China with ITA-1?

Trade theory suggests that all countries would benefit from trade liberalization once they adjust their policies to comply with WTO regulations. However, while China's electronics industry has benefited substantially from ITA-1, in India, the gains from trade liberalization have been overshadowed by (domestically determined) major costs, which are eroding domestic electronic manufacturing and innovation in the context of trade liberalization. What explains the puzzling India–China contrast?

India joined ITA from a position of weakness in electronics manufacturing. In line with Grossman and Helpman (1991), this gap in technological development explains why tariff reductions increased IT imports, but failed to stimulate domestic electronic manufacturing and innovation. In addition, India's thriving IT services industry and its R&D system remained disconnected from electronics manufacturing. By contrast, China joined ITA from a position of strength as the global export factory in electronics manufacturing. But most importantly, it is through a continuous upgrading of its absorptive and innovation capacity that China was able to reap the gains from trade in the electronics industry.[17]

[17]Defined as "a firm's ability to recognize the value of new information, assimilate it, and apply it to commercial ends," absorptive capacity is the necessary prerequisite for developing and improving the firm's innovation capacity (Cohen and Levinthal, 1990).

3.1. *Why India struggles*

India joined ITA-1 early in 1997 with an underdeveloped electronics manufacturing industry, and its main concern was to facilitate the growth of its then still-nascent IT services industry through ICT imports and inward FDI. ITA-1 participation and the resulting price reduction for IT imports did, indeed, facilitate the expansion of India's IT services industry.[18] At the same time, however, domestic electronic manufacturing does not seem to have benefited much from India's participation in ITA-1 and, indeed, was most likely hurt by it.

Electronics manufacturing in India is struggling despite a huge and still-growing domestic market and pockets of world-class capabilities in IT services and chip design.[19] Local production faces cost disadvantages that constrain investment in plants and equipment, technology absorption, and innovation. Cost disadvantages result, for instance, from the lack of a unified tax system, poor infrastructure, the lack of power and water, too many ill-defined and often conflicting regulations, and the lack of a robust ecosystem for components and production equipment. In addition, local production hardly benefits from India's chip design capabilities, which are integrated, instead, into global MNC networks of innovation and production. And there is a huge gap between the rapid growth of domestic demand and nearly stagnant domestic production.

Given the weakness of domestic production, India's growing demand for electronic products generates rising imports of final products and high import dependence for key manufacturing components. These imports have become the third most important contributor, after petroleum and gold, to the country's record current account deficit.

A defining characteristic of India's electronics market is that a handful MNCs dominate without engaging in substantial manufacturing in the country (whether directly or through electronics manufacturing services, or EMSs), with the exception of low value-added final assembly. Those MNCs can rely on their extended GPNs to source the relevant products for

[18] Sen (2014).
[19] Ernst (2014c).

the Indian market from low-cost production sites, primarily in China.[20] Where local Indian firms seek to compete with MNCs for the domestic market, they follow the same pattern of sourcing their products from off-shore production sites in China. Indian firms thus have to compete on the "China price," relying on offshore outsourcing to China-based EMSs.[21]

India's liberalization of information and telecom services has boosted the demand for telecom equipment, but this has not led to the development of a domestic telecom manufacturing industry. Instead, global telecom equipment vendors such as Alcatel, Ericsson, and increasingly Huawei and ZTE, have been the primary beneficiaries. And consumer electronics, the largest segment of India's electronics market, is dominated by Japanese, Korean, and now also Chinese companies. Over the last few years, these companies have substantially decreased domestic production and now rely overwhelmingly on imports from China. As for domestic vendors, they are even less reliant on domestic production — they almost completely source from China.

It is important to emphasize that India's thriving integrated circuit (IC) design sector remains largely disconnected from the domestic market. Most of the design work is done for MNCs, and the design is taken back to their home country where decisions are made on where to locate manufacturing, many times ending up in places like Shenzhen. As a result, India's electronics manufacturing industry is not benefiting from its rich pool of sophisticated IC design engineers. The work of these engineers remains disconnected from the development of the domestic electronics manufacturing industry.

In short, deep integration of electronic design capabilities into global R&D networks is paired in India with almost no integration into the domestic electronics manufacturing value chain. For instance, all major electronic design automation (EDA) tool providers for chip design have large facilities in India. But all of them are 100% focused on export markets (Ernst, 2014c). Hence, little of these capabilities are likely to disseminate within India.

---

[20]In addition to China's substantial cost advantages, what matters most for MNCs is that they can benefit from accumulated capabilities for rapid and low-cost scaling up of sophisticated production lines that exceed India's current capabilities by far.

[21]According to an Indian component supplier, "MNCs … are asking us to be 15% lower than Chinese cost, only then they will source from us" (Ernst, 2014c, p. 30).

Despite India's integration into GINs for chip design, its integration into GPNs of the ICT industry remains very limited. FDI in India's electronics industry has been extremely low, even relative to other sectors — the industry ranks 26 out of 64 sectors in terms of the cumulative FDI received from April 2000 to April 2013. This very low global network integration cuts India off from global knowledge sources, setting it apart from China.

India faces a fundamental challenge — its electronics industry cannot rely so far on a vibrant industrial innovation system. In turn, this constrains the industry's capacity for productivity-enhancing innovation. There is ample evidence that India's R&D system remains stuck at a low level, unable to provide capabilities and innovations that would enable its electronics industry to reap the benefits of ITA liberalization.[22]

For many domestic firms, inadequate size prevents economies of scale and scope, while high costs of doing business, Byzantine regulations, and low domestic value-added constrain profit margins, and hence investment in production and R&D. In addition, larger foreign original equipment manufacturers (OEMs) and EMSs typically conduct only final assembly in India, and are reluctant to invest in full-scale manufacturing and R&D.

In the final analysis, the misery of India's electronics manufacturing industry points to a broader challenge. India's economic institutions, both public and private, were largely designed for a time before the country was opened to the global economy. These institutions are ill-equipped to cope with the requirements of transforming India into an internationally competitive industrial economy that can reap the benefits of ITA-1 participation.[23]

The following data briefly illustrate India's negative experience with ITA-1. Joining ITA-1 led to a reduction in India's tariffs for final products to zero or close to zero, and this led to an acceleration of ITA imports. In 2000, 96 product lines were reduced to zero tariff, while in 2005, it was

---

[22] The most recent Global Innovation Index 2015 provides ample evidence of India's weak industrial innovation capacity — out of a total 143 countries, India is ranked 81 (Cornell University, INSEAD and WIPO, 2015). According to the Battelle Institute, a primary source of international R&D data, India's gross expenditure on research and development (GERD) was 0.85% of gross domestic product (GDP) in 2012 (compared to 1.6% in China), a figure essentially unchanged since 2000. Note that China's GDP is much larger than India's and growing substantially faster.

[23] For an early analysis, see Kelkar (1999).

extended to 121 product lines. While India's ITA imports grew by 18% annually between 1997 and 2000, their growth rate increased to nearly 38% between 2001 and 2005.[24]

An immediate effect has been an increase in the import content of raw material consumption by India's electronics industry in the last seven years, from 50.5% to 55.9%.[25] Further, India's electronics imports under HS code 85 have grown faster than its electronics consumption.[26] Between financial year (FY) 2010–2011 and FY 2012–2013, India's imports grew especially fast for integrated circuits (82.02%), the second largest electronics import category.[27]

In principle, trade deficits are not always negative for economic growth. Empirical research points to the importance of imports in boosting productivity.[28] Yet in India's case, the local value-added of electronics manufacturing is around 7%, while electronics imports account for almost two-thirds of the country's consumption of electronics products (Frost and Sullivan, 2013). In addition, an analysis of the number of Triadic patent families by applicant's country of residence shows that between 1999 and 2005, India recorded merely 26 such patents. During the same period, China recorded 208, Korea 4,862, and the US 37,907.[29]

In short, the Indian electronics industry has failed to reap the potential benefits from ITA-1 participation. Due to the zero tariffs under ITA-1,

---

[24] Kallummal (2012).

[25] Kallummal (2012, p. 36, Figure 19). The import content was growing even faster for two strategically important product groups. In computers, peripherals, and storage devices, from an already very high import content of 61.76% in FY 2005, it became more than 75% in FY 2012. Other electronics (which basically can serve as a proxy for electronic components) saw it move from more than 45% in FY 2005 to almost 53% in FY 2012.

[26] HS code 85 covers electrical machinery and equipment and parts thereof; sound recorders and reproducers, television image and sound recorders and reproducers, and parts. India's consumption of electronics products grew by 14% from FY 2010 to FY 2012. During the same period, its imports of electronics (HS 85) products increased by 31%.

[27] India's imports grew very fast for some electronic components like capacitors (35.9%), rectifiers and inductors (38.4%), and consumer-related products like video recorders and monitors (81.11%), and microphones and loudspeakers (109.1%). For details, see the DGFT website.

[28] OECD (2010).

[29] WTO Secretariat data, based on OECD Stat Extracts (February 27, 2012), quoted in WTO (2012).

India is now an easy target for low-cost electronics imports, especially from China — a country that is far ahead in its electronics manufacturing industry. Some observers fear that with the influx of such imports into the domestic market, "the existing domestic producers may become domestic assemblers and traders in the IT products" (Kallummal, 2012, p. 15).

3.2. *China's ITA-1 experience is very different*

China joined the ITA-1 six years after India in 2003 from a position of strength as the third largest exporter of ITA products and the fourth largest importer. By 2003, China's per capita GDP (US$1,270) was three times higher than India's 1997 per capita GDP (US$427).[30] "China was . . . [already] a leading manufacturer and trader of IT products prior to joining the ITA and deeply engaged in the global IT production chain even before tariff liberalization" (Anderson and Mohs, 2010, p. 7).

In 2004, China expanded its market share, becoming the world's largest exporter of ITA products. The next year, China surpassed both the European Union (EU) and the US to become the leader in overall ITA trade. From 2009 to 2011, China's high-technology exports (in current US$) expanded by almost 48% (from US$309.6 billion to US$457.2 billion (compared to US ICT exports of US$145.3 billion in 2011).[31] Further, China's share of global exports of IT products rapidly expanded from 2.2% in 1996 to 27.5% in 2012.[32]

China's innovation policy has played an important role in the country's rapid rise in the ICT industry. Massive investments in the country's R&D infrastructure and higher education, on a scale never seen before, have been fast-tracking the speed of learning and capability development.[33] Since 2000, China has increased R&D spending roughly 10%

---

[30] See: http://search.worldbank.org/data?qterm=India%20per%20capite%20GDP%20for%201997&language=EN.

[31] See: http://data.worldbank.org/indicator/TX.VAL.TECH.CD, quoting the United Nations Comtrade database. High-technology exports are products with high R&D intensity, such as in aerospace, computers, pharmaceuticals, scientific instruments, and electrical machinery.

[32] See: http://stat.wto.org/StatisticalProgram/WSDBStatProgramHome.aspx?Language=E.

[33] For China's push to upgrade its semiconductor industry, see Ernst (2015). See also Ernst (2016c).

each year — a pace the country maintained during the 2008–2009 recession. This sustained commitment to a rapid expansion of R&D sets China apart from India's weak industrial innovation system.

China's push to upgrade its high-tech export industries through innovation has produced measurable results. The most recent data available from the OECD-WTO TiVA database for 2011 show that, for key export sectors (like ICT, electrical machinery and transport equipment), the domestic value-added (DVA) content of exports has significantly increased, from 25% in 1995 to almost 50%. Specifically for ICT and electronics, the DVA share moved up from less than 30% in 1995 to almost 50% in 2011.

However, two important caveats need to be added to the above headline figures. First, foreign-invested enterprises (FIEs) continue to dominate China's high-tech exports, rising from 79% in 2002 to 82% in 2010.[34] Second, DVA remains very limited — estimates range from 5% to 15%, depending on the complexity of production technology and value chain stage. Strikingly, for high-end smartphones, China's domestic value addition remains stuck below 5%.[35] Recent research documents that while China's reliance on imported intermediates has declined significantly in lower-tech ICT sectors,[36] its "dependence on imported intermediates did not abate in the higher-tech sectors such as electrical and optical equipment, transport equipment or chemicals".[37]

4. Economic structure and global network integration shape benefits and costs of participation in ITA-1

What explains the very different experiences of China and India? China's ITA-1 experience demonstrates that a progressive integration into GPNs and GINs, which was accelerated by China's WTO accession and policy reform, played an important enabling role in the development of its electronics industry. Global network integration facilitated technology diffusion

[34] Congressional Research Service (2015).

[35] China's domestic value-added for the iPhone 4 was reported to be less than 2% (Lin and Tang, 2013).

[36] For example, DVA for computers improved from 20% in 2002 to 33% in 2007 (Lin and Tang, 2013).

[37] Baldwin and Lopez-Gonzalez (2013).

and absorption and enhanced China's capacity to reap the potential gains from ITA-1 trade liberalization. A proxy for China's high degree of GPN integration is that 44% of its exports are produced under so-called "processing trade" arrangements, in which imported inputs are assembled into exports.[38] Another indicator is that two-thirds of China's production of merchandise goods and services are intermediates, which is substantially higher than the world average (Baldwin and Lopez-Gonzales, 2013, p. 13).

As for integration into GINs, China is the largest "net importer" of R&D, and it is the third most important offshore R&D location (after the US and UK) of the 300 top R&D spending multinationals.[39] China is thus deeply integrated, albeit still unevenly, into the international circulation of technological and managerial knowledge needed to enhance its absorptive capacity.

What distinguishes China, however, is that trade liberalization was combined with well-funded and focused support policies for manufacturing and industrial innovation. Even if one factors in massive inefficiencies, this pragmatic policy mix has produced results. As Peter Petri observes, "China is not averse to intervening, but it has done that against the background of a lot of liberalization. It's paying off".[40] This description nicely captures China's approach until quite recently. However, China's new leadership has shifted the balance in favor of industrial upgrading through innovation, as codified in the Strategic Emerging Industries (SEI) plan.[41] As discussed in Sec. 4, this singe-minded focus on industrial innovation may explain China's approach to ITA-2 negotiations.

India's situation is very different. By negotiating a bad deal for its electronics manufacturing industry and delaying necessary regulatory

---

[38] General Administration of Customs, PRC, "2011 Trade by Trade Regime." See: http://www.customs.gov.cn/publish/portal0/tab44604/module109000/info353199.htm.

[39] Ernst (2011).

[40] E-mail to the author from Professor Peter Petri, January 28, 2014.

[41] 国务院关于印发 "十二五" 国家战略性新兴产业发展规划的通知 [The State Council Notification on the Long-term Development Plan for Strategic Emerging Industries during the 12th Five Year Plan], 国发〔2012〕28号, 7 July 2012. Ernst, D., 2016, Advanced Manufacturing and China's Future for Jobs, East-West Center Working Papers: Innovation and Economic Growth Series, no. 8, August 2016, http://papers.ssrn.com/sol3/papers.cfm?abstract_id=2820433 A revised version is forthcoming as a chapter in Paus, Eva (editor), Emerging Dystopias? The New Technological Revolution and the Future of Work, Cornell University Press, Ithaca, NY 14850.

reforms, India's government bears the primary responsibility for the lack of growth and innovation in this industry. At the same time, however, India's experience with ITA-1 indicates that poorly negotiated membership terms may well create additional challenges once a country seeks to reduce investment barriers in its domestic electronics industry.

Raghuram Rajan, the former Reserve Bank of India's governor, emphasizes that the government needs to better align domestic policies to unblock barriers to investment and growth in the electronics industry.[42] There is ample evidence that existing restrictive regulations and the largely dysfunctional implementation of support policies in the past have constrained investment and growth in India's domestic electronics manufacturing.[43]

When the government of Narendra Modi took power in May 2014, bold action was promised in the "Make in India" initiative to overcome the self-imposed strangulation of India's electronics industry. Chief among the necessary first steps is a speedy transition to a unified goods and services tax (GST) system, the single most commonly cited "reform wish" of electronics manufacturers. There was hope that, in addition, drastic simplifications would be implemented in the business regulatory environment, in particular, in dispute resolution for customs conflicts as well as formal and informal penalties, and that long overdue reforms would finally be implemented in taxation, customs, compliance, and inspections, and all the other restrictive regulations. There was also hope that effective regulatory reform will be accompanied by a longer-term and structural industrial-development agenda that would build on the previous Government's National Policy on Electronics (NPE). That policy initiative sought to improve India's international competitiveness through incentives for capability development, cluster formation, R&D, and technology transfer through FDI.

However, very little of these policy announcements have been implemented thus far.[44] In the electronics industry, the main tangible result of

---

[42] Rajan (2013).

[43] As demonstrated in Ernst (2014).

[44] Pradeep S. Mehta, 2017, "A new industrial policy for Bharat India's new industrial policy must focus on improving overall competitiveness and avoid the lure of protectionism.", http://www.livemint.com/Opinion/zNSs9mYJupx0exXBjTcKSN/A-new-industrial-policy-for-Bharat.html. See also D'Costa (2015).

the "Make in India" initiative is the Taiwanese-owned Foxconn's announcement to invest up to $5 billion until 2020 in a new facility in Maharashtra to assembly smartphones from Apple, as well as China's Xiaomi.[45] Foreign investors, who are the main target of this policy, remain unconvinced "that India has a competitive advantage in making anything".[46]

In short, differences in domestic economic structure and global network integration may go a long way in explaining the very different experiences of India and China with participation in ITA-1. But are these domestic determinants also useful to explain why India and China differ in their approaches to the ITA-2 negotiations?

5. Divergence continues — China's role in the making of ITA-2

In May 2012, on the occasion of the 15th anniversary of ITA-1, a group of WTO members, led by the US, decided that the original agreement should be revisited to expand its product coverage to reflect rapid technical change in the ICT industry. Serious negotiations about a possible substantial expansion of the list of products started in July 2013 in Geneva.

5.1. *India withdraws from the ITA-2 negotiations*

The Indian government decided not to join these negotiations. According to the Commerce and Industry Ministry, "India's experience with the ITA-1 has not been encouraging as it has almost wiped out the IT industry from India. After examining the matter in consultation with the nodal Ministry i.e. Department of Electronics and Information Technology and other stakeholders, it has been decided, for the present, not to join the negotiations as it will not be in our national interest".[47] The Indian government believes that a small group of developed country signatories, led by the US, the EU, and Japan, has designed an expanded list with a focus on products that these countries, especially

[45] "How did 'Make in India' fare in 2015?" *EET India*, December 21, 2015.

[46] Sender (2015).

[47] Quoted in "India to Skip Talks on Expanding ITA Scope," *The Hindu*, March 13, 2013.

the US, continue to lead by a wide margin. Behind this nationalist rhetoric, domestic politics looms large. India's refusal to participate in the ITA-2 negotiations may well have reflected the uncertainty before the 2014 election, and a fear of being accused of betraying the nation's strategic interests.

An alternative strategy for India had been proposed by the Associated Chambers of Commerce and Industry of India.[48] ASSOCHAM argued that merely resisting ITA expansion is unlikely to have positive effects for India's electronics industry. Instead, the Indian government should engage in a smart strategy of co-shaping the consolidated product list of ITA-2. "India needs to address the ITA expansion, weighing carefully its long-term as well as short-term objectives in a strategic manner rather than becoming overly influenced by ad hoc approaches and concerns" (ASSOCHAM National WTO Council, 2012, p. 8).

In this view, non-participation in ITA-2 negotiations comes at a heavy cost. India would lose the option of co-shaping the contents of the new expanded ITA product list. Non-participation may also discourage international investors from expanding their presence in India, and it might act as a disincentive for existing FDI manufacturing projects to expand and upgrade their facilities.

The successful conclusion of ITA-2 negotiations in December 2015 shows that India's decision not to participate has been a major strategic blunder. As stated in a note prepared for the World Bank:

> "India should actively participate in ITA-2 talks in order to co-shape negotiations on the expanded product list. As a first step, India should insist on extending the phased-in implementation period for up to five or seven years. US IT industry now seems to be willing to compromise, in order to bring China on board.[49] India should use this opportunity and push for extending the phased-in implementation period. Progress towards zero tariffs on all electronics inputs and outputs is critical for

[48] ASSOCHAM National WTO Council (2012).

[49] "U.S. Tech Industry Mulls Longer Tariff Phase-Outs To Spark ITA Progress." See: http://insidetrade.com/Inside-US-Trade/Inside-U.S.-Trade-04/18/2014/us-tech-industry-mulls-longer-tariff-phase-outs-to-spark-ita-progress/menu-id-710.html.

removing barriers to India's electronics manufacturing industry. The extended time period would make it politically feasible for the new Indian government to agree to it.

As for implementation of this initiative, India should explore scope for joint action with countries that have joined China in pushing for extended phased-in implementation periods, i.e. Malaysia, Thailand, the Philippines, Turkey, and Costa Rica. India should also exploit the fact that the EU has recently pressed to exclude flat panel displays from tariff cuts in a bid to protect production by Korean and Japanese companies in Europe".[50]

5.2. *China's approach again is very different*

China's approach to the ITA-2 negotiations, in fact, followed precisely such a co-shaping negotiation strategy, in combination with persistent delaying tactics. China bargained hard to get an acceptable solution. In the end, China's approach proved to be highly successful.

Before the July 2013 meeting, China presented a list of sensitive products that called for the removal of 106 products rather than asking for an extended implementation period. Under pressure to shorten this list, China, on 17 July, reduced the list of sensitive products that it did not want to be part of ITA-2 to around 90. However, China's revised list of sensitive products continued to include product groups that are among US priorities for ITA-2, i.e., multicomponent integrated circuits (MCOs), [51] light-emit-

---

[50] Ernst (2014d).

[51] MCOs are used in a wide variety of products, including smartphones, tablets, medical devices, household appliances, and in car parts such as braking, steering, and air bag systems. Until the conclusion of ITA-2, MCOs thus can be classified under a wide range of HS subheadings. As a result, no one really knows for sure how important MCOs are for US exports. US-ITC estimates that, in 2011, sales of MCOs accounted for between 1.5% and 3.0% of global semiconductor sales, or an estimated US$1.2 billion to US$2.4 billion. A report of the USITC for the USTR identified five priority sub-sectors of the consolidated draft product list for ITA-2: medical devices; relay and industrial control equipment; optical media, including LED; loudspeakers and handsets; and, most importantly, MCOs. The USITC selected these subsectors to illustrate "the potential for increased market access opportunities for USA firms as a result of ITA expansion" (USITC).

ting diodes (LEDs), and medical devices. While some sources believed that China's long list of sensitive products "…indicated a lack of coordinated domestic consultation…," others said Beijing "… has industrial policy goals in mind".[52]

China faces a fundamental challenge. If ITA-2 would indeed broaden the product list to include zero tariffs for MCOs, LEDs, and medical equipment, this may well have reduced any realistic chances for China to develop significant domestic manufacturing and innovation capabilities in these important ICT sub-sectors that China's government considers to be essential for its industrial upgrading efforts.[53]

During a second round of ITA-2 negotiations, started in October 2013, the impasse continued, culminating in the suspension of talks on November 21, 2013. Only 58 of the 78 current ITA members were participating, with India and Indonesia remaining on the sidelines. This time, China was joined by Malaysia, Thailand, the Philippines, Costa Rica, and Turkey who "also want many products removed from the negotiation".[54] In a sign of the growing complexity of ITA negotiations, the EU has requested to exclude flat panel displays from tariff cuts in a bid to protect production by Korean and Japanese companies in Europe.

China presented a slightly shortened revised list of sensitive products, leaving "59 items it still wants to exclude from the scope of an expanded ITA deal altogether, and around 80 … for which it is requesting tariff staging" to be extended, sometimes beyond five years.[55] But the real sticking point remained advanced semiconductors, the so-called MCOs, where China was adamant "that it will not accept tariff cuts".[56] For the US, this

---

[52] "US, Others Suspend ITA Talks to Pressure China to Soften its Stance," *Inside US Trade*, July 29, 2003, p. 2.

[53] For developments since 2015 in China's push to upgrade its semiconductor industry, see Ernst (2016d). See also Ernst (2016e).

[54] Palmer (2013).

[55] "ITA Talks Extended as China Holds Firm to 'Sensitivities' List," *China Trade Extra*, November 20, 2013. Available at: http://chinatradeextra.com.

[56] "ITA Expansion Talks Suspended Again; No Timeline for Resumption Set," *Inside US Trade*, November 21, 2013. Available at: http://insidetrade.com/Inside-US-Trade/Inside-U.S.-Trade-11/22/2013/ita-expansion-talks-suspended-again-no-timeline-for-resumption-set/menu-id-172.html.

position was unacceptable. Talks broke down in November when the US publicly stated that China's request to join the Trade in Services Agreement (TISA) negotiations would only succeed if it provided "an improved offer in the ITA expansion talks .... Other TISA participants have criticized the US stance on China's participation as setting preconditions that were not applied to other members before they joined. The strongest pressure to include China came from Australia, whose services industry depends on the Chinese market for growth, according to informed sources".[57]

Until the very last moment before the final ITA-2 agreement on December 16, 2015, China successfully used a combination of delay tactics and a strategy of co-shaping the design of an expanded ITA. This reflects China's overriding concern to upgrade its electronics industry through innovation and the development of generic technology platforms like MCOs. The new Chinese leadership has clearly stated, especially since the Comprehensive Reform Blueprint of the Third Plenum,[58] that it is serious about pushing forward is serious about pushing forward important innovation policy initiatives related to advanced manufacturing and services (Guidelines to Promote National IC Industry Development; MIC 2025; Internet Plus Plan).[59]

China's negotiation strategy was based on the assumption that ITA-2 without China would have been an oxymoron. Not only is China the world's biggest smartphone market, it is also by far the most important market for US semiconductor firms. As John Neuffer, a former Senior Vice-President

---

[57] "China Categorically Rejects US Preconditions To Participate in TISA," *Inside US Trade*, November 22, 2013. Available at: http://insidetrade.com/Inside-US-Trade/Inside-U.S.-Trade-11/22/2013/china-categorically-rejects-us-preconditions-to-participation-in-tisa/menu-id-710.html.

[58] US-China Economic and Security Commission (2013).

[59] Ernst, D., 2016, China's Bold Strategy for Semiconductors-Zero-Sum Game or Catalyst for Cooperation?, East-West Center Working Papers: Innovation and Economic Growth Series, no. 9, September. Ernst, D., 2016, Advanced Manufacturing and China's Future for Jobs, East-West Center Working Papers: Innovation and Economic Growth Series, no. 8, August 2016, http://papers.ssrn.com/sol3/papers.cfm?abstract_id=2820433 A revised version is forthcoming as a chapter in Paus, Eva (editor), Emerging Dystopias? The New Technological Revolution and the Future of Work, Cornell University Press, Ithaca, NY 14850.

of Global Policy at the Information Technology Industry Council (ITIC) pointed out in November 2013, "China has got to be part of this. They are too big a player. You can't have an outcome without the Chinese".[60]

A successful conclusion of ITA-2 negotiations thus required first and foremost an agreement between the US and China on how to resolve the tariff treatment of MCOs. That bilateral agreement was reached in November 2014, when President Obama announced success in negotiations with China. However, the agreement prompted fresh opposition from South Korea (about LCD display panels and lithium ion batteries), the EU (about analog car radios), and Taiwan (about flat-panel displays). As the US agreed to further small concessions to China to help South Korea and the EU secure their own deals with China, it seemed that an expanded ITA-2 deal had been reached on July 27, 2015.[61]

However, additional hurdles emerged when China insisted, in late October, on seven-year phase-outs on roughly 20 items (including LEDs), and five-year phase-outs on another 70 items (including most importantly MCOs).[62] By December 11, during an Ambassador-level ITA-2 meeting, it became clear that "WTO members have essentially capitulated to China's demand that it be allowed to phase out tariffs on dozens of ITA products ...[including MCOs and LEDs] ...over five or seven years".[63]

Under the agreed package that was finally reached in Nairobi on December 16, 2015, WTO estimates that more than 65% of tariff lines, accounting for 88% of imports, will be immediately eliminated by July 1, 2016.[64] By 2019, WTO estimates that almost 90% of the tariff lines and 95% of imports will be reached. China's successful negotiation strategy is hidden in the following terse sentence of the WTO Briefing Note: "For a very limited number of sensitive products, tariffs will be progressively eliminated over a period of 5 years, and 7 years for the most exceptional

---

[60]As quoted in Donnan (2013).

[61]"Taiwan Caves, Says It Will Accept ITA Deal without Flat-Panel Displays," *China Trade Extra*, July 27, 2015.

[62]"China Stands Firm in Resisting Calls to Shorten Phase Outs Under ITA Deal," *China Trade Extra*, October 28, 2015.

[63]"China's Lack of Final Tariff Offer Pushes ITA Expansion Talks to Nairobi," *China Trade Extra*, December 11, 2015.

[64]WTO (2015).

cases." It so happens that these delays concern some of the most sensitive products for both the US and China. As for the implementation of ITA-2, The Committee on the Information Technology Agreement (ITA), at a meeting on April 18, 2016, established that the first set of tariff cuts were to be implemented on July 1, 2016, and the second set no later than July 1, 2017, with successive reductions taking place on July 1, 2018, and effective elimination no later than July 1, 2019.

6. Conclusions

This analysis of the contrasting experiences of China and India with ITA demonstrates that the success or failure of trade liberalization is determined by the *domestic determinants* of their ICT industry — i.e., its stage of development, its growth model and policies, its market size and sophistication, the institutional framework, and the managerial and technological capabilities of its firms. Equally important is the country's position in the GPNs and GINs of the ICT industry. These parameters shape the gains from trade for industrial manufacturing and innovation that both countries were able to reap from ITA.

The outcome of the ITA-2 negotiations shows that China needed a successful ITA-2 as much as the US needed it. A viable agreement was impossible without China's participation. India's absence hardly matters, given its insignificant role in the global electronics industry.[65]

The agreement reached in the final ITA-2 package demonstrates that neither the US alone nor with its main allies can dictate any longer the outcome of plurilateral agreements.[66] On the other hand, China's trade negotiators seem to have learned how to compromise without undermining the scope for upgrading the country's semiconductor and solid-state

[65]A particularly telling sign is that India does not even show up in McKinsey's list of top-10 countries in the global value-added electronics industry. See: http://www.slideshare.net/morellimarc/mckinsey-manufacturing-future-2013-22958651 .

[66]The eroding power of the US to shape trade agreements becomes obvious now that the US has withdrawn from the TPP. None of the other 11 members of the TPP wants to engage with the US in a bilateral RTA. After all, the US is openly saying that we'll use our size to get "wins" and force trade surpluses (or reductions in deficits). Even Japan is adamant in opposing this US bilateral mercantilism. See Introductory chapter and Preface.

lighting industries. As China remains by far the largest market for US semiconductor firms, Chinese policy-makers also knew that they had a strong hand of cards to play, strong enough to cope with a possible stalemate or breakdown of ITA-2 negotiations. In contrast, India did not hold anything comparable to China's trump card.

It is necessary, however, to broaden the analysis. Mega-developing countries such as China and India have enough resources to cope with the contingent outcomes of international trade negotiations. For the majority of developing countries, however, a failure to capture the potential gains for innovation from ITA and other plurilateral and megaregional trade agreements could have quite serious consequences, and deprive these countries from speedy access to critical learning opportunities and productivity-enhancing generic technologies.

Bold action is required to avoid zero-sum game or even negative-sum game outcomes and resultant trade conflicts. Thus far, policy debates on the distribution of trade gains and costs have focused on negotiating "special and differential" (SD) arrangements (delays, phase-outs for product coverage, etc.). However, there seems to be limited scope for expanding such SD arrangements.[67]

Once the current disruptive rise of Economic Nationalism will have run its course, future research will then have to address questions like: What changes are necessary in domestic regulations as well as in industrial and innovation policies to reap the potential benefits of ITA trade liberalization? And, equally important, how does a country's innovation capacity in a particular industry affect its approach to and its position in plurilateral and megaregional trade agreements? Most immediately, econometric research and case studies are needed to examine whether participation in ITA-2 might help to accelerate industrial manufacturing and innovation in different types of developing countries, as well as to narrow the gap in productivity and income relative to leading industrialized countries.

In the end, a broad portfolio of diverse policy approaches is required to enable developing countries to increase the gains that they can reap

[67]The scope for compromise has been dramatically reduced by the rise of economic nationalism, especially the US "America First" trade doctrine. See Introductory Chapter.

from ITA participation. The mix of policies will differ across countries and sectors. And the appropriate mix will evolve over time. Hence, flexibility in policy implementation is of the essence.

Acknowledgments

I am grateful for comments and suggestions received from Shigeyuki Abe, Center for Contemporary Asian Studies, Doshisha University, Tokyo; CHEN Tain-Jy, National Taiwan University, Taipei; Susan Helper, Case Western Reserve University, Cleveland; Derek L. Hill, National Science Foundation and the Federal Renewable Energy Laboratory, Arlington; Brian Kahin, University of Washington, Seattle; Konstantinos Karachalios, IEEE-Standards Association, Piscataway; Bill Lazonick, University of Massachusetts at Lowell and The Academic-Industry Research Network; Kate Linton, U.S. International Trade Commission, Washington, D.C.; Doug Lippoldt, HSBC, London; Keith E. Maskus, University of Colorado, Boulder; Michael Plummer, The Johns Hopkins University, SAIS, Bologna; WANG Ping, China National Institute for Standardization (CNIS), Beijing; Mark Wu, Harvard Law School, Cambridge; and XUE Lan, Dean, School of Public Policy and Management, Tsinghua University, Beijing.

References

Anderson, M. and Mohs, J. (2010). "The Information Technology Agreement: An Assessment of World Trade in Information Technology Products," *Journal of International Commerce and Economics,* US International Trade Commission, p. 41. Available at: http://www.usitc.gov/publications/332/journals/info_tech_agreement.pdf.

ASSOCHAM National WTO Council (2012). "Observations on 'Expanding the Information Technology Agreement'," ASSOCHAM Roundtable, December 20, New Delhi. Available at: http://view.officeapps.live.com/op/view.aspx?src=http%3A%2F%2Fwww.assocham.org%2Fdocs%2FITA-Roundtable-Observation-Paper_13.2.13.doc.

Baldwin, R. and J. Lopez-Gonzalez (2013). "Supply-Chain Trade: A Portrait of Global Patterns and Several Testable Hypotheses," NBER Working Paper No. 18957, April, p. 48.

Cohen, W. M. and B. A. Levinthal (1990). "Absorptive Capacity: A New Perspective on Learning and Innovation," *Administrative Science Quarterly*, 35(1), pp. 128–152.

Congressional Research Service (2015). China's Economic Rise: History, Trends, Challenges, and Implications for the United States, October 21, p. 13.

Cornell University, INSEAD and WIPO (eds.) (2015). *Global Innovation Index 2015* (Geneva, Ithaca, and Fontainebleau).

Curtis, J. (2013). "Trade and Innovation: Challenges and Policy Options." Background Paper for the E-15 Expert Group on Trade and Innovation, International Center for Trade and Sustainable Development (ICTSD), Geneva, June 6–7.

D'Costa, A. (2015). "Modi's Mantra to 'Make in India.'" Available at: http://www.eastasiaforum.org/2015/10/25/modis-mantra-to-make-in-india/.

Donnan, S. (2013). "Negotiators Nervously Eye China's Resistance in IT Trade Talks," *Financial Times*, November 19.

Ernst, D. (2011). "Testimony To the US-China Economic and Security Review Commission Hearing on China's Five Year Plan, Indigenous Innovation and Technology Transfers, and Outsourcing," 15 June, note 17. Available at: http://scholarspace.manoa.hawaii.edu/bitstream/handle/10125/20265/wire20110615testimony-PP.pdf?sequence=2.

Ernst, D. (2013). "Is the Information Technology Agreement Facilitating Latecomer Manufacturing and Innovation? India's Experience." East-West Center Working Papers, Economics Series, No. 135, November.

Ernst, D. (2014a). "The Information Technology Agreement, Industrial Development and Innovation - India's and China's Diverse Experiences." The World Economic Forum and the International Center for Trade and Sustainable Development (ICTSD), Geneva, March.

Ernst, D. (2014b). "Trade and Innovation in Global Networks — Regional Policy Implications." In *Can Policy Follow the Dynamics of Global Innovation Platforms?* J. Leijten (ed.), Chapter 3 (The International Innovation Policy Network of the Six Countries Programme, Delft).

Ernst, D. (2014c). *Upgrading India's Electronics Industry — Regulatory Reform and Industrial Policy. An East-West Center Special Study* (East-West Center, Honolulu), 90 pp.

Ernst, D. (2014d). *Fast-Tracking India's Electronics Manufacturing Industry — Trade Policy Related Action Points*, Note prepared for the World Bank New Delhi Office, May 19, p.1.

Ernst, D. (2015). "From Catching Up to Forging Ahead: China's Policies for Semiconductors," East-West Center Special Study, September, 74 pp.

Available at: http://www.eastwestcenter.org/node/35320, http://papers.ssrn.com/sol3/papers.cfm?abstract_id=2744974.

Ernst, D. (2016a). "The Information Technology Agreement, Manufacturing and Innovation — China's and India's Contrasting Experiences." Available at: http://papers.ssrn.com/sol3/papers.cfm?abstract_id=2737082.

Ernst, D. (2016b). *Beyond Value Capture — Exploring Innovation Gains from Global Networks.*

Ernst, D. (2016c). "From Catching Up to Forging Ahead? China's New Role in the Semiconductor Industry," *Solid State Technology*, May.

Ernst, D. (2016d). "China's Bold Strategy for Semiconductors — Zero-Sum Game or Catalyst for Cooperation?" East-West Center Working Papers: Innovation and Economic Growth Series, No. 9, September. Available at: http://www.eastwestcenter.org/node/35798 and http://papers.ssrn.com/sol3/papers.cfm?abstract_id=2836331.

Ernst, D. (2016e). "A New Grand Strategy For Semiconductors," *China Economic Quarterly*, September, pp. 15–22.

Ezell, S. J. (2012). *Boosting Exports, Jobs and Economic Growth by Expanding the ITA* (Information Technology and Innovation Foundation (ITIF), Washington, D.C.) March, pp. 8–9.

Grossman, G. and E. Helpman (1991). *Innovation and Growth in the Global Economy* (The MIT Press, Cambridge, MA).

Kallummal, M. (2012). "Process of Trade Liberalisation under the Information Technology Agreement (ITA): The Indian Experience." Centre for WTO Studies (CWS) Working Paper 200/3, Indian Institute of Foreign Trade, New Delhi, p. 15.

Kelkar, V. (1999). "India's Emerging Economic Challenges," *Economic and Political Weekly*, August 14, pp. 2326–2329.

Lin, F., and H. C. Tang (2013). "People's Republic of China Cracks Electronic Export Sophistication: Fact or Fallacy?" September 17, p. 2. Available at: http://www.aienetwork.org/blog/48/the-peoples-republic-of-china-cracks-electronic-export-sophistication-fact-or-fallacy.

Nakatomi, M. (2013). *Plurilateral Agreements: A Viable Alternative to WTO?* Presentation at the WTO, March 11. Available at: https://www.wto.org/english/res_e/reser_e/wts_future2013_e/nakatomi.pdf.

OECD (2010). *How Imports Improve Productivity and Competitiveness* (OECD, Paris).

Palmer, D. (2013). "US Seeks Expanded Information-Technology Pact," *Politico*, November 11. Available at: http://www.politico.com/story/2013/11/us-pact-trade-info-technology-99631.html.

Rajan, R. (2013). "Why India Slowed," April 30. Available at: http://www.project-syndicate.org/print/the-democratic-roots-of-india-s-economic-slowdown-by-raghuram-rajan

Semiconductor Industry Association (2010). "Doubling Semiconductor Exports Over the Next Five Years." Available at: http://www.sia-online.org/clientuploads/directory/DocumentSIA/ Export/Doubling_Exports_Paper_0610.pdf.

Sen, A. (2014). "IT Sector Exports to Grow 13-15% in FY15, says Nasscom," *LiveMint*, February 11.

Sender, H. (2015). "Modi's 'Make in India' goals are hampered by woeful infrastructure," *Financial Times*, March 4.

USITC, The Information Technology Agreement, Advice and Information on the Proposed Expansion: Part 2: p. VI. Available at: http://www.usitc.gov/publications/332/pub4382.pdf.

US-China Economic and Security Commission (2013). "Third Plenum Economic Reform Proposals: A Scoreboard." November 19. Available at: http://origin.www.uscc.gov/sites/default/files/Research/Backgrounder_Third%20Plenum%20Economic%20Reform%20Proposals--A%20Scorecard%20(2).pdf.

Valdes, R., and M. McCann (2014). "IP Provisions in Regional Trade Agreements: Revision and Update." Staff Working Paper ERSD 2014-14, 24 June. Available at: https://www.wto.org/english/res_e/reser_e/ersd201414_e.pdf.

WTO (2012). *15 Years of the Information Technology Agreement, Trade, Innovation and Global Production Networks* (World Trade Organization, Geneva), p. 107.

WTO (2015). "Briefing Note: The Expansion of Trade in Information Products (ITA Expansion)," December 16, 2015. Available at: https://www.wto.org/english/news_e/news15_e/itabriefingnotes161215_e.pdf

Chapter 17

Promoting Win–Win Development of Global Value Chains*

Susan Helper and Timothy Krueger

Case Western Reserve University, 10900 Euclid Ave, Cleveland, OH 44106, USA

1. Introduction

In recent decades, networks of financially independent companies located around the world have come to account for an increasing share of global production. These global value chains (GVCs) operate differently from other models of international production, such as the export of finished products made largely in a single country, vertically integrated multinational operations, or arm's-length transactions across national borders. The rise of GVCs has significant implications for policies in areas such as international trade, national development, intellectual property (IP), and international standards in areas such as cybersecurity, treatment of investors, workers, and the environment.

*This manuscript was originally prepared for the 2016 East-West Center NSF Workshop titled *Mega-Regionalism: New Challenges for Trade and Innovation.*

According to some researchers, these GVCs offer the potential for complementary growth and development in the firms and nations that participate in them, as well as benefits for consumers. In this view, suppliers locate in nations that have comparative advantages in their particular part of the value chain. Alternatively, GVCs may facilitate a race to the bottom, as multinationals seek suppliers that accept ever-worse terms of trade. A variety of literatures addresses some, but rarely all, of these issues.[1] Policies affecting value chains are often based on implicit assumptions, the validity of which has not been robustly tested. This paper will sketch a conceptual framework and identify empirical research needed to identify sources of complementary growth. It will also discuss how this research can inform policies that promote innovation and development for workers and firms throughout global supply chains.

Below, we outline three perspectives about the impact of trade policy on the development of GVCs:

(1) The "multinational-led development" perspective asserts that GVCs can play a key role in economic development trade policy facilitates the actions of the multinational firms that lead these value chains. By facilitating the flow of goods and services across borders and protecting firms' investments, these policies give workers in less developed countries access to global production systems they otherwise would not come into contact with.[2] In addition to providing better employment in the short term, firms that start at the bottom of GVCs can eventually expand their competencies and learn from production networks, thus capturing wealth and foreign investment.
(2) The "race to the bottom" perspective, in contrast, argues that such multinational-led development creates conditions for sustained poverty and exploitation. Fierce competition to enter GVCs means

[1] Useful studies of the development implications of GVCs include Bruhn (2014), Ravenhill (2014), Stiglitz (2013), OECD, WTO, UNCTAD (2013), Draper and Freytag (2014) and OECD (2013).

[2] This view is often called the "free trade" perspective, and indeed reducing barriers to trade such as tariffs is a key policy goal. However, as we discuss later, other elements of the proposed policies raise barriers to trade, for example by giving protection to intellectual property.

that firms which cut corners will generally win contracts. Even if lead firms or their customers desire high standards for workers as well as investors, GVCs are far too complex for firms to maintain accurate information about all of their suppliers' suppliers, let alone enforce adherence to laws such as labor standards.

(3) Somewhere between these camps is the "reform" perspective, which argues that trade agreements have produced highly inconsistent results, some desirable and some undesirable. There is little reason to believe that participation in GVCs will automatically allow emerging economy firms to capture wealth. However, with oversight and reform, it is possible to craft future trade agreements and strengthen domestic institutions in ways that mitigate or prevent the race to the bottom. In this case, trade could be a catalyst for sustained economic development and could create benefits for supply-chain workers and owners, lead firms, and consumers.

These perspectives have different implications for policy, as seen in the debate over the efficacy of agreements such as the Trans-Pacific Partnership, policy initiatives (such as the High-Level Economic Dialogue between the US and Mexico), and private efforts to establish corporate codes of conduct to govern value chains. As we discuss later, tension exists over which assets are most important to nurture and protect — are intellectual property (often belonging to multinationals), domestic manufacturing eco-systems, and worker and/or environmental rights the keys to future prosperity? Although differences of opinion in part represent differences in values, many of the disagreements reflect different models of how the economy works, so are in principle subject to empirical tests.

2. What are global value chains?

In recent decades, economic activity has become fragmented in a variety of ways, with different phases of production often located far away from each other. A GVC links companies, often in multiple industries and multiple locations, in order to design, produce components for, assemble, and distribute a final product (such as a car or a computer) (Banga, 2013).

A classic example is Apple's 2008 Video iPod digital music player. Apple designed the product largely in the US but did not manufacture it. The iPod's parts came from the US, Japan (hard disc drive from Toshiba), Korea (memory from Samsung), and China (small parts from anonymous companies). The manufacturing was overseen by Taiwanese companies with factories in mainland China. Apple captured about one-third of the output value; the value-added in China from assembly, testing, and packaging was 3% or less of the output price (Dedrick *et al.*, 2008).[3] Other studies of tablets, mobile telephones, and laptops suggest a similar division of activities and value in global production, in which advanced nations specialize in capital and high-skilled labor, capturing most of the value.

Such imbalance often persists despite focused government effort to adjust it. China has established that upgrading its semiconductor industry is a national priority and has unrolled a number of policy initiatives (such as protecting IP and harnessing market forces to incentivize higher value-added production) to advance the industry. However, as Ernst (2015) describes, "China is still playing second fiddle in the industry, because the state's indigenous innovation policy collides with the global technology sourcing needs of Chinese semiconductor firms." These examples illustrate many of the key features of GVCs: the lead firm designs and directs production by multiple tiers of suppliers in many locations; the lead firm directs production but does not own most of these suppliers (Timmer *et al.*, 2013; Baldwin and Lopez-Gonzalez, 2013).

These GVCs operate differently from other models of international production, such as vertically integrated multinational operations, arm's-length transactions across national borders, or the export of finished products made largely in a single country. Vertical integration was epitomized by Henry Ford's production model, in which most parts of the value chain were located close together and were owned by Ford. In the 1920s, 1930s, and 1940s, Ford Motor Company's River Rouge production complex was a mile-and-a-half wide and over a mile long, with 93 buildings and 15.8 million square feet of floor space. It included offices for designing the product, furnaces for making iron and steel, electric power generators, plants for making tires, stamped parts, engines, transmissions, radiators, and tool and die equipment, as well as an assembly plant. Ford

[3] See also Ali-Yrkkö *et al.* (2011); Timmer *et al.* (2013).

also owned the natural resources needed to produce automobiles: 700,000 acres of forests, iron mines and limestone quarries, coal-rich land, and a rubber plantation (U.S. Department of Commerce, 2015).

Most GVCs are different from economists' model of perfect competition, in which transactions between firms are arm's-length and the only information that crosses firm boundaries is price information. In their upper tiers at least, GVCs are characterized by firms making specialized products for a lead firm. Suppliers may have a role in designing the products they provide. And the lead firm often exercises detailed control over the operations of suppliers, specifying quality-control procedures, inventory practices, etc. Lead firms thus find GVCs advantageous over perfectly competitive markets in that they are able to quickly obtain specialized components; GVCs have a disadvantage in that firms are often not able to quickly switch suppliers (U.S. Department of Commerce, 2015).

Much work is underway to quantify the importance of GVCs. Most sources agree that GVCs are increasingly important. However, as the Data Appendix shows, there is a difference of opinion about the level of GVC activity, and how much is foreign versus domestic. In the US, for example, supply chains are an important source of cost for most firms, and offer the potential for competitive advantage. However, these supply chains remain largely domestic. The average US firm buys intermediate inputs that for a multinational comprise about three-fourths of its output value, and for a domestically owned firm, equals about one-half of output value. Only about 10% of these intermediates are imported (Fetzer and Strassner, 2015). (In US manufacturing, imported content is about 20% of the value of output for a typical firm (Nicholson and Noonan, 2015).) Globally, Timmer *et al.* (2013) found that, for 85% of GVCs, the foreign value-added share has increased over the period from 1995 to 2008, and the (unweighted) average share rose from 28% in 1995 to 34% in 2008.

There is significant variation in the structure of GVCs. Gereffi *et al.* (2005) identify three industry characteristics that they see as critical determinants: the complexity of information that is relevant to an industry's transactions, the ease of communicating this information, and the level of supplier capability. For instance, industries with complex transactions, high supplier capabilities, and difficult-to-codify product specifications lend themselves to "relational value chain" governance. Firms in such industries develop interpersonal supplier relationships in order to effectively communicate tacit knowledge and gain access to high-capability suppliers (86). This arrangement is common within the apparel industry. Conversely, the bicycle manufacturing industry, with uniform components and highly specialized supply firms, lends itself to "market-based coordination." Essentially, suppliers need little direction from their customers, and component price governs relationships.[4]

This characterization, while useful, ignores intra-industry variation in the structure of supply chains and does not provide insight into their likely evolution (Coe and Yeung, 2015). Much GVC literature simply assumes that some aspect of "technology," such as the complexity of information to be communicated, exogenously determines parameters for all transactions in that industry. In contrast, Helper and Levine (1992) present a model of endogenous determination of lead firm strategy and supplier capability. That is, product complexity and the nature of the governance mechanism are jointly determined; purchasers in this model can choose either a simple design with static supplier capability and arm's-length governance, or a complex co-design governed by a long-term contract, with mutual learning. A key determinant of what choice the lead firm makes is the amount of market power it has over its customers (Helper and Levine, 1992). For most of the 20th century, General Motors had a great deal of market power in the US, giving the firm an incentive to protect the resulting profits from its suppliers. GM did this by unbundling tasks (e.g., separating design and production, dividing components into smaller sub-components). Dividing work in this way reduces barriers to entry, creating competitive markets for individual tasks and allowing lead firms to protect their final-product market rents from being shared with suppliers. Such strategies were common in US manufacturing in the mid-20th century (Helper, 1991; Helper and Levine, 1992). Recent advances in communication and transportation have offered the possibility of sub-dividing tasks further and opening competition to suppliers around the globe.

To retain bargaining power in this set-up requires maintaining a credible threat to exit, meaning that the lead firm often offers only

---

[4]Gereffi *et al.* (2005) outline three other options for value chains: "modular," "captive," and "hierarchy."

short-term contracts, and neither the supplier nor the purchaser invests much in relationship-specific knowledge or equipment. Maintaining power in this way thus comes at a cost to efficiency. Over time, this separation between tasks can lead to a weakening of productive eco-systems and loss of national competitiveness. Particularly damaging is the geographic separation of innovation and production; the loss of knowledge about how to make a part leads to an erosion of knowledge about potential opportunities for innovation (Pisano and Shih, 2009; Berger, 2005).

In contrast, Toyota, also an automaker, established longer-term, more information-rich relations with suppliers, for example, gaining information about the root causes of quality problems. Toyota helped to transform its suppliers in Japan from small workshops into world-class producers through extensive technical assistance (Nishiguchi, 1994; Smitka, 1991). Toyota too has globalized its supply base; as the company has grown, with new suppliers in dozens of countries problems in maintaining these deep levels of knowledge flow account for a significant part of the automaker's recent quality troubles (MacDuffie and Fujimoto, 2010).

In general, manufacturing clusters or eco-systems are difficult to sustain through private action alone. When firms invest in their suppliers, they do not capture all of the benefits of doing so; firms that do not invest also benefit. Due to this "free rider problem," firms will underinvest in activities to upgrade suppliers by helping them invest in training, new equipment, or innovative processes (U.S. Department of Commerce, 2015). Nations that retain such rich eco-systems typically have a rich set of institutions, such as subsidized research offices, unions, or works councils that promote use of "good jobs" or "high road" strategies, and technical assistance providers (Helper *et al.*, 2012; Ezell and Atkinson, 2011; Lazonick, 1993).

It is difficult for a lead firm to change its procurement strategy. For example, even though decades of competing with Toyota had reduced GM's market power to a point where there was little profit to protect from suppliers, GM still struggles to adopt the collaborative relational contracts that have worked so well for its rival. Changing such contracts is difficult, due to interaction with other corporate incentive structures (such as finance and engineering) and in credibility that collaboration will continue even if times are tough (Helper and Henderson, 2014).

3. Global value chains and economic development

In recent decades, regions and nations have embraced entry into GVCs as a key part of their development strategy. For example, Juarez, Mexico specializes in the production of wiring harnesses — a small part of a car. Harnesses are specified by multinational automakers and "first-tier" suppliers, produced in Juarez facilities owned by the suppliers (or sometimes Mexican sub-contractors), and then shipped from Juarez, in the sequence in which they will be used, to automobile assembly facilities all over North America (Jorge Carrillo). This strategy contrasts with other forms of economic arrangement, such as:

(a) Import-substituting industrialization (ISI), in which a withdrawal from global production is held to promote development by promoting infant industries. The experience of countries such as Mexico, Brazil, and Argentina, which embraced this model in the mid-20th century, shows that while some indigenous capabilities were developed, industries suffered from lack of economies of scale and inefficiencies bred by lack of competition.

(b) Development of domestic supply chains via controlled engagement with foreign technology (e.g., Korea, Japan). Through government action to allocate foreign exchange and coordinate access to foreign technology, these nations developed skilled lead firms of their own (Samsung, Toyota); by seeking to export much of their output, these nations avoided the insularity of ISI strategies (Amsden, 1992; Patrick and Rosovsky, 1976).

As Baldwin (2012) describes, a benefit of the new strategy of insertion into GVC's is that:

> Developing countries can now industrialize by joining GVCs instead of building their own value chain from scratch, as Japan and the Republic of Korea had to do in the 20th century. Developing countries can benefit

from foreign-originated intellectual property, trademarks, managerial and business practices, marketing expertise, and organizational models.

However, while GVCs facilitate entry into manufacturing, they may diminish returns from it because the GVC sector remains disconnected from the rest of the economy (Taglioni and Winkler, 2014).

How can value chains consistently achieve desirable outcomes such as wage growth, high workplace standards, and firm profitability for all participants? Each of our three perspectives has a different view on this question.

The "multinational-led" view points out two main benefits of engaging with corporate engines of growth:

(1) Efficient specialization and access to new and larger customer markets. For example, after NAFTA, Mexico's auto industry moved away from its old focus on the domestic market (making a wide variety of models, each with inefficiently small production runs) to assembling a few models and a few labor-intensive components (such as wiring harnesses), and began to export the vast majority of its production.

Similarly, in agriculture, countries can specialize in a few crops where they have competitive advantage and import other crops. The growth of Peru's asparagus industry demonstrates the way in which free trade agreements can benefit fledgling industries by affording them access to new markets. The Peruvian asparagus industry saw large growth in the wake of the 2009 US–Peru Free Trade Agreement (Abcarian, 2015). According to the US Department of Agriculture's Foreign Agriculture Service, the US imported over 204 million pounds of fresh asparagus from Peru in 2014. The Peruvian asparagus industry is estimated to employ over 60,000 people and has led to low unemployment in the Ica region.

(2) Participation in international networks that facilitate learning, and direct investment from large corporations. Technology transfer, knowledge diffusion, IP leakage, and access to higher quality intermediate inputs can all increase firm productivity. In a study of Colombian firms, Fernandes and Isgut (2005) found that the level of intermediate goods imported is positively correlated with plant

> productivity. Coe and Helpman (1995) examined the underlying mechanism of the knowledge transfer facilitated by imports, finding that firms in a country that experiences an increase in imports tend to borrow product ideas and reverse engineer products, allowing for knowledge transfer even when contact between firms or individuals is minimal.

A substantial amount of research has been done on inter-firm learning, ("knowledge spillovers") as a result of trade (Blalock and Veloso, 2007; Coe and Helpman, 1995). After lowering trade barriers in 1986, Indonesia saw an increase in intermediate imports across many manufacturing sectors. In a study of the period between 1988 and 1996, Blalock and Veloso (2007) found that learning and productivity improvements accompanied the rise in imports, with the strongest improvements in large firms, firms in highly competitive industries, and firms making complex products.

Liu and Buck (2007) identify three mechanisms for knowledge transfer: knowledge diffusion, in which supplier firms learn from multinational enterprises (MNEs); Foreign Direct Investment (FDI) by MNEs, which relocate R&D facilities; and personnel exchange or departure, in which employees and staff leave for other companies in the same industry or form their own companies which then become suppliers to the MNE. They observe that some MNEs have relocated their R&D headquarters to China, with "400 out of the biggest 500 MNEs" having recently established R&D centers there. They find that this results in positive knowledge spillovers for Chinese firms.

Specialized production clusters can benefit from trade agreements because a geographic region or small nation no longer needs to meet its own consumption needs. It can develop production strengths in a small range of activities and import everything else. Israel, which maintains bilateral free trade agreements with seven nations as well as the EU and Mercosur, provides an example (Israel Foreign Trade Administration, 2015). Israel lacks the natural resources to meet many of its own consumption requirements, such as energy. Yet strong relationships with international trading partners have allowed it to focus instead on advanced technology clusters in areas such as communications, biotechnology, and computing (De Fontenay and Carmel, 2001). The proposition that free trade enables clusters may seem to contradict popular notions that clusters are no longer viable or valuable in a world with globalized sourcing. Buciuni and Pisano (2015) confront this paradox by examining centuries-old footwear production clusters in Italy. They argue that access to "tacit" knowledge, as well as co-location among suppliers who use similar technology, continue to explain why some clusters retain "stickiness" despite globalization.

In brief, the "multinational-led" perspective supports the idea that trade promotes long-term improvements for supply chains in developing and developed nations. Increased trade itself, rather than government policy or the many details of trade agreements, is the primary mechanism by which fledgling industries become more profitable and productive over time. Importantly, this process should benefit manufacturers in the US and Europe as well.

In theory, these phenomena can ultimately result in higher productivity and higher wages. Proponents of this perspective sometimes point out that these are long-term processes. While profits at the lower rungs of a global supply chain are initially low, over time firms learn to improve their value-adding processes, and OEMs eventually come under pressure to improve working conditions throughout their industries.

To gain these benefits, policy (both international trade policy and domestic policies) should focus on attracting foreign investment, which is necessary for these outcomes to occur. Richard Baldwin (2012) of the Center for Economic and Policy Research writes:

> "If a high-tech firm is to locate production stages in a developing nation, the nation's government must ensure the necessary free movement of goods, services, information and the protection of tangible and intangible property rights. Old fashioned protection, anti-FDI policies, or lax property rights almost guarantee that the offshored stages will go somewhere else. Developing nations that got the offshored factories became hyper-competitive and wiped out the exports of developing nations that clung to import-substitution industrialisation. In the world of supply-chain industrialisation, protectionism has become destructionism… The justification for SDT [Special and Differential Treatment] also disappears. The

> cooperation helps developing nations credibly commit to policies that are good for them. Allowing a poor nation to not assure protection of the assets that trigger supply-chain trade would harm rather than help."

In contrast, proponents of the "race to the bottom" perspective argue that international trade generally leads to low or stagnant levels of development at the bottom of supply chains. With multitudinous sourcing options and substantial labor cost differentials, the basis of competition between large multinationals becomes cost rather than value-added. Direct investments in foreign suppliers are rare because MNEs switch suppliers frequently based on a supplier's ability to outbid its peer firms. Short-term relationships are the norm (Mayer and Milberg, 2013).

While many MNEs would like to promote living wages, good working conditions, and technology investment throughout their supply chains, the way they have (collectively and individually) structured supply chains constrains their ability to effectively do so (see the IBM example below). Multinationals often use third- or even fourth-party logistics coordinators to manage their supply chains.[5] They often have trouble collecting high-quality information beyond their first-tier suppliers and logistics coordinators. From the other end, managers and workers at lower-tier suppliers sometimes have limited information about which multinationals their products are destined for. Under these conditions, it can be difficult for an OEM to exert consistent influence throughout its supply chain or for it to monitor results.

Moreover, the incentives that lower-tier suppliers face contradict one another. They find it difficult to simultaneously raise wages and invest in high-tech production processes while underbidding their competitors. Race-to-the-bottom proponents argue that the firms cutting corners typically win contracts. Worse, when wages in one developing nation gradually increase, MNEs often shift production to a new lower-cost location. Upwards mobility is fitful, and long-term gains are questionable.

Finally, some authors caution that absolute gains fail to become relative gains. As noted in the win–win discussion, one can find examples of

---

[5]A fourth-party logistics coordinator (4PL), manages numerous third-party logistics coordinators.

low-cost regions like coastal China that gradually experience wage increases and technology transfer. However, wages and productivity remain low in relation to highly developed countries. Likewise, the technology transfer occurs too late in the product's lifecycle for the developing nation to ever control the innovation process. Authors in this camp maintain that the Global North is still the locus of innovation and design, and that globalization has done little to change this. In fact, companies use a variety of strategies to *avoid* transferring technology, as a way of protecting intellectual property. A 2010 *Wall Street Journal* article outlines tactics that some companies have begun using to limit IP leakage within their Chinese value chains. These include:

> Not sharing the most sensitive intellectual property; sending more of their own employees to oversee manufacturing; partnering with a smaller firm that's less able to become a rival; and structuring joint ventures more carefully so that the Western firm has more control (Mattioli, 2010).

Similarly, a Stanford study (Gillai *et al.*, 2014) notes, "It is critical for companies to maintain physical and electronic security to protect trade secrets and other confidential and proprietary information. Trade secrets and other proprietary information should be made available to employees and third parties on a need-to know basis and subject to company procedures and confidentiality agreements. Multiple technological innovations can also be used by companies to safeguard their IP, examples of which include encryption of confidential information; restrictions on electronic documents so that they exist only for a limited amount of time, can only be accessed with a special code, and are restricted from being saved, forwarded, or printed; and more."

Manning and Baines' 2004 study of globalization among poultry industry supply chains finds other reasons for concerns. The authors find that the removal of trade barriers has caused the industry to leave Europe, instead clustering in countries with less environmental oversight and fewer restrictions on animal well-being. Intense competition on the basis of cost has reduced profit margins throughout the industry. Tyson's profit margin shrank from 11.2% in 2000 to 7.7% in 2002, which in turn created a margin-shrinking cascade effect throughout Tyson's supply chain. Globalization of the poultry industry also created a situation in which firms at the top of the industry now constantly shuffle supply chains to plug production gaps and smooth seasonal production variations in an attempt to meet consumer demands throughout North America and Europe. This balancing act depends on a complex web of suppliers, making disease control more difficult and globalizing the reach of disease outbreaks. In short, the poultry industry's experience shows how the removal of trade barriers can restructure an entire industry without creating any obvious benefits for suppliers, and even introduce new risks to consumers.

IBM's quest to monitor its product lifecycle code also illustrates many of these challenges. IBM uses a complex web of 18,000 suppliers in over 100 countries (Susman, 2015). While most of these firms are engaged in production activities, about 300 of these suppliers are downstream waste-disposal firms. These firms fall into two categories: those that collect and repurpose old IBM hardware, and those that dispose of hazardous and non-hazardous waste generated through IBM's manufacturing processes. Although IBM manufactures less than it did in the past, end-of-life repurposing volumes reflect the company's past focus on manufacturing.

These downstream recycling and waste disposal firms are spread across six continents. While IBM strives to enforce high environmental standards for disposal of its hazardous and non-hazardous waste, it has significant difficulty enforcing its corporate code. Because it uses a fourth-party logistics coordinator, many suppliers do not know they are working for IBM. Many of these firms are in parts of the world where IBM's brand carries little cachet. Its leverage also suffers from its declining volumes of waste material due to the aforementioned shift away from manufacturing. Thus, although IBM attempts to schedule regular site visits to monitor standards at these waste disposal facilities, employees report that suppliers often fail to respond to e-mail requests about scheduling such visits. Auditing a network of 300 waste management suppliers is also an unwieldy task for IBM's 26-person Corporate Environmental Affairs staff charged with this task.

Race-to-the-bottom proponents argue that trade has had negative effects on manufacturers in advanced economies as well. Researchers have questioned the assertion that NAFTA has created jobs in North America, let alone lived up to its proponents' lofty initial predictions. The Council on Foreign Relations notes that trends in overall employment, productivity, and international trade do not appear to have changed significantly as a result of NAFTA (Sergie, 2014). The agreement may have contributed to broadening wage inequality and the flight of US manufacturing (Helper and Tucker, 2016). The US lost one-third of its manufacturing jobs in just the ten years from 2000 to 2010. Careful studies (using different methodologies) attribute one-quarter to one-half of this job loss to the entry of China into the World Trade Organization, an event which led to the substitution of Chinese for US suppliers in many value chains (Autor *et al.*, 2013; Pierce and Schott, 2012).

Studies in the "multinational-led" perspective often neglect to raise questions about GVCs' countervailing effects on suppliers in advanced industrial economies. For instance, the aforementioned study on Peru's asparagus suppliers does not explore the magnitude of decreased asparagus production in the US, or the environmental impacts of increased distances traveled. It does not consider the possibility that asparagus quality may vary based on region and transportation distance.

In sum, race-to-the-bottom proponents generally view the long-term effects of trade agreements as inescapably negative. The destructive aspects of free trade are not functions of inadequate policy, but rather increased globalization itself.

The reform view is distinct from the previous approaches in that its supporters see increased levels of international trade as neither inherently productive nor undesirable for industries in developing nations. They note that trade agreements have incredibly diverse and complex outcomes. A developing nation's ability to benefit from free trade depends upon variables like government structure, policy decisions, the industries its suppliers participate in, existing competencies and assets, the other nations it primarily trades with, and even random decisions made by managers. Still, proponents of this perspective feel that, with a good deal of work, future trade agreements can more consistently produce the desirable results past agreements have sporadically achieved.

A central tenet of this perspective is the inconsistency of outcomes. For instance, Frankel (1997) finds positive and statistically significant effects on trade flows due to Mercosur, but no statistically significant effects of the Andean pact. In instances when trade agreements do not produce higher absolute levels of international trade, it is difficult to see how lower-tier suppliers in developing nations can capture the theorized benefits of international trade.

If one accepts that trade agreements do increase international trade in many instances, the question then becomes what outcomes this has produced for firms and workers. Barrientos and Smith (2007) ask this question of the South African fruit industry, the Indian garment industry, and the Vietnamese garment and footwear industries. They find that as supply chains have become more globalized, firms have instituted Ethical Trading Initiatives (ETIs) intended to improve conditions for workers in the lower rungs of value chains. These have produced visible benefits in some categories but not others. For instance, ETIs tended to be effective at increasing health and safety standards, as well as leading to the creation of maternity benefits and some reductions of working hours. They were ineffective at resolving harsh worker treatment or discrimination based on gender, ethnicity, and caste. The authors attribute this to management's tendency to focus on "outcomes" (working hours and workplace safety) instead of complex "processes" (discrimination and management style). They find it easier to influence outcome variables and may be more accustomed to thinking about these variables as part of their jobs.

The outcomes of ETIs vary between regions and industries as well. Barrientos and Smith (2007) find that the South African fruit industry was already strongly unionized, and ETIs produced no impact on collective bargaining power. In Vietnam, where unions are affiliated with government, ETIs did result in greater collective bargaining power. In India, ETIs were ineffective at overcoming workers' common perception that union affiliation would threaten their jobs.

Factors such as the level of supply-chain complexity, duration of supplier relationships, and the level of interdependence contributed to this variance in ETI efficacy. Suppliers who depended upon a single customer for a large percentage of their sales were more likely to adhere to that corporation's formal standards. Firms filling numerous contracts, or short-term contracts, were less aware of their customers' ETIs. Similarly, supply chains full of indirect relationships, such as sub-contractors and "networks of agents" who source work on a project-by-project basis, were less likely to adhere to an ETI. Suppliers were most likely to comply when they depended upon several different customers with similar formal standards; managers sometimes perceived a "critical mass" of fairly similar ETIs. Barrientos and Smith (2007) find that ETIs have produced major health and safety improvements[6] in Vietnam and India, where such standards were previously low. In South Africa, national legislation had already led to health and safety improvements, so the effect of ETIs was trivial.[7]

Liu and Buck's work echoes the idea that preexisting political and workforce conditions strongly influence free trade's ability to generate positive externalities. A country needs a preexisting critical mass of scientists and engineers in order for MNEs to want to locate R&D facilities there. They conclude that the process whereby China captures positive high-tech spillovers "is not automatic" — it is related to the government's strategy of attracting R&D facilities for just this purpose. This may cast doubt upon whether this process is realistic in all developing nations. Liu and Buck (1993) also caution that FDI can have a negative effect on supplier learning, as it can allow foreign suppliers to monopolize a market for high-tech production.

Taiwan's experience developing and upgrading its IT industry tells both stories. In many ways, Taiwan is a Cinderella story of economic development. Ernst (2013) explains how the poor, underdeveloped nation of the mid-20th century suffered from a number of size-related disadvantages. Taiwan's small domestic markets limited its ability to support "sophisticated lead user" firms. In larger nations, such firms can support robust domestic supply networks or grow internationally competitive without foreign support. Taiwan's small size also constrained its ability to absorb large fluctuations in international demand. Moreover, the country's

[6]This is defined as "clearer procedures, information and training, fire safety, personal protective equipment, safer use of chemicals, lighting and ventilation, toilets and drinking water."

[7]Some work on industrial organization of fair trade coffee finds that in the long run, workers' wages do not increase on average, due to entry and the high cost of certification (see De Janvry *et al.*, 2015).

lack of industrial diversity and depth limited its domestic knowledge base, forcing it to rely heavily on foreign "technology, tools, and ideas."

Yet Taiwan employed novel strategies to compensate for these deficits. It joined a diverse and constantly evolving array of industrial networks. This exposed Taiwanese firms to more learning opportunities than static relationships in a limited number of industries would have. Taiwan was also very effective at blending formal and informal learning relationships. FDI, venture capital, and other contract-based relationships formed the basis of integration into formal networks, whereas a constant international exchange of students and workers formed the basis of informal network integration. Finally, Taiwan seems to have benefited by choosing a single, primary learning partner — in this case, the US.

Implementing these strategies allowed Taiwan to achieve what Ernst labels "fast-follower" status. That is, Taiwanese firms upgraded faster than the North American firms they supplied, narrowing the competency gap over time. However, Ernst does not anticipate that the same strategy will be available in the future. The new landscape of GVCs is characterized by shrinking margins and "decreasing returns of export-led industrialization" (Ernst, 2003). Taiwanese firms and policy-makers are attempting to respond by finding ways to increase the amount of R&D done in domestic firms, but this type of upgrading is both costly and difficult. The Taiwanese experience therefore indicates that while appropriate development strategy can allow fledgling firms to capture wealth from GVCs, conditions change from one decade to the next and past success does not guarantee continued success.

These findings support the idea that global supply chains can create channels for sharing knowledge, spreading legal compliance, and improving outcomes for workers. They also underscore how haphazard these results can be. Given the complexity and internal incentives of these supply chains, corporate intent is not always sufficient for producing desirable outcomes. Even when trade agreements or ethical trading agreements require better conditions at the lower rungs of supply chains, corporations find it difficult to enforce compliance (Locke, 2013).

These difficulties can be seen in the aftermath of the disastrous 2013 Rana Plaza collapse in Bangladesh in which 1,200 people were killed. Many (though not all) lead firms operating in Bangladesh signed up for extensive efforts to monitor suppliers. In the succeeding two years, international monitoring agencies have inspected 3,425 factories. However, only eight of these factories have been remediated such that they passed the final inspection. In addition, sub-contractors have moved a significant amount of production to unregulated indirect suppliers, factories that are often unknown to regulators. An in-depth NYU survey found that only 27% of Bangladesh garment factories in the export sector were subject to the accord. The survey found 3 million more workers in the garment export sector than had been realized; these workers are not subject to the protections of the agreement (Labowitz and Baumann-Pauly, 2015).

In the reform view, trade agreements alone are not sufficient for generating desirable outcomes (GAO, 2014). They should be coupled with growth-oriented government investments as well as effective enforcement measures that maximize a trade agreement's efficacy. Rodrik (2015) characterizes the most successful complementary policies as:

> pragmatic, opportunistic, often "unorthodox" and promotes domestic manufacturing industries — such as protection of home market, subsidisation of exports, managed currencies, local-content rules, development banking, special investment zones...

Note that many trade agreements forbid items on this list, especially local content rules.

Some studies have focused more on private governance rather than the art of crafting effective trade agreements. Locke *et al.* (2009), observing working conditions in the global apparel industry, claim that compliance programs have had a lesser effect than a "commitment-oriented" approach where lead firms train suppliers in joint problem-solving, information exchange, and diffusion of best practices. These programs require significant upfront investment but provide benefits to lead firms (such as Nike) in reduced inventory and higher quality. The greater management and worker capability required to operate factories in this manner reduces the ability of suppliers to evade regulation by moving production to shadowy indirect contractors.

Similar efforts can lead to improved environmental performance. Plambeck *et al.* (2011) note that the right incentives and collaborative efforts can help suppliers achieve better environmental performance. The experience of suppliers in China supports this idea: "Rather than simply monitoring Chinese suppliers' compliance with local environmental, health and safety (EHS) standards, leading companies are giving suppliers tools and incentives to independently improve environmental performance."

Gereffi and Mayer hypothesize that four factors affect the success of private governance:

(1) the structure of the particular global value chain in which production takes place;
(2) the extent to which demand for a firm's products relies on its brand identity;
(3) the possibilities for collective action by consumers, workers, or other activists to exert pressure on producers; and
(4) the extent to which commercial interests of lead firms align with social and environmental concerns.

Thus, the most effective options to better global working conditions require the linking of public, private and social forms of governance (Coe and Yeung, 2015).[8] Gereffi and Mayer (2010) argue that trends in the dynamics of global supply chains mean that the power and information

---

[8]This blended public–private logic echoes Albert Hirschman's seminal ideas on the most effective forms of development. Hirschman was deeply influenced by his time as a World Bank employee in Colombia during the early 1950s. The era's prevailing logic assumed successful development efforts should fit into an economy-wide vision, crafted by governments with the help of international financial institutions. These macro-level plans envisaged "balanced" development, meaning simultaneous initiatives to develop all industries that would eventually constitute the nation's new economy. This logic was predicated on the assumption that underdevelopment in some areas would undermine investments in others; that a nation must essentially jump from a pre-industrial equilibrium to an industrial equilibrium, moving as quickly as possible through the unsustainable middle-ground (Alacevich, 2007). In contrast, Hirschman argued that imbalances in "backward and forward linkages" could be effective at catalyzing further development as firms sought to remedy bottlenecks or find uses for excess capacity (Hirschman, 1958).

asymmetries that characterized what were referred to as 'buyer-driven' chains have diminished. The principal countries in which these supply chains are concentrated, such as China, India, Brazil, Indonesia, and South Africa, now have greater bargaining power to pressure foreign companies for changes to benefit local interests.

Cambodia is a case in which a trade agreement, firms' desire to avoid egregious human rights violations in their supply chains, and the home government's desire for upgrading came together to generate positive outcomes. In less than a decade, Cambodia's garment sector was transformed from a "cluster of sweatshops where human rights violations were pervasive, into hundreds of monitored factories that are more likely to comply with labor law than they are to engage in forced overtime." In 2010, the International Labour Organization (ILO) mechanism found that 98% of the surveyed factories did not engage in anti-union activity, and union membership in the sector increased ninefold, from 39,000 to 337,000 from 1998 to 2004. (Despite the advances in the garment industry, Cambodia's overall indicators of social protection remain among the world's worst.)

In 1999, Cambodia and the US (its largest trading partner) signed a Bilateral Textile Agreement that offered Cambodia an increased quota of exports to the US if Cambodia could show that producers substantially complied with Cambodian labor law. The US convinced the ILO to oversee a monitoring mechanism, and the precursor to what is now known as "Better Factories Cambodia" (BFC) program began operation in 2001. The Cambodian government required all garment factories to participate in order to receive permission to export, thus ensuring widespread participation. The formal agreement expired with the end of the international quota system (the MultiFiber Agreement) in 2005 but the monitoring program continues today under the multistakeholder BFC program. Alisa DiCaprio (2011) writes:

> Both the government of Cambodia and the manufacturing association supported this because they recognize the benefits to be gained from being able to verify labour standards for branded apparel companies that are sensitive to working conditions in their suppliers. It is this ability that

attracts companies like Disney and GAP, which explicitly brand their clothing as socially responsible, to buy from Cambodia.[9]

4. Conclusion

We have reviewed three perspectives on the impact of trade agreements on development of GVCs:

(1) the "multinational-led development" perspective, that stakeholders in all parts of the supply chain do best if trade agreements and national policies focus on ensuring lead firms a return on their investment in building out supply chains, by promoting a smooth flow of goods and services, low tariffs, and strong protections for intellectual property and other investments
(2) the "race-to-the-bottom" perspective, that lead firms do not have incentives to promote development of firms or workers at lower levels of their supply chains
(3) the "reform" perspective, that while GVCs do not automatically promote development, the right type of trade agreements can protect worker rights and allow space for national development. This requires complementary domestic policies that help establish and upgrade productive eco-systems.

Our brief review of evidence suggests that it is unlikely that multinationals on their own have sufficient incentive to invest in all socially beneficial supply-chain collaborative activities. On the other hand, evidence shows that consumer pressure and the private benefits of innovation do mean that lead firms have some incentives to invest in supply-chain upgrading. On balance therefore, the evidence favors the intermediate, "reform" perspective. But much work needs to be done to understand what kinds of international and domestic institutions are most effective in different circumstances, and how these institutions affect, and are affected by, supply-chain dynamics.

[9] See also Kolben (2015).

We need serious empirical research on the gains from global network integration and on its costs, especially in terms of barriers to domestic upgrading. Many empirical questions remain. While proponents of the different perspectives may differ in some of their value judgments (e.g., the weight to be given to income equality), many of the questions are at least, in principle, susceptible to empirical analysis. Some thoughts that may guide this research:

(1) *Counterfactual.* It is very important in estimating impacts of a policy to clearly specify a counterfactual. Would people employed as part of GVCs be unemployed if not part of a GVC? In regular employment in a domestic firm? Employed precariously in the informal sector? Participation in GVCs may have large impacts in affecting the choice set available to workers and firms. Particularly in agriculture, the increased value of cash crops from participating in a GVC may create incentives to overturn traditional land tenure arrangements, leading to displacement (sometimes violently) of small holders.
(2) *Nature of impacts.* It is unlikely that trade agreements will have large and lasting impacts on overall employment levels — the economy tends to full employment, after a period of unemployment caused by the dislocation due to changed trade arrangements (Petri).[10] Instead, we should look for impacts on productivity (due to specialization within supply chains, technical assistance by multinationals, incentives for innovation) and wage levels (which could rise due to increased productivity, or fall due to reduced bargaining power).
(3) *Nature of inter-firm ties.* Much of the GVC literature focuses only on the fact of outsourcing and does not examine the impacts (on innovation, job quality, etc.) of the collaborativeness of the ties. The impacts of Toyota's technical assistance to suppliers (both domestically and abroad) suggest that these ties matter a great deal. What is the relative

[10]In a time of "secular stagnation" with unemployment in equilibrium, it is possible that increased net exports could lead to employment gains (Bernstein and Baker, 2016). However, the stimulative effect of reducing the current account deficit would be offset by reduced foreign investment in the capital account. See Ikenson (2013), Bernstein (2014), and Benedetto (2014).

effectiveness of policies that aim to create productive suppliers in contrast to those that aim to allow firms to quickly switch partners? That is, are good suppliers and customers "made," or are they simply "born," endowed with a special asset that makes them uniquely productive?

(4) *Interaction of different institutions.* Earlier, we looked at several efforts to improve working conditions in the global garment industry: Cambodia's Better Work Program, Nike's supplier training in Mexico, and the Rana Plaza Accords in Bangladesh. The first two efforts are regarded by their authors as successes while little improvement has been seen in Bangladesh. By what standards are these efforts successful, and what factors account for these differential outcomes?

(5) *GVC dynamics.* What forces internal and external to firms affect the evolution of GVCs — the incorporation of new countries and suppliers, and the squeezing or upgrading of existing suppliers?

(6) *Infant clusters?* Can productive, spillover-rich clusters be established and grow in the context of trade agreements that prohibit domestic content requirements? Or can such prohibitions increase the dynamism of clusters by reducing their insulation from outside ideas and competitors?

(7) *Data harmonization.* As the data appendix suggests, there is much we do not know even about the size of GVCs at a point in time. To what extent are GVCs composed of a lead firm from an industrialized country and suppliers from an emerging market? This is the case focused on by the literature, but other data suggest that supply chains are often very heavily domestic or involve mostly rich nations.

Data Appendix

Much work is underway to quantify the importance of GVCs. Most sources agree that GVCs are increasingly important. However, there is a difference of opinion about the level of GVC activity, and how much is foreign versus domestic.

For example, in its World Investment Report 2013, the United Nations Conference on Trade and Development (UNCTAD, 2013) estimated that

80% of international trade was organized through global production networks coordinated by lead firms investing in cross-border productive assets and trading inputs and outputs with partners, suppliers, and customers. Similarly, Johnson and Noguera (2012) found that intermediate inputs play an outsize role in international trade. They estimate that as much as two-thirds of gross trade is composed of intermediate goods that pass a given customs frontier multiple times before becoming final goods and reaching their ultimate consumer.

Other figures are significantly lower. Timmer *et al.* (2013) found that, for 85% of the GVCs, the foreign value-added share has increased over the period from 1995 to 2008. The (unweighted) average share rose from 28% in 1995 to 34% in 2008 (Timmer *et al.*, 2013). Baldwin and Lopez-Gonzalez (2013) report that

> "About half of the world's output of goods and services are sold as intermediate inputs. An important misconception, stoked by a few examples of truly globalised production, concerns the relative openness of final and intermediate markets. The world is still more globalised for final goods than it is for intermediates; the domestic-sales-to-export split is about 60-40 for final manufactures while it is about 70-30 for intermediates (i.e. 70% sold locally and 30% exported). Overall, world production is not yet very internationalised: The imported intermediates share of total world manufacturing is only 16%; for the production of all goods and services, it is just 8%."

It is important to understand the sources of these differences. Do Baldwin and Lopez-Gonzalez do less double-counting than other sources? Do they not truly take into account the work using the new Trade in Value Added (TIVA) frameworks? Their numbers are consistent (the author thinks):[11]

> "The share of intermediate goods in world non-fuel exports has hovered around the 50 per cent mark since 2000, dropping during the 2009 crisis and reaching its highest point (55 per cent) in 2011. The European Union was the biggest exporter of intermediate goods (36 per cent of all

[11] See https://www.wto.org/english/res_e/statis_e/its2013_e/its13_highlights4_e.pdf

intermediate goods) in 2011. One-third of these goods were exported outside the European Union, and two-thirds within the EU. China and the United States were the second and third largest exporters of intermediate goods with a share in world trade of 10.0 per cent and 9.8 per cent, respectively. The six largest exporters of intermediate goods — the European Union, China, the US, Japan, the Republic of Korea and Hong Kong, China — represent nearly two-thirds of world trade in intermediate goods."[12]

In the US, supply chains are an important source of cost for most firms and offer the potential for competitive advantage. However, these supply chains remain largely domestic. The average US firm buys intermediate inputs that for a multinational comprise about three-fourths of their output value, and for domestically owned firm, equal about one-half of their output value. Only about 10% of these intermediates are imported (Fetzer and Strassner, 2015; Nicholson and Noonan, 2014). (In US manufacturing, imported content is about 20% of the value of output for a typical firm; see Nicholson and Noonan (2014).) These figures are probably somewhat underestimated, due to the difficulties involved in properly setting prices for imported inputs (Houseman *et al.*, 2010; Houseman and Mandel, 2015).

Clearly, more effort is needed in this area. A very useful roadmap (with particular focus on understanding the effects of outsourcing on job quality) is provided by Appelbaum *et al.* (2015).

References

Abcarian, L. (2015). "Impact of Free Trade Agreement a win-win," *The Produce News*, August 6. Available at: http://www.producenews.com/9-news-section/story-cat/16585-impact-of-free-trade-agreement-a-win-win.

Ahmad, N. and J. Ribarsky (2014). "Trade in Value Added, Jobs and Investment." *33rd General Conference of the International Association for Research in Income and Wealth*, Rotterdam, August. Available at: http://unstats.un.org/

[12] See also Ahmad and Ribarsky (2014).

unsd/trade/events/2014/mexico/documents/session1/IARIW%20-%20 Session%204C%20-%20Ahmad%20Paper.pdf.

Alacevich, M. (2007). "Early Development Economics Debates Revisited." The World Bank, December.

Ali-Yrkkö, J., P. Rouvinen, T. Seppälä, and Ylä-Anttila, P. (2011). "Who Captures Value in Global Supply Chains? Case Nokia N95 Smartphone," *Journal of Industry*, 11, pp. 263–278.

Amsden, A. H. (1992). *Asia's Next Giant: South Korea and Late Industrialization* (Oxford University Press).

Appelbaum, E., R. Batt, A. Bernhardt, and S. Houseman (2015). "Domestic Outsourcing in the U.S.: A Research Agenda to Assess Trends and Effects on Job Quality." Prepared for *The Future of Work Symposium*, U.S. Department of Labor, Washington DC, December, version dated November 10.

Autor, D., D. Dorn, and G. H. Hanson (2013). "The China Syndrome: Local Labor Market Effects of Import Competition in the United States," *American Economic Review*, 103(6), pp. 2121–2168.

Baier, S. and J. Bergstrand (2011). "On the Endogeneity of International Trade Flows and Free Trade Agreements," *Researchgate*.

Baldwin, R. (2012). "WTO 2.0: Thinking Ahead on Global Trade Governance." Center for Economic and Policy Research, December. Available at: http://www.voxeu.org/article/wto-20-thinking-ahead-global-trade-governance.

Baldwin, R. and J. Lopez-Gonzalez (2013). "Supply-Chain Trade: A Portrait of Global Patterns and Several Testable Hypotheses." NBER Working Paper No. 18957, April.

Banga, R. (2013). "Measuring Value in Global Value Chains." *UCTAD*. May. Available from: http://unctad.org/en/PublicationsLibrary/ecidc2013misc1_bp8.pdf

Barrientos, S. and S. Smith (2007). "Do Workers Benefit from Ethical Trade? Assessing Codes of Labour Practice in Global Production Systems," *Third World Quarterly*, 28(4).

Benedetto, J. B. (2014). "Who Financed Recent U.S. Trade Deficits?" *Journal of International Commerce and Economics*, May 2014. Available at: https://www.usitc.gov/publications/332/journals/volume_vi_u_s_trade_deficit.pdf.

Berger, S. (2005). *How We Compete: What Companies Around the World Are Doing to Make it in Today's Global Economy* (Crown Business).

Bernstein, J. (2014). "Full Employment, Trade Deficits, and the Dollar as Reserve Currency. What are the Connections?" *On The Economy*, October 7. Available at: http://jaredbernsteinblog.com/full-employment-trade-deficits-and-the-dollar-as-reserve-currency-what-are-the-connections/.

Bernstein, J. and D. Baker (2016). "Why Trade Defticts Matter." *The Atlantic*. December 8. Available from: https://www.theatlantic.com/business/archive/2016/12/trump-trade-deficit/509912/

Blalock, G., and F. Veloso (2007). "Imports, Productivity Growth, and Supply Chain Learning," *World Development*, 35(7), 2007.

Bruhn, D. (2014). "Global Value Chains and Deep Preferential Trade Agreements: Promoting Trade at the Cost of Domestic Policy Autonomy?" German Development Institute Discussion Paper. Available at: https://www.die-gdi.de/uploads/media/DP_23.2014_01.pdf.

Buciuni, G. and G. P. Pisano (2015). "Can Marshall's Clusters Survive Globalization?" Harvard Business School Working Paper 15-088. Available at: http://www.hbs.edu/faculty/Publication%20Files/15-088_cd4f1da8-4988-408b-b7eb-8c0e5798a398.pdf

Carrillo, J. (2014). "Transnational Strategies and Regional Development: The Case of GM and Delphi in Mexico." Industry and Innovation 11:1/2, pp. 127–153. March/June. Available from: http://www.academia.edu/12302043/Transnational_Strategies_of_GM_and_Delphi_in_Mexico

Coe, D. and E. Helpman (1995). "International R&D Spillovers," *European Economic Review*, 39, pp. 859–887.

Coe, N. M. and H. W. Yeung. (2015). *Global Production Networks: Theorizing Economic Development in an Interconnected World*. Oxford University Press.

De Fontenay, C., and E. Carmel (2001). "Israel's Silicon Wadi: The Forces Behind Cluster Formation." Stanford Institute for Economic Policy Research, June.

De Janvry, A., C. McIntosh, and E. Sadoulet (2015). "Fair Trade and Free Entry: Can a Disequilibrium Market Serve as a Development Tool?" *The Review of Economics and Statistics*, 97(3), pp. 567–573.

Dedrick, J., K. L. Kraemer, and G. Linden (2008). "Who Profits from Innovation in Global Value Chains? A Study of the iPod and notebook PCs." *Alfred P. Sloan Foundation Industry Studies Annual Conference*, May.

DiCaprio, A. (2011). "The Demand Side of Social Protection: Lessons from Cambodia's Labour Rights Experience." World Institute for Development Economics Research Working Paper No. 81. Available at: http://www.econstor.eu/bitstream/10419/53975/1/678365709.pdf.

Draper, P. and A. Freytag (2014). "Who Captures the Value in the Global Value Chain? High Level Implications for the World Trade Organization." E15 Initiative Think Piece, September. .Available at: http://e15initiative.org/

publications/who-captures-the-value-in-the-global-value-chain-high-level-implications-for-the-world-trade-organization/.

Ernst, D. (2003). "Pathways to Innovation in the Global Network Economy: Asian Upgrading Strategies in the Electronics Industry." East-West Center No. 58, June.

Ernst, D. (2013). "Industrial Upgrading through Low-Cost and Fast Innovation — Taiwan's Experience." East-West Center No. 133, September.

Ernst, D. (2015). "From Catching Up to Forging Ahead: China's Policies for Semiconductors." East-West Center.

Ezell, R. D. and S. J. Atkinson (2011). "International Benchmarking of Countries' Policies and Programs Supporting SME Manufacturers." The Information Technology & Innovation Foundation, September. Available at: http://www.itif.org/files/2011-sme-manufacturing-tech-programs.pdf.

Fernandes, A. and A. Isgut (2005). "Learning-by-exporting, and Productivity: Evidence from Colombia." World Bank Policy Research Working Paper No. 3544.

Fetzer, J. and E. H. Strassner (2015). "Identifying Heterogeneity in the Production Components of Globally Engaged Business Enterprises in the United States." Bureau of Economic Analysis, U.S. Department of Commerce, Washington, D.C., June.

Frankel, J. A. (1997). *Regional Trading Blocs* (Institute for International Economics, Washington, D.C.).

Gaston, N. and D. Trefler (1997). "The Labour Market Consequences of the Canada-U.S. Free Trade Agreement," *The Canadian Journal of Economics*, 30(1).

Gereffi, G., J. Humphrey, and T. Sturgeon (2005). "The Governance of Global Value Chains," *Review of International Political Economy*, 12(1), pp. 78–104.

Gillai, B., S. Rammohan, and H. Lee (2014). "Managing Supply Chain Sustainability and Intellectual Property: Are They More Similar than Different?" Stanford Initiative for the Study of Supply Chain Responsibility, March. Available from: https://www.gsb.stanford.edu/sites/gsb/files/publication-pdf/white-paper-vcii-sustainability-ip-supply-chains.pdf.

Government Accountability Office, Report to Congressional Requesters (2014). "U.S. Partners are Addressing Labor Commitments, But More Monitoring and Enforcement Are Needed." November. Available at: http://www.gao.gov/assets/670/666787.pdf.

Helper, S. and D. I. Levine (1992). "Long-Term Supplier Relations and Product-Market Structure," *Journal of Law, Economics, & Organization*, 8(3), pp. 561–581.

Helper, S. (1991). "Strategy and Irreversibility in Supplier Relations: The Case of the U.S. Automobile Industry," *The Business History Review*, 65(4), pp. 781–824.

Helper, S., T. Krueger, and H. Wial (2012). "Why Does Manufacturing Matter? Which Manufacturing Matters?" Brookings Institution, February. Available at: http://www.brookings.edu/research/papers/2012/02/22-manufacturing-helper-krueger-wial.

Helper, S. and R. Henderson (2014). "Management Practices, Relational Contracts, and the Decline of General Motors." NBER Working Paper No. 19867, January.

Helper, S. and T. Tucker. "We Need New Trade Policies, But Not Trump's." Real Clear Politics, December 19, 2016. Available from: http://www.realclearpolicy.com/articles/2016/12/19/we_need_new_trade_policies_but_not_trumps_110121.html

Hirschman, A. (1958). *The Strategy of Economic Development* (Yale University Press, New Haven), p. 108.

Houseman, S. *et al.* (2010). "Offshoring Bias in U.S. Manufacturing: Implications for Productivity and Value Added." Federal Reserve, International Finance Discussion Paper Number 1007, September. Available at: http://www.federalreserve.gov/pubs/ifdp/2010/1007/ifdp1007.pdf.

Houseman, S. and M. Mandel (2015). "Measuring Globalization: Better Trade Statistics of Better Policy." Available at: http://research.upjohn.org/up_press/232/.

Ikenson, D. (2013). "Trade Deficit Oped In The New York Times Is Heavy On Fallacy And Wrong On Economics," *Forbes*, November 8. Available at: http://www.forbes.com/sites/danikenson/2013/11/08/trade-deficit-oped-in-the-new-york-times-is-heavy-on-fallacy-and-wrong-on-economics/.

Israel Foreign Trade Administration (2015). "Trade Agreements and Bilateral Issues 2015." Available at: http://www.moit.gov.il/NR/exeres/00E158D1-2672-4B9F-B35C-29D4A6D56009.htm.

Johnson, R. and G. Noguera (2012). "Accounting for intermediates: Production sharing and trade in value added." *Journal of International Economics*, 2012, 86:2, pp. 224–236.

Kolben, K. (2015). "Dialogic Labor Regulation in the Global Supply Chain," *Michigan Journal of International Law*, 36(3).

Labowitz, S. and D. Baumann-Pauly (2015). "Beyond the Tip of the Iceberg: Bangladesh's Forgotten Apparel Workers." NYU Stern Center for Business and Human Rights, December. Available at: http://people.stern.nyu.edu/twadhwa/bangladesh/downloads/beyond_the_tip_of_the_iceberg_report.pdf.

Lazonick, W. (1993). *Business Organization and the Myth of the Market Economy* (Cambridge University Press).

Liu, X. and T. Buck (2007). "Innovation Performance and Channels for International Technology Spillovers: Evidence from Chinese High-tech Industries," *Research Policy*, 36.

Locke, R., M. Amengual, and A. Mangla (2009). "Virtue out of Necessity? Compliance, Commitment, and the Improvement of Labor Conditions in Global Supply Chains," *Politics and Society*.

Locke, R. (2013). *The Promise and Limits of Private Power: Promoting Labor Standards in a Global Economy* (Cambridge University Press).

MacDuffie, J., and T. Fujimoto (2010). "Why Dinosaurs Will Keep Ruling the Auto Industry," *Harvard Business Review*.

Manning, L. and R. N. Baines (2004). "Globalisation: A Study of the Poultry-meat Supply Chain," *British Food Journal*, 106(10/11).

Mayer, F. and G. Gereffi (2010). "Regulation and Economic Globalization: Prospects and Limits of Private Governance." *Business and Politics*, 12(3).

Mayer, F. and W. Milberg (2013). "Aid for Trade in a World of Global Value Chains: Chain Power, The Distribution of Rents and Implications for the Form of Aid." Capturing the Gains Working Paper 34, June.

Mattioli, D. (2010). "In China, Western Firms Keep Secrets Close," *The Wall Street Journal*, August 30. Available at: http://www.wsj.com/articles/SB10001424052748704913704575453612099883050.

Morash, E. and D. F. Lynch (2002). "Public Policy and Global Supply Chain Capabilities and Performance: A Resource-Based View," *Journal of International Marketing*, 10(1).

Nicholson, J. R. and R. Noonan (2014). "Digital Economy and Cross-Border Trade: The Value of Digitally-Deliverable Services." U.S. Department of Commerce, ESA Issue Brief #01-14, January 27. Available at: http://www.esa.doc.gov/sites/default/files/digitaleconomyandcross-bordertrade.pdf.

Nishiguchi, T. (1994). *Strategic Industrial Sourcing: The Japanese Advantage* (Oxford University Press).

OECD, WTO, UNCTAD (2013). "Implications of Global Value Chains for Trade, Investment, Development and Jobs." G-20 Leaders Summit, September 2013. Available at: http://unctad.org/en/PublicationsLibrary/unctad_oecd_wto_2013d1_en.pdf.

OECD (2013). "Interconnected Economies: Benefiting from Global Value Chains." Synthesis Report. Available at: http://www.oecd.org/sti/ind/interconnected-economies-GVCs-synthesis.pdf

Patrick, H. T. and H. Rosovsky (eds.) (1976). *Asia's New Giant: How the Japanese Economy Works* (Brookings Institution, New York).

Pierce, J. R. and P. K. Schott (2012). "The Surprisingly Swift Decline of U.S. Manufacturing Employment." National Bureau of Economic Research Working Paper 18655, December. Available at: https://www.usitc.gov/research_and_analysis/documents/Pierce%20and%20Schott%20-%20The%20Surprisingly%20Swift%20Decline%20of%20U.S.%20Manufacturing%20Employment_0.pdf

Pisano, G. P. and W. C. Shih (2009). "Restoring American Competitiveness," *Harvard Business Review*. Available at: https://hbr.org/2009/07/restoring-american-competitiveness/ar/1

Plambeck, E., H. L. Lee, and P. Yatsko (2011). "Improving Environmental Performance in Your Chinese Supply Chain," *MIT Sloan Management Review*, December. Available at: http://sloanreview.mit.edu/article/improving-environmental-performance-in-your-chinese-supply-chain/

Ravenhill, J. (2014). "Global Value Chains and Development," *Review of International Political Economy*, 21(1), pp. 264–274.

Rodrik, D. (2015). "The Future of Economic Transformation in Developing Countries." Report for Overseas Development Institute. Available at: http://set.odi.org/wp-content/uploads/2015/07/Dani-Rodrik_The-future-of-economic-transformation-in-developing-countries_Event-report_23-June-2015.pdf

Sergie, M. A. (2014). "NAFTA's Economic Impact." Council on Foreign Relations, February. Available at: http://www.cfr.org/trade/naftas-economic-impact/p15790

Smitka, M. (1991). *Competitive Ties: Subcontracting in the Japanese Automotive Industry* (Columbia University Press, New York).

Stiglitz, J. E. (2013). "South Africa Breaks Out." Project Syndicate. Available at: http://www.project-syndicate.org/commentary/joseph-e--stiglitz-on-the-dangers-of-bilateral-investment-agreements

Susman, G. (2015). "Selecting and Evaluating Environmentally Responsible Suppliers of Hazardous, Nonhazardous Special Waste and Product-End-Of-Life Management Services." Prepared by Gerry Susman for The Second Smeal Sustainability Case Competition, Fall, Pennsylvania State University.

Taglioni, D. and D. Winkler (2014). "Making Global Value Chains Work for Development." The World Bank, Economic Premise, No. 143, May. Available at: https://openknowledge.worldbank.org/bitstream/handle/10986/18421/880530BRI0EP140d0VC0knowledge0notes.pdf?sequence=1

Timmer, M. P., A. A. Erumban, B. Los, R. Stehrer, G. J. de Vries (2013). "Slicing Up Global Value Chains." Groningen Growth and Development Centre and The Vienna Institute for International Economic Studies, May.

Tinbergen, J. (1962). *Shaping the World Economy* (The Twentieth Century Fund, New York).

United Nations Conference on Trade and Development (UNCTAD) (2013). "World Investment Report 2013." Available at http://unctad.org/en/PublicationsLibrary/wir2013_en.pdf

U.S. Department of Commerce (2015). "Supply Chain Innovation: Strengthening America's Small Manufacturers." The Executive Office of the President and the U.S. Department of Commerce, March. Available at: https://www.whitehouse.gov/sites/default/files/docs/supply_chain_innovation_report.pdf.

Chapter 18

Labor Implications of TPP: A Game Changer?*

Ronald C. Brown

University of Hawai'i Law School, USA

1. Introduction

The Social Dimension Labor Provision of the Trans-Pacific Partnership (TPP) indeed represents the "Next Generation" in Free Trade Agreements (FTAs) and is possibly a "game changer" in the several Asian countries with TPP-side agreements with the US. As one of the few countries in the world that includes labor provisions in FTAs with Asia-Pacific countries, the US, in the TPP, has built on its past FTA labor provisions and through

*This chapter is based on a paper presented at the NSF Workshop on *Mega-Regionalism: New Challenges for Trade and Innovation* (MCTI) at the East-West Center, Honolulu, Hawaii, on January 20–21, 2016.

its side agreements on labor with Vietnam, Malaysia, and Brunei to advance new and higher labor standards.[1] The significance of the Side Agreements cannot be over-emphasized but, as always, the proof is in the actual enforcement. In the author's opinion, future scrutiny should be on making enforcement meaningful.

This chapter places discussion of the TPP in the context of addressing the growth of FTAs and the limited use of social dimension provisions on labor,[2] particularly with many Asia-Pacific countries, thus highlighting the progressive measures on labor protections undertaken by the TPP and how it fits with emerging developments in the region.[3] This discussion is followed by identifying a new research agenda implicated by TPP labor provisions and its Side Agreements and providing analyses to the challenges and opportunities they present. Finally, in the conclusion, summaries, suggestions, and challenges to other disciplines will be made.

[1]The Side Agreements of the TPP that the US has with Vietnam, Malaysia, and Brunei, by express terms, require their labor laws to be changed and improved *prior* to their implementation of the TPP so as to allow independent labor unions, strikes, proper treatment of immigrants, anti-discrimination provisions, labor inspections, and the basic labor standards affecting working conditions. That is, *before* they are allowed to export goods duty-free to the US and otherwise use the provisions of the TPP, the required legislative labor reforms must be concluded. The side agreements are very detailed in their obligations. TPP Full Text. Available at: https://ustr.gov/trade-agreements/free-trade-agreements/trans-pacific-partnership/tpp-full-text.

[2]Social dimension provisions in FTAs include environmental and labor provisions. The ILO explains it as follows: "labour provisions are defined as: '(i) any labour standard which establishes minimum working conditions, terms of employment or worker rights, (ii) any norm on the protection provided to workers under national labour law and its enforcement, as well as (iii) any framework for cooperation in and/or monitoring of these issues'. This chapter provides an overview of the social and economic rationale for labour provisions in trade agreements as well as the various approaches that have been taken in different country contexts." See ILO (2015).

[3]Not all agree the TPP is "progressive." The AFL-CIO retains many of its pre-signing reservations. See Drake (2015). Also see USTR (2015).

2. Emerging Mega-FTAs; Prior US FTA labor protections and shortcomings; and the TPP — A new generation?

2.1. *Mega-FTAs currently emerging*[4]

In the last decade there has been a blossoming of trade liberalization initiatives resulting in FTAs.[5] As of January 2013, the number of ratified FTAs involving at least one country from the Asian region is 109. This is more than triple the number in 2002. There are another 148 FTAs at various stages of development, bringing the total to 257.[6] Nearly three-quarters of the FTAs, or 189 out of a total of 257, were bilateral (i.e., involving two countries); only 68 were multilateral.[7]

---

[4]This material is more fully discussed in Brown (2016). Leading the way in numbers are the FTAs involving the ten members of the Association of Southeast Asian Nations (ASEAN) and their six trade partners: Australia, China, India, Japan, South Korea, and New Zealand. FTAs involving these countries have increased from 27 in 2002 to 179 by January 2013, a 600% increase. The ADB report states that ASEAN is leading negotiations with its six partners to establish the Regional Comprehensive Economic Partnership (RCEP) by 2015, the deadline ASEAN set to achieve the ASEAN Economic Community. This will have a market size of 3.4 billion people and a combined GDP of $21.4 trillion or about 30% of world GDP. Currently, the ASEAN+6 countries account for 70% of the total FTAs in Asia. See ADB (2013c). Negotiations on RCEP, an initiative involving the ten ASEAN members and its partners — Australia, China, India, Japan, New Zealand, and South Korea — started in 2013, seeks an economic partnership which would form the world's largest economic bloc. See *ASEAN Affairs* (2012).

[5]See discussion in ADB (2013c, pp. 17–26). See also ARIC (n.d.). The potential for negative effects of FTAs because of the complexity and inconsistency of some regulations has been described as the "noodle bowl" (or "spaghetti bowl") effect arising from different rules of origin (ROOs) (Zhang and Shen, 2011). See summary at ADB (2013a). Two proposals have been advanced to disentangle the Asian free trade noodle bowl: consolidation — which creates a regional FTA to harmonize bilateral FTAs; and multilateralization — which grants non-discriminatory preferences to non-members, eliminating preference discrepancies. See ADB (2013b).

[6]See ADB (2013c, p. 17, 21). See also ARIC (n.d.).

[7]See ADB (2013c, p. 18). "Two leading participants in Asian FTAs—Japan and Singapore—strongly favor a WTO-plus approach to FTAs. All of Japan's agreements and

More recently, *megaregional* trade deals have come into vogue.[8] Asia's own megaregional trade deal — the Regional Comprehensive Economic Partnership (RCEP) — was launched in 2012. This trade deal could create the world's largest trading bloc and have significant implications for the world economy.[9] The objective is to achieve an economic partnership agreement among the ASEAN member states and ASEAN's FTA Partners"[10] and the deal envisages a highly competitive single market and production base that is fully integrated into the global economy.[11] At the same time, a key purpose is to reconcile two long-standing proposals into a large region-wide trade agreement: the East Asian Free Trade Agreement,[12] which included ASEAN, China, Japan, and South Korea, and the RCEP, which added Australia, India, and New Zealand. RCEP bridges the two proposals by adopting an open accession scheme. It permits external countries to join later and does not prohibit members from acceding to other free-trade groupings, such as the TPP.[13] Negotiations among the 16 parties began in early 2013 and are predicted to conclude in 2016.[14] It does not include a social dimension provision protecting labor.

---

most of Singapore's are WTO-plus. Likewise, Brunei Darussalam, Indonesia, the Republic of Korea, Malaysia, the Philippines, and Viet Nam largely follow a WTO-plus format" (*Ibid.*, p. 19). Singapore's primary issues were found to be newer additions to the scope of a FTA agreement and included investment, competition, intellectual property, and public procurement ("the four Singapore issues") (*Ibid.*, p. 2). These go beyond the scope of the WTO requirements and thus are sometimes incorporated to FTAs and characterized as "WTO-plus." These newer issues were economically focused, without inclusion of a social dimension. See Kawai and Wignaraja (2010).

[8] See Wignaraja (2013). "*Megaregional* trade deals" is a term used to describe multicountry trade agreements made between (1) a large number of countries; (2) countries covering a large geographical area; or (3) very large countries.

[9] See ASEAN (n.d.) and Wallar (2014).

[10] See ASEAN (2012) and Wallar (2014).

[11] ADB (2013d).

[12] ARIC (2012).

[13] Hiebert and Hanlon (2012).

[14] *Economic Times* (2015) and Wignaraja (2013). "Because the RCEP will contain three of the largest economies in the world — China, India and Japan — it is globally important. The bloc represents 49 per cent of the world's population and accounts for 30 per cent of world GDP. It also makes up 29 per cent of world trade and 26 per cent of world FDI

In May 2016, President Obama, in promoting the need for the TPP, underscored that China and 15 other nations recently met in Australia with a goal of finalizing their deal, the RCEP, before the end of 2016 (now slated for 2017). He added, "[t]hat trade deal won't prevent unfair competition among government-subsidized, state-owned enterprises. It won't protect a free and open Internet. Nor will it respect intellectual property rights in a way that ensures America's creators, artists, filmmakers and entrepreneurs get their due. And it certainly won't enforce high standards for our workers and our environment".[15] In fact, China does not include labor protection provisions in its FTAs.

Besides RCEP, the newest FTA in world trade these days is the TPP. "Hailed as a state-of-the-art free trade agreement (FTA), it will unite 11 countries — Australia, Brunei, Canada, Chile, Malaysia, Mexico, New Zealand, Peru, Singapore, the US, and Vietnam — with a combined GDP of almost $21 trillion (about 30 percent of world GDP) and $4.4 trillion in exports of goods and services, or about a fifth of total world exports. If Japan and South Korea are added (they are actively exploring entry later this year), TPP would cover 40 percent of world GDP and nearly a third of world exports".[16] The TPP countries already have a number of existing FTAs, as seen in Fig. 1.[17]

---

inflows. Conservative estimates using various computable general equilibrium models suggest that if the RCEP were implemented it would bring large income gains to the world economy of between US $260–644 billion in a decade or so" (*Ibid.*). A critique on the trajectory of the RCEP and its relationship with regional FTAs in East Asia is found in a recent ADB report. See generally, Fukunaga and Isono (2013). Talks appear to be advancing. India, backed by South Korea and China, has made a two-tier proposal at the Regional Comprehensive Economic Partnership (RCEP), offering wider duty eliminations to the 10 ASEAN countries in the trade bloc and a lower market access to the five non-ASEAN members. The exact categories that will be given duty-free status will be decided in further deliberations in April, with an end-2015 deadline to conclude negotiations. See Seth (2015). Other regional developments are also moving forward. In March 2014, China, Japan, and South Korea held their fourth round of negotiations over a trilateral free trade agreement (China–Japan–South Korea FTA) (Tiezzi, 2014).

[15] Obama (2016).

[16] Kotschwar and Schott (2013).

[17] Fergusson *et al.* (2015, p. 9).

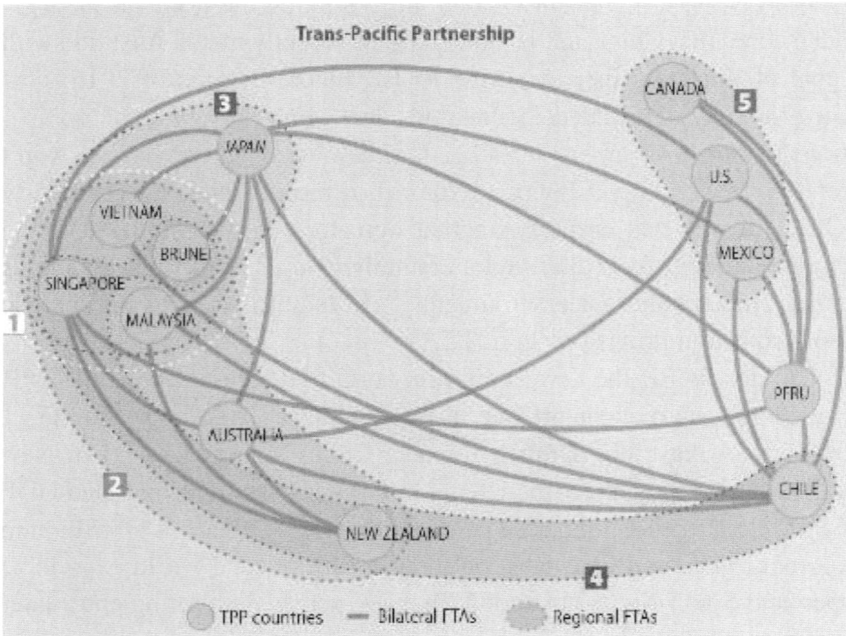

Figure 1. Existing FTAs among TPP countries.
Source: WTO FTA database and websites of TPP countries' trade Ministries (renumbered). Trade data from IMF.

All current TPP participants are members of the Asia-Pacific Economic Cooperation (APEC) Forum[18]; and, many see their prospective agreement as a step toward APEC's long-standing goal to create a Free Trade Area of the Asia-Pacific (FTAAP) among APEC members. At the same time, other trade integration arrangements are in place in the Asia-Pacific area.[19] These are ASEAN,[20] of which four of its ten

[18]These countries include Australia, Canada, Chile, China, Indonesia, Japan, Mexico, Peru, and the US (Global Affairs Canada, n.d.).
[19]These include ASEAN, RCEP, and the Latin America trade bloc, the Pacific Alliance (including Chile, Columbia, Mexico, and Peru). See Ramirez (2013).
[20]ASEAN members are Brunei, Cambodia, Indonesia, Laos, Malaysia, Myanmar, the Philippines, Singapore, Thailand, and Vietnam.

members are not signatories to APEC; Pacific Alliance[21]; and, in current negotiations, is the RCEP. China thus far has not agreed to a social dimension provision in any of its FTAs relating to the protection of labor.[22] One could imagine that if the TPP were not ratified by the US Congress, the RCEP with China as a leading member would be poised to be concluded without labor protection provisions.[23]

Of the six countries (China, India, Japan, South Korea, Australia, and New Zealand), all are ASEAN FTA *partners*, and several are also members of the TPP.[24] The RCEP, with the relatively poorer Asian nations included (contrasted with TPP's relatively wealthier nations) and less ambitious agenda, may be seen as advantageous way for some countries to reform more gradually and serve as a possible pathway to harmonization and an eventual FTAAP.[25]

In addition to the TPP, other mega-FTAs that may come into being are mentioned in Fig. 2 and include the US–EU FTA (i.e., Transatlantic Trade and Investment Partnership (TTIP)), the FTAAP, the RCEP, and the Japan–EU Economic Partnership Agreement (EPA).

---

[21] The Pacific Alliance is a Latin American trade bloc aimed toward integration. It currently has four member states — Chile, Colombia, Mexico, and Peru, which all border the Pacific Ocean. Costa Rica began the process of joining the Alliance in February 2014. The four founding nations of the Pacific Alliance represent nearly 36% of the Latin American GDP. If counted as a single country, this group of nations would be the sixth largest economy in the world, with a PPP GDP of more than US$3 trillion. Pacific Alliance (2013); see also: http://en.wikipedia.org/wiki/Pacific_Alliance.

[22] Ongoing negotiations of the US–China Bilateral Investment Treaty (BIT), which has US-proposed labor provisions, has not concluded at this time, and if agreed to, would be the first agreement in which China accepts ILO standards in an FTA or BIT. See discussion in Brown (2015).

[23] See, for example, Dennis (2015).

[24] New Zealand Ministry of Foreign Affairs and Trade (n.d.). The seven countries presently included in both the TPP and RCEP are Malaysia, Singapore, Vietnam, Brunei, Japan, Australia, and New Zealand. South Korea already qualifies for the TPP but has not applied to join the negotiations. Indonesia, Thailand, and China are potential future members (Frost, 2014, pp. 8–9).

[25] Aggarwal (2016).

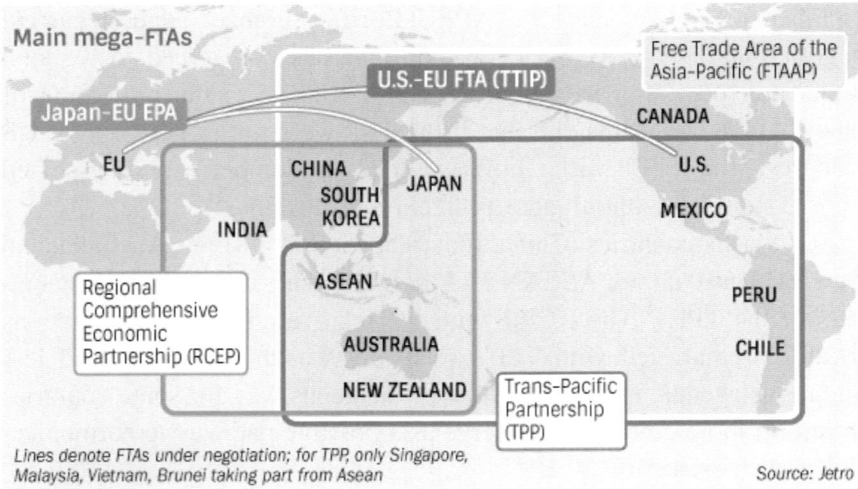

Figure 2. Main mega-FTAs.

2.2. *Prior US FTA labor protections and shortcomings*

In 2007, the US adopted the so-called "May 10 Standard" or "model" labor protection language in its FTA social dimension provisions to advance international labor standards.[26] It seeks to embrace and promote commitments to the four ILO core labor standards as embodied in the 1998 ILO Declaration on Fundamental Principles and Rights at Work[27] and provide enforcement mechanisms, with sanctions, for states and investors for alleged trade or labor violations. The US model bilateral trade and investment frameworks are significantly distinguishable from EU FTAs in two important ways.[28] First, alongside trade obligations, the

[26] For full discussion, see Brown (2016); also discussed in Brown (2015b). As will be seen, these US frameworks are not fully efficacious to do so, and recommendations to bolster them, including comments on incorporation or utilization of Corporate Social Responsibility (CSR) and International Framework Agreements (IFA) provisions are provided in Brown (2016b). For additional discussion, see Brown (2016c). See also, Bolle (2014).
[27] ILO Declaration on Fundamental Principles and Rights at Work. See: http://www.ilo.org/declaration/lang--en/index.htm. Accessed December 2, 2014.
[28] The "EU perspective" may provide insights about the relative strengths and weaknesses of US FTAs. See generally, European Commission — Trade. "Dispute Settlement."

FTAs and Bilateral Investment Treaties (BITs) include labor obligations that incorporate the ILO Declaration, not the Conventions as does the EU. Second, unlike the EU, US FTAs, though not the BITs, *wed* bilateral trade with these labor obligations under a new substantive legal procedure and provides an unprecedented unitary enforcement mechanism for both sets of obligations, wherein trade sanctions may also be brought for labor violations.[29]

While the US is ahead of Asia in implementing labor protection in trade agreements through its social dimension provisions, there are still shortcomings in its furthering ILO Standards. Without question, the US model bilateral trade and investment frameworks are an improvement in that they attempt to wed free trade with international labor norms in an effort to halt the "race to the bottom" in terms of labor regulations, but, it can be argued, its efforts to advance international labor standards are inadequate for several reasons.[30] First, while the US model frameworks integrate international labor norms, it is clear that the ILO commitments are to the ILO Declaration and not the ILO Conventions and likely are not sufficiently substantive to be enforceable.[31] Second, the dispute settlement mechanisms for enforcing the labor obligations are somewhat vague on what constitutes a violation and how violations would likely be remedied.

Available at: http://ec.europa.eu/trade/policy/accessing-markets/dispute-settlement/. Accessed August 15, 2014. The US has adopted only two of the core International Labor Organization Conventions, although it has approved the ILO Constitution and Declaration *vis-à-vis* its membership in the ILO.

[29] For further discussion, see Brown (2016d).

[30] See discussion in Brown (2016b) *supra* n. 25. See also, Brown, *Fostering Labor Rights in a Global Economy*, *supra* n. 25. Though the US model has agreement on the "conventions," by a footnote in the treaty it explicitly limits its definition to the Declaration. (*Ibid.*)

[31] "Where labour provisions refer to ILO conventions the parties can rely on the reports of the ILO supervisory bodies, which provide guidance on the interpretation of labour standards. By contrast, the 1998 Declaration is, as such, not subject to the supervision of the ILO's supervisory bodies although some guidance on the 1998 Declaration may be drawn from the comments of the ILO supervisory bodies on the respective fundamental conventions" (International Institute for Labour Studies, 2013). The EU, on the other hand, commits to the Conventions, but its enforcement mechanism is not as strong as that of the US See discussion on the draft EU–Columbia FTA at TUC (2010). For the US standard language in its FTA's Social Dimension provisions, see KORUS FTA, USTR (n.d.).

Third, while the FTA's dispute settlement mechanisms allow for wedded enforcement of trade and labor obligations,[32] only states and individual investors may directly bring claims that can lead to sanctions[33]; third parties such as non-governmental organizations, unions, and labor groups, and/or workers cannot directly bring similar challenges, except in some FTAs through cumbersome administrative requests to the state's discretion, thus inhibiting enforcement. And, fourth, while the frameworks place an onus on the states to enforce labor standards, they do not *require companies* to adhere to basic corporate social responsibility norms (CSRs) in their operations, and multinational corporations (MNCs) can easily evade compliance through such activities as sub-contracting.[34]

2.3. *TPP — A new generation?*

Highlights of the TPP labor provision found in Chapter 19 (see appendix for a summary), with its US model language promoting ILO core labor standards, includes emphasis on not using labor standards for protectionist purposes (19.2:2) or derogating from its labor standards (19.4:b).[35] The TPP also calls for "impartial and independent tribunals that are fair and

[32]However, the European Union (EU) in its FTAs does not wed the labor and trade enforcement mechanisms. Each have their own mechanism, with the former not having trade sanctions available. For more detailed discussion, see Brown, *Fostering Labor Rights in a Global Economy*, *supra* n. 25.

[33]Some FTAs permit third parties to present allegations to institutionalized contact points. See: http://www.ilo.org/global/standards/information-resources-and-publications/free-trade-agreements-and-labour-rights/lang--en/index.htm. Accessed July 19, 2017; ILO (2013, p. 32). In a few cases, the only sanction available is the modification of development cooperation. This is the case of the Canada–Costa Rica Trade Agreement and, to some extent, of the EU–CARIFORUM Economic Partnership Agreement (*Ibid.*, p. 32 n. 30). Also, see the Side Agreement of NAFTA, The North American Agreement on Labor Cooperation (NAALC) that authorizes third-party complaints on labor disputes, but which is reported ineffective and not used (Bieszcat, 2008). It is argued that the system is flawed due to the fact that it follows an adversarial and litigious approach toward solving labor issues (Studer, 2010; see also Kay, 2003)

[34]For further discussion, see Brown (2016b), *supra* n. 25.

[35]TPP Full Text of labor provisions. Available at: https://ustr.gov/trade-agreements/free-trade-agreements/trans-pacific-partnership/tpp-full-text.

transparent" (19.8:2–3). A new provision is to "encourage" enterprises to "voluntarily" adopt CSR initiatives (19.7). Provisions for cooperative activities, including technical assistance, are very wide in scope and encompass 21 very specific categories of improving labor standards and conditions in a list from a–u (19.10:6). These activities can be accomplished through a variety of modes from workshops to exchange of technical expertise and assistance (19.10:7). Cooperative dialogue is authorized through contact points under the labor ministries (19.11), and a Labour Council is established to oversee all processes (19.12) and through labor consultations, use of experts and panels, and other means, resolve labor issues that are raised (19.15). If there is no resolution by consultation within 60 days, the establishment of a panel can be requested and pursuant to the Dispute Resolution (Chapter 28) the case may proceed.[36] Under the Dispute Resolution provisions, labor violations affecting trade can proceed through the mechanism ultimately resulting in trade sanctions.

But by far, the most dramatic breakthrough on labor protections is found in the Side Agreements of the TPP that the US has with Vietnam, Malaysia, and Brunei, i.e., those countries with the poorest record of ratification of ILO labor standard conventions. By express terms, their labor laws must be newly established, changed, and improved to allow independent labor unions, strikes, proper treatment of immigrants, anti-discrimination provisions, labor inspections, and the basic labor standards affecting working conditions, *before* they are allowed to export goods duty-free to the US and otherwise benefit from the provisions of the TPP. The Side Agreements are very detailed in their obligations.[37]

In the Side Agreement with Vietnam (Plan for the Enhancement of Trade and Labour Relations), Vietnam agrees to eight categories of detailed legislative labor reforms, including the rights of union organization and collective bargaining, strikes, prohibitions on forced labor, discrimination, and the creation of administrative institutions to effectively enforce these obligations.[38] Oversight, review, cooperation, and

---

[36] See Appendix for a summary of dispute resolution in the TPP.

[37] TPP, *ibid.*

[38] United States–Vietnam Plan for the Enhancement of Trade and Labour Relations (TPP, *ibid, n.* 34).

technical assistance are also provided. The Agreement states these obligations to be enacted "prior to the entry into force of the TPP Agreement between the United States and Vietnam".[39]

The Side Agreement (Labor Consistency Plan) with Malaysia creates certain targeted legislative labor reforms to protect labor unions, union membership, collective bargaining and strikes, as well as access to remedies.[40] Specific provisions on prohibiting forced labor, child labor, and discrimination are included, as well as the creation of administrative institutions to effectively enforce these obligations. Oversight, review, cooperation, and technical assistance are also provided. The Agreement states these obligations to be enacted "prior to the entry into force of the TPP Agreement between the United States and Malaysia".[41]

The Side Agreement with Brunei (Labor Consistency Plan) obligates Brunei to legislative labor reforms addressing specific rights and obligations under the rights of labor unions to organize, collectively bargain, strike and to provide meaningful tribunals to resolve labor disputes.[42] It also targets forced labor, child labor, employment discrimination, and acceptable conditions of work. It obligates Brunei to have in place meaningful institutions to implement the legal reforms, including a labor inspectorate. Oversight, review, cooperation, and technical assistance are also provided. The Agreement states these obligations to be enacted "prior to the entry into force of the TPP Agreement between the United States and Brunei".[43]

The Labor Advisory Committee (LAC), composed of 19 American labor union leaders, strongly opposes the TPP.[44] Its conclusion is that the TPP does not meet labor objectives and lists 14 deficiencies,[45] fails to advance the economic interests of the US, and does not fulfill all of the negotiating objectives identified by Congress in the Trade Priorities and Accountability Act of 2015.

---

[39] *Ibid.*, Art. VII 1.

[40] Malaysia–United States Labour Consistency Plan (TPP, *supra, n.* 34).

[41] *Ibid.*, Art. VII 1.

[42] Brunei–United States Labour Consistency Plan (TPP, *supra, n.* 34).

[43] *Ibid.*, Art. VII 1.

[44] USTR (2015) (hereinafter, LAC Report).

[45] *Ibid.*, pp. 15–17. See also, Warren (2015).

It continues that while the TPP's Labor Chapter broadly meets the standards of the "May 10" Agreement, "it fails to include the vast majority of the improvements that the LAC recommended to build upon "May 10" to better ensure that member nations actually protect worker rights. The TPP also clearly falls short of "May 10" for a major reason — a critical component of "May 10" was that member nations needed to bring their laws into compliance with "May 10" standards before a congressional vote on implementing legislation for the agreement. It appears that some TPP member nations' laws will not be in compliance with the Labor Chapter's mandates before the agreement takes effect. For example, under the side understanding with Vietnam, Vietnam will have five years after the TPP enters into force to allow independent unions to form *beyond the enterprise level* [emphasis added]".[46] While the TPP also includes side letters/consistency plans to improve labor rights laws in three partner countries (Brunei, Malaysia, and Vietnam), by focusing on legal changes to the exclusion of implementation and enforcement benchmarks, the plans adopt the same failed approach as the Colombia Labor Action Plan.[47]

The International Trade Union Confederation (ITUC) released its list of labor issues that it advocates but are unaddressed or are under-addressed by the labor provisions in the TPP.[48] The critique does not address the three TPP Side Agreements with Vietnam, Malaysia, and Brunei, but only the TPP. More specifically, it raises issues about continued ambiguities in the TPP provisions. For example, it highlights the vagueness of obligations, such as referencing the Declaration, rather than Conventions; "acceptable conditions of work" and non-derogation, where minimum standards of the laws are met though conditions that are still not acceptable; the uncertain location of when a labor violation "affects trade" so as to trigger remedial action; and continued inaccessibility of non-state parties to trigger the Investor–State Dispute Settlement (ISDS) mechanism.

[46] LAC Report *supra* n. 42, p. 50.

[47] *Ibid.*, p. 52. The LAC Report also evaluates the three Side Agreements. (*Ibid.*, pp. 71–72 (Brunei); pp. 76–79 (Malaysia); p. 87 (Singapore); and pp. 87–90 (Vietnam)).

[48] ITUC (2015). The AFL-CIO (2014) shares the reluctance of the ITUC to rely on the agreements' promises, based on the US–Columbia LAP experience.

Some progress was noted regarding treatment of migrants and on forced labor.

The Office of the US Trade Representative provided the following summary of the concluded, relevant provisions of the TPP.[49]

> On October 4, 2015, Ministers of the 12 Trans-Pacific Partnership (TPP) countries — Australia, Brunei Darussalam, Canada, Chile, Japan, Malaysia, Mexico, New Zealand, Peru, Singapore, US, and Vietnam — announced conclusion of their negotiations. The result is a high-standard, ambitious, comprehensive, and balanced agreement that will promote economic growth; support the creation and retention of jobs; enhance innovation, productivity and competitiveness; raise living standards; reduce poverty in our countries; and promote transparency, good governance, and enhanced labor and environmental protections. We envision conclusion of this agreement, with its new and high standards for trade and investment in the Asia Pacific, as an important step toward our ultimate goal of open trade and regional integration across the region.

In March 2016, the Peterson Institute for International Economics, in its assessment of the TPP, concluded that the TPP is the "most ambitious of US FTAs" [50] and while it does not address "all the concerns of labor rights advocates, it does move the labor agenda along".[51]

It concludes that the "the text of Chapter 28 [disputes between governments] reflects real learning from past mistakes in setting up FTA dispute settlement systems and offers genuine hope for a strong, efficient, and effective system for resolving TPP disputes".[52] It "is designed to be a

[49] See: https://ustr.gov/about-us/policy-offices/press-office/press-releases/2015/october/summary-trans-pacific-partnership. It specifies that the TPP is open to accession by members of the Asia-Pacific Economic Cooperation Forum and other States or separate customs territories as agreed by the Parties, again after completing applicable legal procedures in each Party. (*Ibid.*)

[50] Cimino-Isaacs (2016a, p. 2). See full text in Vol. 2 of PIIE Briefing (2016); and see Cimino-Isaacs (2016b).

[51] Cimino-Isaacs (2016b).

[52] Hillman, (2016).

faster, more transparent and more comprehensive hybrid of the WTO's Dispute Settlement Understanding and the dispute settlement provisions in a number of US bilateral and regional trade agreements".[53]

The TPP has two separate dispute settlement chapters: Chapter 2 applies to disputes between governments and Chapter 9 covers disputes between private parties. The assessment of the Peterson Institute for International Economics on the latter is that it "is designed to be broader, deeper, faster, and more transparent than either the WTO's Dispute Settlement Understanding or any prior bilateral or regional free trade agreement. It covers more chapters and issues than prior dispute settlement systems (including systems on labor, the environment, cross-border data flows, and state-owned enterprises)".[54]

"[T]he ISDS provisions in the TPP are a significant improvement over those in previous agreements; … it ensures the transparency of dispute proceedings … [and] echoing recent EU agreements, the TPP more narrowly defines "fair and equitable treatment" than the definition in previous such accords. It also eliminates forum shopping — i.e., the attempt by companies to litigate in the most sympathetic places — and enables the assessment of costs against a losing party (a provision that discourages frivolous cases)".[55]

---

[53] *Ibid.*, p. 109.

[54] However, the Peterson Institute Report points out that "it leaves out some key issues, including the side agreement on currency manipulation, trade remedies, and many of the new issues included in the TPP itself, such as capacity building, competitiveness and business facilitation, and regulatory coherence" (Hufbauer, 2016). TPP Chapter 9 also clarifies that the investor bringing the case against the government bears the burden to prove all elements of its claims. It makes proceedings fully open and transparent and allows civil society organizations and others not party to the dispute to participate. In addition, Chapter 9 will, for the first time, clarify important concepts in the non-discrimination and minimum standard of treatment obligations. For example, it clarifies the definition of legitimate public welfare objectives as a ground for defending a country's regulations and makes clear that frustrated investor expectations are not in and of themselves sufficient to overturn such regulations. TPP countries will also establish a code of conduct for ISDS arbitrators to guide arbitrator independence and impartiality (*Ibid.*, p. 110).

[55] *Ibid.*

3. Implications for new research agenda

The overall economic impact of the potential TPP agreement depends on a number of factors, including the extent of the trade liberalization achieved in the agreement, as well as the current level and potential growth of trade and investment among TPP members. [56] On both measures, the TPP appears significant, given that the TPP would be the largest US FTA by trade flows ($727 billion in US goods exports and $882 billion in imports in 2014).[57]

The potential and actual economic impacts of the TPP on the living standards, institutional structures, employment, job creation, etc., in all countries, including the US, would seem to present a worthwhile challenge and endeavor for future research and analysis. For example, the LAC expects the TPP to drive down wages, cost jobs, and have a negative impact on the US economy as a whole.[58] Economists might also solve the perennial riddle of how to measure the impacts of "labor violations affecting trade," which can trigger the Dispute Settlement mechanism.

Labor market implications of trade and trade agreements[59]

Trade also has notable impacts on labor markets, many of them a direct result of the classic gains from trade in terms of increased productivity and innova-

---

[56] Fergusson *et al.* (2015).

[57] *Ibid.*

[58] The position of the Labor Advisory Committee on Trade Negotiations and Trade Policy (LAC) is as follows. "The primary measure of the success of our trade policies should be increasing jobs, rising wages, and broadly shared prosperity, not higher corporate profits and increased offshoring of America's jobs and productive capacity. Trade rules that enhance the already formidable economic and political power of global corporations — including ISDS, excessive monopoly rights for pharmaceutical products, and deregulatory financial services and SPS rules — will continue to undermine worker bargaining power, here and abroad, as well as weaken democratic processes and regulatory capacity across all 12 TPP countries" (LAC, 2015). Opposing views by economists are also expressed. "On the other side of the ledger, while the TPP is not likely to affect overall employment in the US, it will involve adjustment costs as US workers and capital move from less to more productive firms and industries. Section 4 estimates that 53,700 US jobs will be affected i.e., that number is both eliminated in less productive import-competing firms and added in exporting and other expanding firms — in each year during implementation of the TPP" (Petri and Plummer, 2016).

[59] White House, Office of the President (2015).

tion. US businesses that expand in response to the increased foreign market access due to US trade agreements support additional jobs. The importance of such export-led job growth for the Nation's income is reinforced by the fact that wages in export-intensive manufacturing industries tend to be higher than wages in non-export-intensive manufacturing industries.

Of course, while the aggregate benefits of trade may be large, they may also be unevenly distributed. Trade, therefore, can also have adverse effects for some workers. Domestic policies that the Administration supports, such as investment in infrastructure, worker training, and education, can help our labor force take advantage of the considerable opportunities that trade opens up. For displaced workers and their families, effective policies can help smooth the adjustment into new, potentially higher-paying jobs. The changes will need to be addressed by data collection and research analysis.

3.1. *Increased economic growth*[60]

When countries specialize in goods and services they can produce relatively efficiently and trade for the rest, world production and consumption increase as the existing global resources are more effectively utilized. Simple international trade theory, therefore, suggests that increased international trade can boost incomes. However straightforward this may seem, it is actually quite difficult to discern empirically a causal relationship between trade and income. For instance, countries may trade more because they are richer. Richer countries have better trading infrastructure, such as good ports, and better access to information about opportunities abroad. The fundamental challenge for statistical inference, then, is that while trade may affect income, income also affects trade.[61]

An extensive body of work examines the empirical association between trade openness and per capita income growth. Frankel and Romer (1999) were among the first to attempt to establish a causal picture of the impact of trade on income, arguing that the geographic placement of countries provides a way to isolate a component of trade that is unrelated

[60] *Ibid.* However, economists assert that econometric work can control for these and still find a positive relationship. See, for example, OECD, ILO, World Bank, and WTO (2010).
[61] *Ibid.*

to income except through its effect on trade. The authors use the distance between countries to predict the amount of their bilateral trade, relying on the well-established relationship that proximity is a strong predictor of trade flows (the so-called "gravity model"). Because a component of trade flows isolated in this manner is highly correlated with income, the authors conclude that trade has a positive causal effect on income.[62]

The OECD and World Bank Group, in an October 6, 2015, report, urged consideration of expanding inclusiveness of Lesser Developing Countries (LICDs) into Global Value Chains through arrangements lowering the trade barriers.

> "In November 2014, the Group of Twenty (G-20) Leaders concluded that trade and competition are powerful drivers of growth, increased living standards and job creation. They also acknowledged that one important way for countries to connect to the global economy and develop is through global value chains (GVCs); a recognition that GVCs provide opportunities to empower the local economy with sophisticated imported technology, know how, and a richer skill-set. G20 leaders at the Brisbane Summit stated that 'we need policies that take full advantage of global value chains and encourage greater participation and value addition by developing countries'".[63]
>
> "G20 countries are the key trading partners for LIDCs, with around 70% of imports of LIDCs originating from G20 countries and close to 80% of LIDCs exports directed to the G20. Thus, policy reforms that would lower trade costs in G20 economies and LIDCs can have important implications for GVC participation. GVC trade is particularly affected by trade barriers: when goods and services cross borders multiple times, as both imports and exports, trade costs are compounded. This is especially problematic for firms in LICDs and for SMEs, which are less able to absorb these costs. Moreover, foreign direct investment (FDI) is the most common vehicle for countries to participate in global value chains. The US government, for example, estimates that intrafirm transactions constitute close to 50% of US imports and around 30% of US exports. According to UNCTAD, an estimated 80% of global trade now occurs within international production networks of

[62] *Ibid.*

[63] OECD (2015).

multinational companies. And it is these companies that are responsible for more than 1 trillion dollars of global FDI flows annually. FDI are vital for SMEs, as their natural predisposition is to join GVCs indirectly as upstream supplier to exporters".[64]

Further, the Report indicates that

"[O]ne of the most important issues is about assessing the costs of complying with international standards as compared with the opportunity cost of serving regional markets. This is the case because adopting higher standards involves greater difficulties and challenges for SMEs and LIDCs than for larger firms and rich countries. Increases in production and trade costs can originate from the required compliance with a multitude of standards and technical regulations. For this reason … sequencing may be important. … LIDCs and SMEs may first need to expand into the regional market to gain scale and learning economies and then adopt stringent international standards".[65]

3.2. *The potential impact of the TPP on US trade policy*[66]

The US pursuit of the TPP and the possible non-approval by the US Congress raise other questions regarding its possible impact on the status and shape of current and future US trade policy. For example, how will the TPP agreement relate to the recently launched TTIP negotiations, or RCEP or the FTAAP? If all negotiations conclude successfully, could they eventually be merged in the FTAAP, for example?

Similarly, the TPP raises the issue of the US and the future of the WTO as a major force for trade liberalization. Some may argue, for example, that the US has signaled the death knell of future rounds of multilateral agreements in favor of regional pacts. Others might assert that the TPP could serve as a building block for a more viable multilateral trade system that responds to trade challenges of the 21st century. Some may even say that the TPP may become the predominant force for trade liberalization going forward, that is, if it can be agreed to by the current parties.

[64] *Ibid.*, p. 8.
[65] *Ibid.*, p. 77.
[66] Fergusson *et al.* (2015).

Another issue for possible consideration is: What would be the impact on US trade policy if the TPP negotiations are not completed successfully or are delayed indefinitely? Some could argue that such an outcome would indicate that it is not feasible to negotiate a comprehensive set of rules with a diverse group of countries and that the US would have to tailor its ambitions. In addition, some might assert that such an outcome would signify a temporary, if not permanent setback to the notion of an FTAAP. Still others may conclude that such result could force the US to retreat from negotiating trade agreements altogether.

4. Conclusion and recommendation

The observation of this chapter is that the TPP represents a "new generation" of social-dimension labor provisions of FTAs that can actually protect more workers than are currently being protected by in-place domestic labor laws. In the author's opinion, it is a "game changer" and the labor provisions of the TPP, and especially its three Side Agreements, should be embraced. The TPP is not perfect[67]; clarity of language could be improved and as well as access to the dispute resolution process, but it is progressive and moving toward securing greater labor protections for workers.[68] It should also be noted that American workers can also benefit from this FTA, as the protections for labor unions, strikes, and working conditions can also be raised regarding US workers.

Of course, the articles of the TPP require close monitoring, cooperation, consultation, and enforcement that are all part of the recipe for it to work. And the continuing research and findings of economists and other specialists are needed to actually measure the economic implications of the TPP's labor provisions.

It is significant to note that the alternative mega-FTAs in this region, whether it be the RCEP or the FTAAP do not have, nor is there any certainty

[67] See a more negative appraisal in ITUC (2015).

[68] This view is supported in the Peterson Institute's Report cited earlier, where its conclusion regarding the labor provisions was that they "can play a constructive role through added pressure to upgrade standards and to enforce binding TPP commitments" (Cimino-Isaacs, 2016a, p. 9, 2016b, pp. 41–65)."

they will have, provisions for labor protection given the nations involved and their track record of not having such labor protection.[69]

An alternative approach, perhaps for future FTAs, could be to require state party agreement to enforce *binding* CSRs, rather than *voluntary* as in Article 19.7 in the TPP, with worker rights, grievance procedures, and remedies within the governance document; and, in certain industries be part of an International Framework Agreement (IFA)[70] This "privatization" addresses the labor issues by enforceable contractual obligations between the non-"Parties." The dispute resolution procedures could include private agreement on mediation and arbitration,[71] with the scope of arbitrable items including alleged violations of contract or domestic or ILO labor law obligations, with venue to be determined.[72]

The "new" approach of the TPP's Side Agreements conditioning use of the TPP on prior compliance with "on-the-books" legislative reforms is, in the author's opinion, a big step forward in gaining closer compliance with meeting the obligations of the ILO labor standards. But observers note they look similar to the US–Colombia Agreement, where the glittery promises did not come into actuality.

The practical difficulties of affording workers enforceable labor rights and achieving these improved labor protections under international trade agreements, besides the issue of government *will,* is finding the right balance with the MNCs who benefit from locating in areas with low labor standards. Currently, where there is lax law enforcement of labor standards, even if employers and MNCs have and adhere to codes of conduct,

[69]President Obama's position on this is: "[T]he United States, not countries like China, should write them. Let's seize this opportunity, pass the Trans-Pacific Partnership and make sure America isn't holding the bag, but holding the pen" (Obama, 2016).

[70]A global framework agreement, also referred to as an international framework agreement, is "an instrument negotiated between a multinational enterprise and a Global Union Federation (GUF) in order to establish an ongoing relationship between the parties and ensure that the company respects the same standards in all the countries where it operates." The framework agreements cover the same ILO conventions on which the labor principles are based. United Nations Global Impact, https://www.unglobalcompact.org/Issues/Labour/Global_Framework_Agreements.html (last visited Jan. 31, 2015).

[71]Though litigation could be considered, issues for exhaustion of domestic remedies and deference to sovereignty would first need to be resolved.

[72]These remedies and others are discussed in Brown (2016a). Also see, Brown (2015b).

OECD guidelines, and CSRs, they are mostly non-binding and do not restrict the use of contractors and sub-contractors who fall outside these "obligations".[73]

In addition to the above approaches, at a future appropriate time, adding or amplifying third-party access to the enforcement mechanisms through more direct and expeditious means would address some of the national and international labor unions' concerns and reservations,[74] perhaps also clarifying the specific facts necessary to prove a violation of applicable labor standards and, more specifically, defining how and when "the labor violation affects trade" so as to trigger the Dispute Settlement machinery. And, perhaps thereafter expediting and mandating advancement of the complaint, and granting arbitration panels increased authority to implement remedies, including suspending benefits or some FTA mechanisms, such as the dispute resolution mechanism, that could facilitate the effectiveness of the dispute resolution mechanisms, and in the judgment of some, to actually make them work. A new proposal keyed into the ISDS mechanism and more "worker-friendly," has been issued by the AFL-CIO, its Australian and New Zealand counterparts, and other

---

[73] For an excellent discussion regarding privatizing the obligations of MNCs, see discussion in van der Heijden and Zandvliet (2014) or Zandvliet and van der Heijden (2014). The same authors argue the post- Rana Plaza disaster and its resulting remedial accords made Bangladesh a "policy laboratory for new ways to enforce fundamental labor rights. These responses, which can be characterized as a network approach, involve many stakeholders cooperating in different coalitions to pursue a variety of goals. The network approach aligns with the idea that improving labor rights in global supply chains is not the sole responsibility of the state in which production takes place. A collaborative effort is required. … [The Authors] argue that businesses with transnational supply chains should cast their labor commitments in a contractual form, following the successful example of the Bangladesh Accord for Fire and Building Safety. Institutionally, the ILO should engage more directly with businesses and use its authoritative role to strengthen supply-chain bargaining" (Heijden and Zandvliet, 2014).

[74] Third-parties could include unions, NGOs, or even the ILO. See Brown (2016c), *supra* n. 25. The ability of *stakeholders'* access to enforcement machinery could be better clarified. By merely allowing investor or state complaints, third-parties are left to petition their state though administrative apparatus, such as NAOs or departments of labor, that may leave too much discretion in the state's decision whether to proceed (USTR, 2006). See also, Government of Canada (2014). This point is reiterated in ITUC (2015). "II. Dispute Settlement."

unions, and "is based, in part, on the labour and dispute resolution chapters of the U.S.-Peru Free Trade Agreement (FTA), two of the Parties to the current [TPP] negotiations".[75]

Of course it is easier for an academic to state the above, than for a union, who over the years sees and lives with broken promises and the frustrations of lack of enforcement in FTAs. There certainly are valid worries about the enforcement provisions regarding labor; however, the author supports the optimistic view, that chipping away is better than doing nothing or being rendered irrelevant on the labor issues by the RCEP or the FTAAP. And, just maybe it is the best deal in town; and there are avenues other than FTAs to fight the unions' fight. As to the overall value of the non-labor provisions of the TPP, the author leaves that to the trade experts.

In sum, in the author's view, the implications of the TPP labor provisions, and particularly on the basis of its three TPP Side Agreements, are generally positive, encouraging, and progressive, building on past US approaches. It brings the promise of improved and enforceable labor standards among the TPP parties and the certainty of needed additional research on its legal, economic, political, and social consequences.[76]

---

[75] See: http://www.trungtamwto.vn/sites/default/files/tpp/attachments/Final%20Official%20ITUC%20TransPacificPartnership%20Labor%20Chapter%202b29%20%20TPP%20labor%20rights.pdf.

[76] President Trump withdrew the US from the TPP, however the other TPP partners are moving toward its final approval. On November 11, 2017, at a side meeting at an APEC meeting in Vietnam, members of the TTP agreed to continue working on the proposed trade deal. According to a draft due to be released on Saturday November 11, 2017 in Vietnam, "the 11 countries committed to 'the core elements' of a deal while making clear more work remains. At a speech in Danang, Trump set out a strong message making clear he was only interested in bilateral deals in Asia which would never put the United States at a disadvantage. China's President Xi Jinping, meanwhile, used the same forum to stress multilateralism and said globalization was an irreversible trend." Takenaka, K. and M. Tostevin. "TPP Countries Agree to Keep Trade Deal Alive, Much Work Remains". Available at https://in.reuters.com/article/apec-summit/trans-pacific-trade-deal-advances-without-united-states-idINKBN1DA2PP. See also, Brown, R. C. (2018). "Chinese Style Global Trade: New Leader in Asia: New Rules; No Labor Protections in its Free Trade Agreements". *UCLA Pacific Basin Law Journal*.

References

ADB (2013a). "Asian Free Trade Agreements: Untangling the Noodle Bowl." August 8. Available at: http://www.adb.org/features/free-trade-untangling-asia-s-noodle-bowl

ADB (2013b). "Asian Economic Integration Monitor — March 2013." Available at: http://www.adb.org/publications/asian-economic-integration-monitor-march-2013. Accessed 16 August.

ADB (2013c). "Regional Cooperation and Integration in a Changing World." Available at: https://www.adb.org/sites/default/files/publication/30224/regional-cooperation-changing-world.pdf

ADB (2013d). "Asian Development Outlook 2013 Asia's Energy Challenge." Available at: http://www.adb.org/sites/default/files/pub/2013/aem-201303.pdf

AFL-CIO (2014). "Making the Columbia Action Plan Work." Available at: http://www.aflcio.org/content/download/123141/3414471/April2014_Colombia Report.pdf;

Aggarwal, V. K. (2016). "Mega-FTAs and the Trade-Security Nexus: The Trans-Pacific Partnership (TPP) and Regional Comprehensive Economic Partnership (RCEP)." Available at: http://www.eastwestcenter.org/system/tdf/private/api123_0.pdf?file=1&type=node&id=35584.

ARIC (n.d.) "Bilateral FTAs by Geographic Area, WTO Notification and Status 2000 and 2015." Available at: http://aric.adb.org/fta. Accessed 16 August 2014.

ARIC (2012). "East Asia Free Trade Area (ASEAN+3)." Available at: http://aric.adb.org/fta/east-asia-free-trade-area-%28asean3%29. Accessed 11 May 2015.

ASEAN (n.d.) "ASEAN Regional Comprehensive Economic Partnership." Available at: http://asean.org/?static_post=asean-framework-for-regional-comprehensive-economic-partnership

ASEAN Affairs (2012). "ASEAN plus Six Likely to Form World's Largest Economic Bloc." November 17. Available at: http://www.aseanaffairs.com/asean_news/trade/asean_plus_six_likely_to_form_world_s_largest_economic_bloc.

ASEAN.org (2012). "Guiding Principles and Objectives for Negotiating the Regional Comprehensive Economic Partnership." August 30. Available at: http://www.asean.org/images/2012/documents/Guiding%20Principles%20and%20Objectives%20for%20Negotiating%20the%20Regional%20Comprehensive%20Economic%20Partnership.pdf

Bieszcat, F. H. (2008). "Labor Provisions in Trade Agreements: From to Now," *Chicago-Kent Law Review*, 86, pp. 1387–1388. Available at: http://scholarship.kentlaw.iit.edu/cgi/viewcontent.cgi?article=3695&context=cklawreview.

Bolle, M. J. (2014). "Overview of Labor Enforcement Issues in Free Trade Agreements," January 29. Available at: https://www.fas.org/sgp/crs/misc/RS22823.pdf.

Brown, R. C. (2015a). "Fundamental Labour Rights in China — Legal Implementation and Cultural Logic." In *China – U.S. Implementation of ILO Standards by BITs and Pieces (FTAs)*, U. Liukkunen and Y. Chen (eds.), (Springer), p. 169.

Brown, R. C. (2015b). "Asian and US Perspectives on Labor Rights under International Trade Agreements Compared." In *Protecting Labor Rights in a Globalizing World*, A. Marx *et al.* (eds.) (Edward Elgar, Cheltenham).

Brown, R. C. (2016a). "FTAs in Asia-Pacific: "Next Generation" of Social Dimension Provisions on Labor?" *Indiana International and Comparative Law Review.*

Brown, R. C. (2016b). "FTAs that Also Protect Workers: Expanding the Reach of Social Dimension Provisions on Labor to Promote, Compel, and Implement ILO Core Labor Standards." In *Employment Relations And Transformation of The Enterprise In The Global Economy*, E. Ales *et al.* (eds.), Part II, Giappichelli-MBF Book Series, p. 109.

Brown, R. C. (2016c). "The Efficacy of the Emergent US Model Trade and Investment Frameworks to Advance International Labor Standards in Bangladesh," *International Labour Review.*

Brown, R. C. (2016d). "China – U.S. Implementation of ILO Standards by BITs and Pieces (FTAs)." In *Fundamental Labour Rights in China — Legal Implementation and Cultural Logic*, U. Liukkunen and Y. Chen (eds.), Chapter 6 (Springer).

Brown, R. C. *Fostering Labor Rights in a Global Economy: The Efficacy of the Emergent US Model Trade and Investment Frameworks to Advance International Labor Standards in Bangladesh.*

Cimino-Isaacs, C. (2016a). "Labor Standards in the TPP." Available at: https://piie.com/system/files/documents/cimino-isaacs20160328ppt.pdf, p. 2. Peterson Institute for International Economics, March 28.

Cimino-Isaacs, C. (2016b). Chapter 4. In *Labor Standards in the TPP*, pp. 41–65, March. Available at: https://piie.com/system/files/documents/piieb16-4.pdf

Dennis, S. T. (2015). "Obama Faces a Tough Road with TPP Trade Deal," November 12. Available at: http://fortune.com/2015/11/12/barack-obama-tpp-trade-deal/

Drake, C. (2015). "TPP Daily Debunk #1: The 'Most Progressive Trade Agreement Ever,'" November 11. Available at: http://www.aflcio.org/Blog/Political-Action-Legislation/TPP-Daily-Debunk-1-The-Most-Progressive-Trade-Agreement-Ever.

Economic Times (2015). "RCEP Set to Miss 2015 Deadline, Talks to Spill Over to Next Year," October 23. Available at: http://economictimes.indiatimes.com/news/international/business/rcep-set-to-miss-2015-deadline-talks-to-spill-over-to-next-year/articleshow/49499225.cms

Fergusson, I. F., McMinimy, M. A., and Williams, B. R. (2015). "The Trans-Pacific Partnership (TPP) Negotiations and Issues for Congress." March 20, p. 9. Available at: https://www.fas.org/sgp/crs/row/R42694.pdf.

Frost, E. I. (2014). "Rival Regionalisms and Regional Order: A Slow Crisis of Legitimacy," pp. 8–9. Available at: http://www.nbr.org/publications/special-report/pdf/free/021115/SR48.pdf. Accessed 27 November 2015.

Fukunaga, Y. and I. Isono (2013). "Taking ASEAN+1 FTAs towards the RCEP: A Mapping Study." ERIA Discussion Paper Series, ERIA-DP-2013-02. Available at: http://www.eria.org/ERIA-DP-2013-02.pdf

Global Affairs Canada (n.d.) "History and Membership of APEC." Available at: http://www.international.gc.ca/apec/map-carte.aspx?lang=eng. Accessed 9 May 2015.

Government of Canada (2014). "Canada-European Union: Comprehensive Economic and Trade Agreement (CETA), Canada-E.U., arts. 24 and 33." September 26. Available at: http://www.international.gc.ca/trade-agreements-accords-commerciaux/agr-acc/ceta-aecg/text-texte/33.aspx?lang=eng, or http://trade.ec.europa.eu/doclib/docs/2014/september/tradoc_152806.pdf.

Hiebert, M., and L. Hanlon (2012). *ASEAN and Partners Launch Regional Comprehensive Economic Partnership*, Center for Strategic and International Studies, December 7. Available at: http://csis.org/publication/asean-and-partners-launch-regional-comprehensive-economic-partnership.

Hillman, J. (2016). "Dispute Settlement Mechanism." In *Assessing the Trans-Pacific Partnership: Market Access and Sectoral Issues*, Vol. 1, Chapter 9, p. 101. Available at: https://piie.com/system/files/documents/piieb16-4.pdf.

Hufbauer, G. C. (2016). "Investor-State Dispute Settlement." In *Assessing the Trans-Pacific Partnership*: *Market Access and Sectoral Issues*, Vol. 2, Chapter 10, p. 109. Available at: https://piie.com/publications/briefings/piieb16-1.pdf

ILO (2013). "Social Dimensions of Free Trade Agreements," Available at: http://www.ilo.org/wcmsp5/groups/public/---dgreports/---inst/documents/publication/wcms_228965.pdf.

ITUC-CSI (2015). "Trans Pacific Partnership Labour Chapter Scorecard Fundamental Issues Remain Unaddressed." November 27. Available at: http://www.ituc-csi.org/IMG/pdf/trans_pacific.pdf.

Kawai, M. and G. Wignaraja (2010). "Asian FTAs: Trends, Prospects, and Challenges."Asian Development Bank Working Paper No. 226, p. 1, 4. Available at: http://www.un.org/esa/ffd/msc/regionalcooperation/ADB_WPs.pdf.

Kay, T. (2003). "Analysis of the Labour Aspects of NAFTA." Preliminary document for commentaries of Work Group I of the Inter-American Conference of Ministers of Labour, ILO, June Available at: white.oit.org.pe/cimt/DocumentosWord/ingles/EDITADO%20NAFTA.doc.

Kotschwar, B. and J. J. Schott (2013). "The Next Big Thing? The Trans-Pacific Partnership & Latin America: Latin America Goes Global," *Americas Quarterly*, 8. Available at: http://americasquarterly.org/next-big-thing-trans-pacific-partnership.

LAC (2015). "Report on the Impacts of the Trans-Pacific Partnership," December 2, pp. 38–42. Available at: https://ustr.gov/sites/default/files/Labor-Advisory-Committee-for-Trade-Negotiations-and-Trade-Policy.pdf.

New Zealand MFAT (n.d.). "Regional Comprehensive Economic Partnership (RCEP)." Available at: http://www.mfat.govt.nz/Trade-and-Economic-Relations/2-Trade-Relationships-and-Agreements/RCEP/. Accessed 10 May 2015.

Obama, B. (2016). "The TPP Would Let America, Not China, Lead the Way on Global Trade," *Washington Post*, May 2. Available at: https://www.washingtonpost.com/opinions/president-obama-the-tpp-would-let-america-not-china-lead-the-way-on-global-trade/2016/05/02/680540e4-0fd0-11e6-93ae-50921721165d_story.html.

OECD (2015). "Inclusive Global Value Chains Policy Options in Trade and Complementary Areas for GVC Integration by Small and Medium Enterprises and Low-income Developing Countries." October 6, p. 7. Available at: http://www.oecd.org/trade/OECD-WBG-g20-gvc-report-2015.pdf.

OECD, ILO, World Bank, and WTO (2010). "Seizing the Benefits of Trade for Employment and Growth." Prepared for the 2010 G-20 Summit. Available at: https://www.oecd.org/trade/benefitlib/46353240.pdf.

Pacific Alliance (2013). "The Pacific Alliance VII Summit." May 23. Available at: http://alianzapacifico.net/documents/abc_eng.pdf.

Petri, P. A. and M. G. Plummer (2016). "The Economic Effects of the Trans-Pacific Partnership: New Estimates," January, p. 3. Available at: https://piie.com/publications/wp/wp16-2.pdf.

PIIE (2016). "Briefing: Assessing the Trans-Pacific Partnership." March 28.

Ramirez, S. (2013). "Regionalism: The Pacific Alliance," *Americas Quarterly*, Spring. Available at: http://www.americasquarterly.org/content/regionalism-pacific-alliance

Seth, D. (2015). "India Offers Wider Duty Cuts at Regional Comprehensive Economic Partnership." *The Economic Times*, February 18. Available at: http://articles.economictimes.indiatimes.com/2015-02-18/news/59268923_1_regional-comprehensive-economic-partnership-asean-rcep

Studer, I. (2010). "The NAFTA Side Agreements: Toward a More Cooperative Approach?" *Wake Forest Law Review*, 45, pp. 469–490.

Tiezzi, S. (2014). "China-Japan-South Korea Hold FTA Talks Despite Political Tension," *The Diplomat*, March 5. Available at: http://thediplomat.com/2014/03/china-japan-south-korea-hold-fta-talks-despite-political-tension/.

TUC (2010). "Text of Draft EU-Colombia Trade Deal Offers Nothing for Threatened Workers," May. Available at: http://www.tuc.org.uk/international-issues/countries/colombia/human-rights/text-draft-eu-colombia-trade-deal-offers

USTR (2006). "U.S. – Peru Trade Promotion Agreement, U.S.-Peru, art. 21." April 12. Available at: https://ustr.gov/trade-agreements/free-trade-agreements/peru-tpa/final-text

USTR (2015a). "KORUS FTA." Available at: http://www.ustr.gov/trade-agreements/free-trade-agreements/korus-fta/final-text. Accessed 31 January 2015.

USTR (2015b). "Report on the Impacts of the TPP by The Labor Advisory Committee on Trade Negotiations and Trade Policy," December 2. Available at: https://ustr.gov/sites/default/files/Labor-Advisory-Committee-for-Trade-Negotiations-and-Trade-Policy.pdf

van der Heijden, P., and R. Zandvliet (2014). "Enforcement of Fundamental Labor Rights, The Network Approach: Closing The Governance Gaps In Low-Wage Manufacturing Industries." Policy Brief, September, p. 12. Available at: http://thehagueinstituteforglobaljustice.org/cp/uploads/publications/policypaper_12-enforcement-of-fundamental-labor-rights_1409068554.pdf

Wallar, J. (2014). "Achieving the Promise of the ASEAN Economic Community: Less Than You Imagine, More Than You Know." National Bureau of Asian Research, 20. Available at: http://www.nbr.org/downloads/pdfs/ETA/wallar_paper_072814.pdf.

Warren, E. (2015). "Broken Promises." Available at: http://big.assets.huffingtonpost.com/WarrenReport.pdf

White House, Office of the President (2015). "The Economic Benefits of U.S. Trade," May, p. 10. Available at: https://www.whitehouse.gov/sites/default/files/docs/cea_trade_report_final_nonembargod_v2.pdf

Wignaraja, G. (2013). "Why the RCEP Matters for Asia and the World." East Asia Forum, May 15. Available at: http://www.eastasiaforum.org/2013/05/15/why-the-rcep-matters-for-asia-and-the-world/

Zandvliet, R. and van der Heijden, P. (2014). "The Rapprochement of ILO Standards and CSR Mechanisms: Towards a Positive Understanding of 'Privatization.'" February 5. Available at: http://dx.doi.org/10.2139/ssrn.2391295

Zhang, Y. and M. Shen (2011). "The Status of East Asian Free Trade Agreements." Asian Development Bank Working Paper No. 282, p. 3. Available at: http://www.econstor.eu/handle/10419/53698

Appendix

Labour[77]

All TPP Parties are International Labour Organization (ILO) members and recognize the importance of promoting internationally recognized labour rights. TPP Parties agree to adopt and maintain in their laws and practices the fundamental labour rights as recognized in the ILO 1998 Declaration, namely freedom of association and the right to collective bargaining; elimination of forced labour; abolition of child labour and a prohibition on the worst forms of child labour; and elimination of discrimination in employment. They also agree to have laws governing minimum wages, hours of work, and occupational safety and health. These commitments also apply to export processing zones. The 12 Parties agree not to waive or derogate from laws implementing fundamental labour rights in order to attract trade or investment, and not to fail to effectively enforce their labour laws in a sustained or recurring pattern that would affect trade or investment between the TPP Parties. In addition to commitments by Parties to eliminate forced labour in their own countries, the Labour chapter includes commitments to discourage importation of goods that are produced by forced labour or child labour, or that contain inputs produced by forced labour, regardless of whether the source country is a TPP Party. Each of the 12 TPP Parties commits to ensure access to fair, equitable and transparent administrative and judicial proceedings and to provide effective remedies for violations of its labour laws. They also agree to public participation in implementation of the Labour chapter, including establishing mechanisms to obtain public input.

The commitments in the chapter are subject to the dispute settlement procedures laid out in the Dispute Settlement chapter. To promote the rapid resolution of labour issues between TPP Parties, the Labour chapter also establishes a labour dialogue that Parties may choose to use to try to resolve any labour issue between them that arises under the chapter. This dialogue allows for expeditious consideration of matters and for Parties to mutually agree to a course of action to address issues. The Labour chapter

[77] *Ibid.* Also see, Cimino-Isaacs (2016b).

establishes a mechanism for cooperation on labour issues, including opportunities for stakeholder input in identifying areas of cooperation and participation, as appropriate and jointly agreed, in cooperative activities.

Dispute settlement[78]

The Dispute Settlement chapter is intended to allow Parties to expeditiously address disputes between them over implementation of the TPP. TPP Parties will make every attempt to resolve disputes through cooperation and consultation and encourage the use of alternative dispute resolution mechanisms when appropriate. When this is not possible, TPP Parties aim to have these disputes resolved through impartial, unbiased panels. The dispute settlement mechanism created in this chapter applies across the TPP, with few specific exceptions. The public in each TPP Party will be able to follow proceedings, since submissions made in disputes will be made available to the public, hearings will be open to the public unless the disputing Parties otherwise agree, and the final report presented by panels will also be made available to the public. Panels will consider requests from non-governmental entities located in the territory of any disputing Party to provide written views regarding the dispute to panels during dispute settlement proceedings.

Should consultations fail to resolve an issue, Parties may request establishment of a panel, which would be established within 60 days after the date of receipt of a request for consultations or 30 days after the date of receipt of a request related to perishable goods. Panels will be composed of three international trade and subject matter experts independent of the disputing Parties, with procedures available to ensure that a panel can be composed even if a Party fails to appoint a panelist within a set period of time. These panelists will be subject to a code of conduct to ensure the integrity of the dispute settlement mechanism. They will present an initial report to the disputing Parties within 150 days after the last panelist is appointed or 120 days in cases of urgency, such as cases related to perishable goods. The initial report will be confidential, to enable Parties to offer comments. The final report must be presented no later than

[78] PIIE Briefing (2016); *ibid.*

30 days after the presentation of the initial report and must be made public within 15 days, subject to the protection of any confidential information in the report.

To maximize compliance, the Dispute Settlement chapter allows for the use of trade retaliation (*e.g.*, suspension of benefits), if a Party found not to have complied with its obligations fails to bring itself into compliance with its obligations. Before use of trade retaliation, a Party found in violation can negotiate or arbitrate a reasonable period of time in which to remedy the breach.

Index